I0038141

Henning Höppe
Rare-Earth Elements

Also of Interest

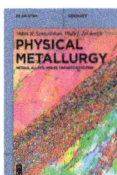

Physical Metallurgy
Metals, Alloys, Phase Transformations
Vadim M. Schastlivtsev and Vitaly I. Zel'dovich, 2022
ISBN 978-3-11-075801-6, e-ISBN (PDF) 978-3-11-075802-3

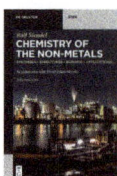

Chemistry of the Non-Metals
Syntheses - Structures - Bonding - Applications
Ralf Steudel, 2020
ISBN 978-3-11-057805-8, e-ISBN (PDF) 978-3-11-057806-5

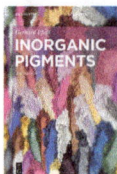

Inorganic Pigments
Gerhard Pfaff, 2023
ISBN 978-3-11-074391-3, e-ISBN (PDF) 978-3-11-074392-0

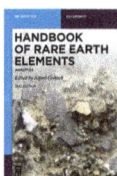

Handbook of Rare Earth Elements
Analytics
Edited by: Alfred Golloch, 2022
ISBN 978-3-11-069636-3, e-ISBN (PDF) 9783110696455

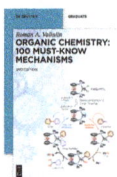

Organic Chemistry: 100 Must-Know Mechanisms
Roman Valiulin, 2023
ISBN 978-3-11-078682-8, e-ISBN (PDF) 978-3-11-078683-5

Henning Höppe

Rare-Earth Elements

Solid State Materials: Chemical, Optical and Magnetic Properties

DE GRUYTER

Author
Prof. Dr. Henning Höppe
Universität Augsburg
Institut für Physik
Universitätsstraße 1
86159 Augsburg
Germany
henning@ak-hoeppe.de

ISBN 978-3-11-068081-2
e-ISBN (PDF) 978-3-11-068082-9
e-ISBN (EPUB) 978-3-11-068088-1

Library of Congress Control Number: 2023948754

Bibliographic information published by the Deutsche Nationalbibliothek
The Deutsche Nationalbibliothek lists this publication in the Deutsche Nationalbibliografie;
detailed bibliographic data are available on the Internet at http://dnb.dnb.de.

© 2024 Walter de Gruyter GmbH, Berlin/Boston
Cover image: Thibault Renard / iStock / Getty Images Plus
Typesetting: VTeX UAB, Lithuania
Printing and binding: CPI books GmbH, Leck

www.degruyter.com

Abbreviations

A	electron affinity
aq	water
bcc	body centered cubic packing
c	cubic close packed layer
c	speed of light
c. n.	coordination number
CTS	charge-transfer states
Equ.	equation
F	Faraday constant
Fig.	figure
g	gaseous
h	Planck constant
h	hexagonal close packed layer
hc	double hexagonal close packed layer
$\mathcal{H}$	Hamilton operator
HOMO	highest occupied molecular orbital
H_2O	dissolved in water
I	ionization energy
IC	internal conversion
ISC	intersystem crossing
IVCT	intervalence charge-transfer transition
k_B	Boltzmann constant
LMCT	ligand-to-metal charge-transfer transition
Ln	any lanthanoid element
LUMO	lowest occupied molecular orbital
MAPLE	Madelung Part of Lattice Energy
n	principal quantum number
n. a.	not available
org	dissolved in organic phase
R	any rare-earth element
r_i	ionic radius
ref.	reference
refs.	references
T	absolute temperature
T_C	Curie temperature
T_c	critical temperature
T_N	Neél temperature
Tab.	table
u. v.	ultraviolet (400–200,nm)
VUV	vacuum ultraviolet (< 200 nm)
χ	electronegativity
η	chemical hardness
ξ	nephelauxetic ratio

https://doi.org/10.1515/9783110680829-201

Preface

Rare-earth elements are maybe the most exciting group in the periodic table featuring a great diversity of properties. Unfortunately, these elements in many chemistry and physics courses only play a minor role. Rarely, a concise overview is given enabling students and researchers to quickly enter this field. Therefore, I intend with this textbook to give an overview to provide the readers a basic understanding of the chemistry and of the materials properties. All this is based on simple and as good examples as possible, so this textbook will and cannot deliver a complete overview about rare-earth metal based compounds and materials. It wants to deliver the ability to assess literature, to provide the foot-in-the-door. Regarding the applications and examples, it is restricted to the topics where rare-earth elements are crucial and on which their properties can be demonstrated well. Further, colors were used, which should be appropriate for color-blind readers.

Most of the more sophisticated chemical and physical basics are discussed, so with a certain base on chemical and physical knowledge, you will be able to follow and understand the topics herein. You are invited to actually *work* with this book. You may read the book from page one until the (bitter) end; or you may pick out certain topics that you either find in the Table of Contents or the Index, and then you may work through considering the cross references in the text. You may choose to follow the recommendations for further reading given in the text.

This textbook is divided in two parts. In the first part, the basics of rare-earth chemistry, the basic data, the basic compound classes will be described and discussed. The second part focuses on the optical and magnetic properties of the elements and especially their solid compounds. Both properties feature clear general trends, which can be understood very well based on part one—understanding is in any case better than memorizing detailed data. While in the case of the magnetic properties the general trends are sufficient to understand the properties of most of the materials, the optical properties are different; here, the general trends yield distinct individual properties. Accordingly, in these chapters the individual ions are discussed. Eventually, I wish all readers to have fun!

March 2024   Henning Höppe

https://doi.org/10.1515/9783110680829-202

Contents

1 The Rare-Earth Elements

1.1 An Introduction

Let us start with a simple question, a question heavily debated in the 18th century:

What is a (rare-earth) element?

Element derives from the latin word *elementum,* which means something like a *basic unit.* Boyle[1] stated in his famous book, *The Sceptical Chemist* of 1661:

Those distinct substances, which concretes generally either afford, or are made up of, may, without very much inconvenience, be called the elements or principles of them.

The great Lavoisier[2] later defined an element as a matter, which cannot be deconstructed by chemical means like heat, light, mechanical or electric energy. Further, as we will see in chapter 4 on natural resources, most of the rare-earth elements are neither actually *rare* nor they are *earths.* Earths in this sense were considered elements until the 19th century, before the English chemist, Davy,[3] showed that via electrolysis, metals can be extracted from *earths,* nowadays known to be oxides as shown in Figure 1.1. The very first publication on a rare earth, yttria, was published in 1794 by Gadolin[4] reporting its discovery [1]. The history following this first report is subject of the second part of this Introduction.

Perhaps almost no one is aware that rare-earth elements are omnipresent in our daily life. They play a crucial role for fighting global warming—from the very beginning of energy generation based on solar or wind energy through lighting and electromobility, heating and air-conditioning. They are present here as the converter of solar radiation, as permanent magnets in wind turbines and electric engines, as energy saving white light emitting diodes or as hydrogen storage material. There is a great diversity of further applications reaching from catalysis, telecommunications and mobile phones through medical applications. Accordingly, many of these elements are publicly considered as critical elements by the European Union and the United States [2, 3]. This may be one reason, and a very good reason to study rare-earth chemistry and physics. Another reason may be to acquire the knowledge and know-how to become a researcher in this field—and maybe it is altogether, and you intend to teach students. In any case, our journey through the exciting world of rare-earth elements starts right now.

---

1 *Robert Boyle,* Irish natural scientist (*1627 †1692).

2 *Antoine Laurent de Lavoisier,* French natural scientist, chemist, lawyer and economist (*1743 †1794).

3 *Humphry Davy,* British chemist (*1778 †1829).

4 *Johan Gadolin,* Finnish chemist, mineralogist and physicist (*1760 †1852).

https://doi.org/10.1515/9783110680829-001

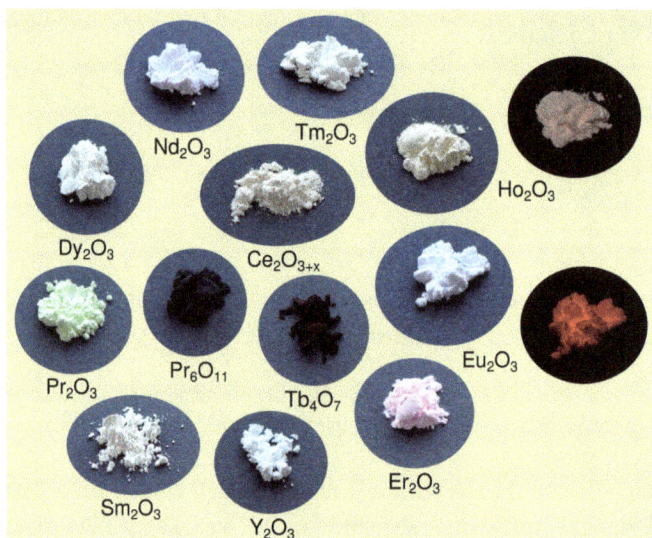

Figure 1.1: Photographs of the rare earths under daylight; Ho_2O_3 and Eu_2O_3 are also shown under a fluorescent and an u. v. lamp, respectively.

Table 1.1: The rare-earth elements with name, atomic number Z, symbol, atomic masses M in g mol^{-1} of the stable elements only, typical R_2O_3 oxide contents of selected minerals in wght-% as well as the year of discovery; data: [4–7].

element	Z	symbol	mass	gadolinite	cerite	samarskite	discovery
scandium	21	Sc	44.956(1)	0.00	0.00	5.50	1879
yttrium	39	Y	88.906(1)	26.86	2.26	46.56	1794
lanthanum	57	La	138.91(1)	2.34	18.01	0.28	1839
cerium	58	Ce	140.12(1)	12.65	43.84	0.88	1804
praseodymium	59	Pr	140.91(1)	3.44	5.58	0.24	1885
neodymium	60	Nd	144.24(1)	25.74	23.67	2.92	1885
promethium	61	Pm		0	0	0	1947
samarium	62	Sm	150.36(2)	11.13	3.65	8.01	1879
europium	63	Eu	151.96(1)	0.00	0.00	0.21	1896
gadolinium	64	Gd	157.25(3)	12.85	2.48	14.75	1886
terbium	65	Tb	158.93(1)	0.41	0.00	2.24	1843
dysprosium	66	Dy	162.50(1)	3.81	0.30	9.19	1886
holmium	67	Ho	164.93(1)	0.43	0.03	1.01	1879
erbium	68	Er	167.26(1)	0.01	0.15	3.28	1843
thulium	69	Tm	168.93(1)	0.00	0.00	2.39	1879
ytterbium	70	Yb	173.05(2)	0.00	0.01	2.18	1878
lutetium	71	Lu	174.97(1)	0.00	0.00	0.35	1907

The rare-earth elements are listed in Table 1.1. By the way, the rare-earth elements will be abbreviated by R in this book. Among these, the elements cerium through lutetium, resembling roughly many of the properties of lanthanum, are normally named *lanthanide* elements. Also in this table, the actually accepted atomic masses are given except that of promethium as this is a radioactive element. In Table 1.1, the relative weight-% of the rare-earth oxides R_2O_3 contained in *gadolinite* and *cerite* from Bastnäs, Sweden, and *samarskite* from Piława Górna, Poland are given; the contents of rare-earth oxides vary even locally considerably. You may notice that generally the elements with even numbers are more abundant than those with odd numbers. This will be addressed in Chapter 4. With these sometimes very small figures in mind, the discovery history of the rare-earth elements becomes even more impressive.

1.2 A Short History of Discovery

The part of Robert Boyle's phrase on substances, which are made up of chemical elements, is an especially important one during the discovery history of the rare-earth elements. For a better overview, you may consult Figure 1.2. The story started in the small Swedish town of *Ytterby* near Stockholm. There, Arrhenius[5] found a black stone in 1787, later named *gadolinite*. Chemically, this is a quite complicated composition with $R_2FeBe_2O_2(SiO_4)_2$, where mainly the iron atoms cause the dark hue. Seven years later, from this mineral Johan Gadolin extracted *yttria* and so discovered the first *rare earth* [1]. This was a quite challenging task as the black stone is known to be a silicate, and silicates are usually hard to digest and dissolve. Many years later, the rare-earth metal *yttrium* was obtained from yttria. Also in Sweden, near the iron and copper

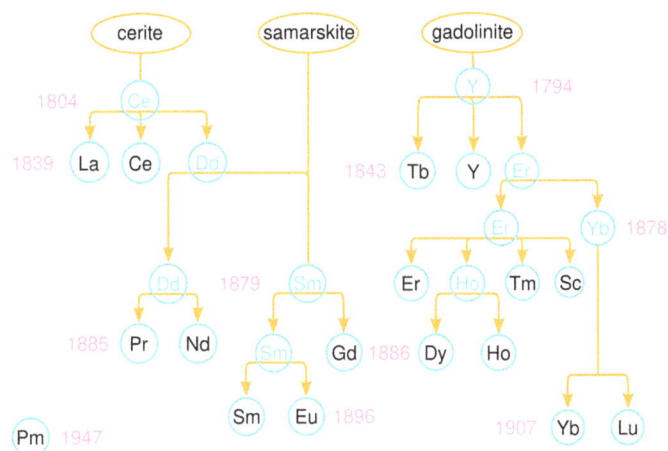

Figure 1.2: Historic development of the discoveries of the rare-earth elements with the important years in red and intermediate steps in blue; Dd = didymium.

---

5 *Carl Axel Arrhenius*, Swedish soldier with an interest in chemistry and mineralogy (*1757 †1824), not to be confused with the chemist and physicist, Svante Arrhenius.

mine *Bastnäs*, on the edge of the 18th century, the discovery of another mineral was reported—*cerite* with a typical composition of $(R,Ca)_9(SiO_4)_3(SiO_3(OH))_4(OH,F)_3$. Its behavior was puzzling and it took decades before Martin Klaproth,[6] Wilhelm Hisinger[7] and Jöns Berzelius[8] independently identified the earth *ceria* and the metal *cerium* in 1804 [8, 9]. Hisinger and Berzelius named the new element after the goddess *Ceres* of agriculture who also was the namesake of the then, recently discovered, dwarf planet.

Another twenty years later, Mosander[9] shed more light on the first two earths, yttria and ceria. He succeeded in isolation of "cerium" metal by reducing its chloride with the aid of gaseous potassium. Further experiments in this way obtained metal samples that revealed different colors of obtained oxides upon oxidation and different weights leading to the conclusion that further metals are hidden here. Careful experiments then yielded a slightly more basic colorless oxide containing *lanthanum* in 1839, named after the Greek term for *hidden*. Still, oxidation of different metal samples yielded slightly different colors leading to the discovery of *didymium*, named after the Greek word for "twin" [10]. In his publication, Mosander described the taste of lanthanum salts as "sweet, slightly astringent." Moreover, didymium oxide should be responsible for the amethyst color hue of the other oxides and that these turn dark brown under red heat. It would take another fifty years to elucidate that didymium comprises in total at least five elements.

Mosander then also reinvestigated yttria and the ominous black stone mineral *gadolinite* closer. After purifying the rare-earth oxide content, he could discriminate three elements with decreasing solubility in aqueous oxalate solutions: first, *erbium* oxalate precipitates, then *terbium* oxalate and finally the *yttrium* compound [10]. After yttrium, also erbium and terbium were named after the location of the black stone, Ytterby. So, until 1843, six rare-earth elements had been identified, some of which still containing further elements as we know today.

To dig deeper in the group of rare-earth elements, new analytical tools were helpful as until then only pure chemical methods were available: dissolution and precipitation behavior, taste, heat treatment as well as color and crystal shape. By the end of the 1850s, spectral analysis was developed. Moreover, the first ideas on the modern periodic table were published by Mendeleev in 1869. Known elements were classified according their atomic weight and many further elements were postulated. Certainly, gaps in the rows of atomic masses suggested missing elements. In the following years, the so far known rare-earth elements were scrutinized, their masses were determined more and more precisely. To obtain cleaner samples, separation methods like the frequently repeated recrystallization and fractional precipitation, were developed. Finally, the existence of

---

6 *Martin Heinrich Klaproth*, German chemist (*1743 †1817).

7 *Wilhelm Hisinger*, Swedish chemist and physicist (*1766 †1852).

8 *Jöns Jacob Berzelius*, Swedish chemist (*1779 †1848).

9 *Carl Gustav Mosander*, Swedish chemist and surgeon (*1797 †1858).

all elements discovered so far was confirmed, but some of the names became mixed up and it is impossible to state today, who actually discovered terbium, for instance. In Table 1.1, I therefore chose the first documented discoveries as long as the existence of these elements was confirmed later.

A chemist participating in many of these controversies was Delafontaine.[10] During his investigations on samples from *samarskite*, i. e., ABO_4 ($A = R$, U, Th, Sn, Fe, Ti; $B = Nb$, Ta), he stated remarkable differences in the spectral properties of didymium with other samples of didymiuim from *cerite* like the absence of certain spectral lines and differing intensities. Lecoq de Boisbaudran[11] repeated Delafontaine's experiments and objected to his findings that the didymium samples differed. Nevertheless, in the fraction containing also didymium he identified unknown spectroscopic lines and separated another element, which could be precipitated as oxalate or sulphate from this fraction. It is now known as *samarium*. He chose the name after the aforementioned mineral *samarskite*.

Meanwhile, Marignac[12] investigated the mineral *gadolinite* further, the black stone. He stated that the earth *erbia* cannot be homogeneous as a colorless portion of the respective chloride solution does not precipitate with sulfoxylic acid, H_2SO_2, while the contained pink ions do. Accordingly, *erbia* could be separated into a colorless oxide— for which no visible absorption could be found—and a pink oxide with the known absorption spectrum of erbium. Because of the similar chemical properties to yttrium and erbium, Marignac named the new element *ytterbium*. This occurred in 1878.

1879 was an extraordinary year for rare-earth chemistry as the hitherto known *erbium* was proven to consist of four elements. The lucky guys conducting once again the very careful separation chemistry were Cleve[13] and Nilson.[14] Nilson found in *gadolinite*—the black stone—another new element. He finally separated a basic nitrate from this mineral and named it *scandium* after his home region, Scandinavia. Scandium was confirmed in the same year by Cleve—he isolated less than a gram of scandium oxide from four kilograms of *gadolinite*. Cleve also again thoroughly investigated the erbium oxide after removal of the ytterbium discovered the year before. Here, he separated the erbium oxide fraction into three fractions—taken during the fractioning process. Based on spectroscopy, he postulated two more elements. Scandinavia acted as namesake already, so he chose the ancient Greek name for the same region, Thule, and named the first element *thulium*. For the second name, he took Sweden's capital Stockholm, and *holmium* was born. The discovery of holmium was also claimed by Swiss chemist, Charles Soret, as element "X" from 1878. Cleve and Lecoq de Boisbaudran agreed that holmium was indeed the same element, and accepted the priority for Soret. Seven years

---

10 *Marc Abraham Delafontaine*, Swiss chemist (*1838 †1911).
11 *Paul-Émile Lecoq de Boisbaudran*, French chemist (*1838 †1912).
12 *Jean Charles Marignac*, Swiss chemist (*1817 †1894).
13 *Per Teodor Cleve*, Swedish chemist (*1840 †1905).
14 *Lars Fredrik Nilson*, Swedish chemist (*1840 †1899).

later, Lecoq de Boisbaudran was successful to prove after an even more elaborate separation process of more than 50 steps—comprising precipitations with ammonia and oxalate—that the assumed element holmium actually contained a second element and named it *dysprosium* after the Greek word for "inaccessible". Although this may sound like frustration, he eventually found a pure element.

By the end of 1879, eleven rare-earth elements were known, perhaps not yet pure. In 1880, Marignac combined several separation methods developed before and applied them to *samarskite* and found a colorless nitrate that behaved quite differently to the others. It took another six years, now based on *gadolinite* samples, until its existence and unequivocal discrimination from the other new elements was confirmed. It was named *gadolinium*.

A new era in rare-earth chemistry started in 1885, when von Welsbach[15] entered the stage. Since the separation of samarium from didymium, doubts persisted that didymium is indeed a pure element. Auer modified the separation processes of *gadolinite* by giving the highly time-consuming fractional crystallization a higher priority. By this approach, he was able to enrich certain fractions of didymium, and finally held two different samples in his hands—different in color, greenish and pink. The element behind the greenish oxide he named *praseodidymium* using the Greek word for "green," the element behind the pink one he named *neodidymium*, i. e., "new twin." Later, however, the shortened names *praseodymium* and *neodymium* were assigned. Auer von Welsbach noticed the luminescence of many rare-earth salts. He combined this finding with the knowledge acquired in Bunsen's laboratory during his studies there. He invented the *incandescent gas mantle* where luminescence of rare-earth oxides is excited by a gas heat source. He set up a business, became rich and invented the brand *Osram*. Most importantly, for rare-earth research, though, was the triggering of further research driven by possible applications.

Still, there were debates whether samarium is already pure or whether there is another hidden element. Demarçay[16] conducted many careful separation experiments by fractional crystallization of nitrates on samarium samples and, finally, in 1896 presented another new nitrate. This was slightly less soluble in water than samarium nitrate, but slightly better soluble than the gadolinium compound. Moreover, it showed different spectra than the other two—and was named *europium* after Europe.

On the brink of the 20th century, Auer von Welsbach spent again more time for rare-earth chemistry and checked the purity of ytterbium with further refined separation processes, now based on the fractional crystallization of rare-earth double oxalates with ammonium. And indeed, there was indication for another element. In 1907, he submitted a manuscript reporting experimental evidence that ytterbium consisted of two

---

15 *Carl Auer von Welsbach*, Austrian chemist and businessman (*1858 †1929).
16 *Eugène-Anatole Demarçay*, French chemist (*1852 †1903).

elements and suggested the nice names *aldebaranium* and *cassiopeium*. Forty days earlier, Urbain[17] had submitted a manuscript reporting the same, but suggesting the names *neoytterbium* and *lutetium*, honoring his home city, Paris, with its latin name Lutetia. Urbain had developed a novel separation process employing the auxiliary element bismuth. Bismuth behaves similar to rare-earth elements and for the crystallization of several salts its solubility lies between those of two adjacent rare-earth elements. Therefore, the separation of the rare-earth elements is enhanced. The dispute on the names continued for several years, and until World War II in German literature the name cassiopeium instead of lutetium was frequently used.

The increasingly thorough understanding of the theory behind the chemical elements and the systematics of the periodic table driven by researchers like Nils Bohr finally solved the important question whether further rare-earth elements exist somewhere in-between the ones known so far. The answer was yes, and it was solely element no. 61, which was missing. First hints that between neodymium and samarium a further element should exist were given in 1902 by Bohuslav Brauner[18] who noted the largest differences of properties along the whole rare-earth series between these two elements. Eventually, it became clear that element 61 only exists in traces in nature as there is no stable isotope for this element. Therefore, it is radioactive (Chapter 4). In 1947, *promethium* was identified during investigations of fission products of uranium in the course of the Manhattan Project [11]. The first two isotopes were $^{147}_{61}$Pm and $^{149}_{61}$Pm with half-times of 3.7 years and 47 hrs, respectively. Presumably, because of its special background, it was named after Prometheus, a titan in Greek mythology, who stole the fire from the Olympian gods to give it to mankind. Zeus was not happy about this incident and punished Prometheus. The titan was tied to a rock, and every single day his liver was eaten by Zeus' eagle—since a new liver grew every night.

The first rare-earth element was discovered following the works in 1787, the last identified in 1947. So, it took 160 years. You may read a more detailed history of all discoveries with even more hot debates on the existence of certain elements, dead ends and some more details on the chemical background of each controversy in [7]. The cited chapter is one of meanwhile 329 chapters in 63 volumes of the Handbook on the Physics and Chemistry of Rare Earths, initiated by Mr. Rare Earth, Karl Gschneidner.[19] In this book, I will occasionally reference to the chapters therein.

And now, it is time to look into the general aspects of rare-earth elements and understand them from their atomic structure leading to physical and chemical properties of the atoms and the rare-earth ions. This is crucial to understand, why certain ions form, why the rare-earth elements behave physically and chemically as they do and why certain rare-earth elements are suited for optical and magnetic properties.

---

17 *Georges Urbain*, French chemist (*1872 †1938).

18 *Bohuslav Brauner*, Bohemian chemist (*1855 †1935).

19 *Karl A. Gschneidner*, American metallurgist and chemist (*1930 †2016).

Part I: **General Aspects of Rare-Earth Elements**

2 Basic Aspects and General Properties

During data collection for this book, I came across a chemistry tutorial, which stated: "The lanthanides and actinides possess more complicated chemistry that does not generally follow any trends. Therefore, noble gases, lanthanides and actinides do not have electronegativity values" [12]. This is an interesting statement which, presumably, will prove wrong in the next chapters.

2.1 The Rare-Earth Elements in the Periodic Table

Within the Periodic Table, the rare-earth elements have either a rather prominent or easy-to-oversee position. Scandium, yttrium and lanthanum are sorted into group 3 as they have in common the valence electron configuration $ns^2 (n-1)d^1$. Moreover, the lanthanide elements are sometimes ignored in teaching as they are physically hidden on some inflatable classroom periodic tables. Certainly, this somewhat hidden position is due to the fact that according to Table 2.1 here the 4f states are subsequently filled and— for reasons to be discussed later—the elements indeed belong also to group 3. Occasionally, the old controverse discussion where to put the lanthanide elements properly is warmed up [13, 14]. Some alternative representations of the Periodic Table, like the one developed by Benfey,[20] displayed in Figure 2.1, give the lanthanide elements the prominent place they actually deserve, though. I like Benfey's idea because it empha-

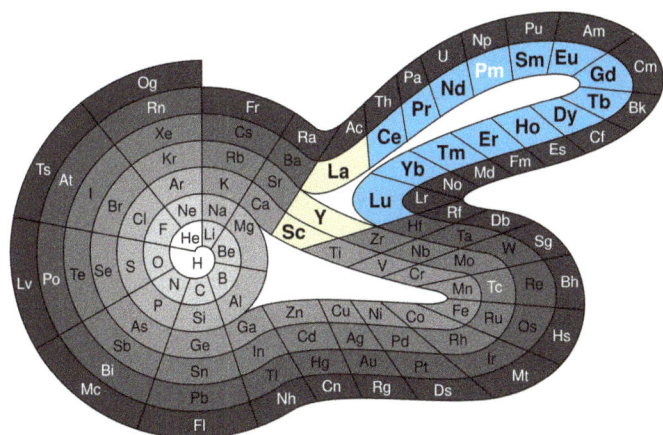

Figure 2.1: A modified representation of the Periodic Table, which was developed by Otto Theodor Benfey; the lanthanide elements are highlighted in blue, and together with the yellow ones, they form the rare-earth elements.

---

20 *Otto Theodor Benfey*, German-British-American chemist and historian (*1925).

https://doi.org/10.1515/9783110680829-002

sizes the periodically evolving elements [15]. I made some changes especially to the borders between the rare-earth elements and the other transition metals as lanthanum and lutetium indeed should share a single place in group 3.

Maybe the most important figure in chemistry is the electronegativity; it is crucial for the basic understanding of the chemical bonding and physical properties of any chemical moiety—like gases, liquids, solids or metals, insulators, semiconductors. Since this central figure will also help us to understand a manifold of specific features found for the lanthanide elements and their ions, a basic discussion will help the readership of this textbook. Regardless of the differing physical definition, electronegativity depends from the screening of the outermost electrons from the nuclear charge. The actually relevant remaining nuclear charge after screening by core electrons is the *effective nuclear charge* targeting the electron under consideration. This screening determines its ionization energy as well as the electron affinity for a further electron to be trapped by a vacant state. An elaborate scrutiny of this concept with a focus up to the outer transition metals may be found in [16]. In the following paragraph, I will employ this concept to derive relevant properties of the rare-earth elements and their ions. For a better understanding of this discussion, you may imagine yourself standing somewhere in a valley, i. e., the electron or vacant state under consideration. Your view will be constricted by more or less transparent hills representing the shielding electron clouds as shown in Figure 2.2. Depending on the problem, your view will be directed toward the nucleus or outward seeking neighboring atoms. The more diffuse the hill becomes, the better you will recognize your target of desire and its charge. In the first part, I will focus the discussion on the view toward the nucleus.

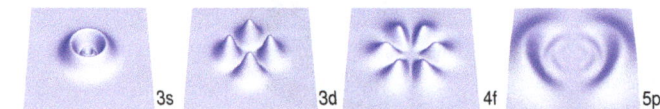

Figure 2.2: Relative radial probability functions of 3s, 3d, 4f and 5p orbitals.

Simply spoken, the higher the effective nuclear charge affects an electron in an atom, the larger its *ionization energy* will be. The ionization energy relates to a valence electron, which is removed. Analogously, one can also consider an electronic vacancy and estimate its attraction for trapping an additional electron present in the energetic continuum. Hence, the *electron affinity* is released. Such vacancies are also called *holes*. Electrons are situated within an atom as electron clouds or *electron probability density* in orbitals of different shape. Electron clouds concentrated closely around the nucleus shield the nuclear charge better than diffuse electron clouds—recall the more or less transparent hill image from above. For this estimation, the diffuseness of both, the screening electron clouds and the screened electrons, have to be considered. The more diffuse an orbital is the less efficient the screening effect of electrons therein will be.

Let us now illustrate what determines this diffuseness. First, it is governed by the orbital type (azimuthal quantum number l), and, second, by the principle quantum number n. Each orbital features l *angular nodes*, i. e., planar nodes containing the nucleus. The shielding efficiency declines with an increasing number of angular nodes as s orbitals have zero, p orbitals one, d orbitals two and f orbitals three of them. This is well illustrated in Figure 2.2 since every angular node cuts two valleys in the shielding electron probability density and enables a better view through. Considering the same orbital type, its shielding effect decreases with rising quantum number n as each additional shell provides a further *radial node*.[21] Thus, the electron clouds become more and more diffuse as depicted in Figures 2.2 and 2.3.

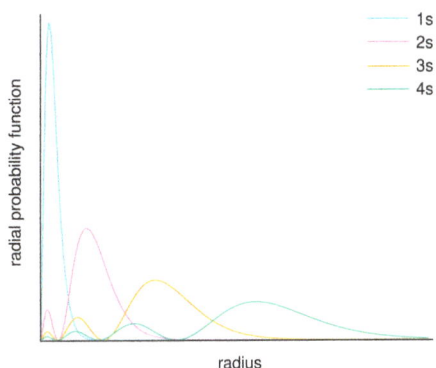

Figure 2.3: The radial probability density of selected s orbitals illustrating the increasing number of radial nodes.

Accordingly, the diffuseness increases with the number of radial nodes. The 1s orbital being the very first occurring s orbital lacks any node while with each subsequent shell an additional radial node emerges. Each shifts the orbital cloud outwards with the relatively strongest effect upon introduction of the first radial node between 1s and 2s; this is reflected in the strikingly different properties of the isoelectronic elements hydrogen (insulator) and lithium (metal). Insulators are characterized by well-localized electron clouds while conduction requires diffuse orbitals with long-range interactions. This basic finding holds for all orbital types, for 2p and 3p, for 3d and 4d, and of course also for the 4f states, which are the first occurring f orbitals comprising no radial node at all and, therefore, being relatively localized (nondiffuse).

Further, the screening of electrons by those comprising the same spin is worse compared with those of opposite spin caused by the Pauli rule [17, Chapter 2.9]. For instance, this explains why half and fully occupied degenerated orbitals are of particular stability. First, such half-occupied orbitals like $4f^7$ feature a symmetric charge distribution. Second, these electrons screen the eighth electron very well as its spin is opposite. And finally, the seventh electron is shielded worse so its ionization energy is increased.

---

21 spherical nodes around the nucleus.

Employing this approach, the octet rule or the relatively low ligand-field splitting of 3d orbitals with respect to 4d are concluded. 4d and 5d orbitals contain one and two radial nodes, respectively. Thus, the 4d and even more the 5d states are significantly more diffuse than the 3d ones. The interaction with approaching ligands will be stronger with the more diffuse orbitals causing a larger ligand-field splitting (see also Chapter 8.5). For this latter example, you will turn and look outwards recalling the hill image from above.

Focusing on the lanthanide elements, we ought consider the 4f states first. These are relatively contracted as they do not contain a radial node. Moreover, these elements possess filled $5s^2p^6$ states (Figure 2.2) and in the case of the nonionized form $6\,s^2$ shells (Figure 2.4). These aspects yield two main consequences. 4f states are badly screened against the nuclear charge and they are well shielded by the $5s^2p^6$ shell against any neighboring atom—ligand or whatever. Since they experience a quite strong effective nuclear charge, their ionization energies are high, and consequently the 4f states are only rarely ionized. Viewing outwards, the excellent shielding against adjacent atoms prevents them from strong interactions with these leading to a specific behavior regarding their optical and magnetic properties; this will be discussed in Chapters 7.8 and 9.

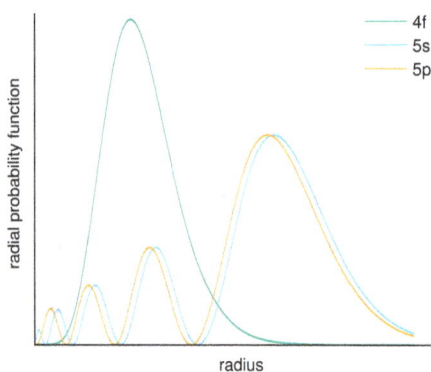

Figure 2.4: The radial probability densities of 4f orbitals versus the well against neighboring atoms shielding $5s^2p^6$ shell in Eu^{3+}.

Based on Allen's[22] approach of averaged valence electron energies, the electronegativities of the elements hydrogen through curium (Cm, no. 96) were calculated in a slightly refined manner [18–20]. The advantage of this approach is that the majority of the necessary data can actually be measured, so that the resulting electronegativity scale is based on experimental data. This does not hold for the 4f and 5f elements where calculations were still necessary; unfortunately, these calculations cannot be exact due to spin-orbit coupling effects, which could not treated thoroughly. Thus, these values show reliable qualitative trends only. In [20], a detailed discussion where to discriminate between *valence* and *core* electrons is conducted; with regard to the lanthanide

---

22 *Leland Cullen Allen*, U.S. chemist and physicist (*1926 †2012).

elements, valence electrons are considered to be found in the outermost 6s, 5d and 4f orbitals. The obtained average valence electron energies are listed in Table 2.1, together with the scaled values to become comparable to Pauling's[23] scale, which will be used for the discussion, and they show a clear trend. From scandium to promethium, the electronegativities range between 1.0 and 1.2, but rise strongly toward gadolinium, where a maximum of 2.3 is achieved. Then, between terbium and erbium, again more or less the same values between 1.3 and 1.4 are found before the electronegativities steeply rise toward 1.7 for ytterbium; lutetium then again plays in the league of the first half. A further discussion follows in Chapter 3.5 where the electronegativities are also depicted in Figure 3.7.

Table 2.1: Electronic configurations of the gaseous rare-earth atoms and the most relevant ions; consider the underlying noble gas configurations in parentheses; recently published electronegativities [20] as configurational energies / eV (RZH) and scaled to the *Pauling* scale (RZH, P).

name	gaseous atom	Ln^{2+}	Ln^{3+}	Ln^{4+}	χ_{RZH}	$\chi_{RZH, P}$
scandium	$3d^1 4s^2$		[Ar]		7.0	1.2
yttrium	$4d^1 5s^2$		[Kr]		6.3	1.1
lanthanum	$5d^1 6s^2$		[Xe]		6.0	1.0
cerium	$4f^1 5d^1 6s^2$		$4f^1$	[Xe]	7.3	1.2
praseodymium	$4f^3 6s^2$		$4f^2$	$4f^1$	6.7	1.1
neodymium	$4f^4 6s^2$		$4f^3$		7.2	1.2
promethium	$4f^5 6s^2$		$4f^4$		7.4	1.2
samarium	$4f^6 6s^2$	$4f^6$	$4f^5$		8.3	1.4
europium	$4f^7 6s^2$	$4f^7$	$4f^6$		9.4	1.6
gadolinium	$4f^7 5d^1 6s^2$		$4f^7$		13.8	2.3
terbium	$4f^9 6s^2$		$4f^8$	$4f^7$	7.7	1.3
dysprosium	$4f^{10} 6s^2$		$4f^9$		8.4	1.4
holmium	$4f^{11} 6s^2$		$4f^{10}$		8.3	1.4
erbium	$4f^{12} 6s^2$		$4f^{11}$		7.6	1.3
thulium	$4f^{13} 6s^2$	$4f^{13}$	$4f^{12}$		9.0	1.5
ytterbium	$4f^{14} 6s^2$	$4f^{14}$	$4f^{13}$		10.2	1.7
lutetium	$4f^{14} 5d^1 6s^2$		$4f^{14}$		6.4	1.1

A further famous consequence of the screening model is based on the scenario that the nuclear charge increases by fourteen units between lanthanum and lutetium while the fourteen respective electrons populate rather badly screening 4f orbitals (Figure 2.2). Obviously, this causes a contraction of the next shells' orbitals, and thus the whole atoms,

---

23 *Linus Pauling*, American chemist, Nobel Prize in Chemistry 1954 and Nobel Peace Prize 1962 (*1901 †1994).

respectively. This consequence is commonly known as *lanthanide contraction* and is accordingly responsible for the increasing densities observed from scandium via yttrium to lanthanum and lutetium (Chapter 3).

2.2 Electronic Configuration

The electronic configurations of the rare-earth elements and their most relevant ions are depicted in Table 2.1. The latter ones are those present in nature and in relevant materials. Here, I add a special remark on *promethium* for which nature did not find a physically stable combination of protons and neutrons; it therefore only forms radioactive isotopes—nonetheless, it is also present in the table as it is occurring naturally in traces (Chapter 4). The gaseous lanthanide element atoms adopt the general configuration $[Xe]4f^{n-1}5d^16s^2$ starting from cerium ($n = 1$).

The configurations given for the gaseous atoms suggest a possibly stable divalent oxidation state for most of the lanthanide elements according to a configuration $[Xe]4f^{n-1}5d^1$ after the ionization of the 6s electrons. But certainly the diffuseness of the 5d states and their accordingly very broad bands in the metals' band structure, well overlaying the narrow 4f bands, should be considered. Then it is obvious that the third electron exhibits most likely 5d character in the solid, which determines the chemical properties. Hence, this third electron is also quite easily ionized leading to the most stable oxidation state R^{3+}. Dorenbos analyzed the lanthanide states' energies with regard to different oxidation states and his findings confirm these arguments [21].

3 General Trends of Physical and Chemical Behavior

In this chapter, I will discuss basic general trends within the series of rare-earth elements. Several properties develop rather continuously with distinct exceptions. After starting with basic data such as the atomic radii, ionic radii and the densities, we will shed light on melting and boiling points as well as the structure chemistry of the elements. Here, a short story is included on cerium as an example for the complexity of the pure elements, which cannot be thoroughly discussed in this book. Then we will slowly turn to more chemistry related properties like electronegativities and chemical hardness before looking at the chemistry in solution and oxidation states. A discussion on simple complexes will complete this chapter.

3.1 Atomic, Ionic and van der Waals Radii

Figure 3.1 shows the evolving van der Waals and atomic radii as well as the ionic radii of eightfold coordinated trivalent rare-earth ions. Certainly, scandium and yttrium comprise fewer electronic shells and, therefore, cause an almost linear increase for all presented radii. Apart from that, the *atomic radii* decrease smoothly with an increasing nuclear charge, and due to the quite ineffective shielding of the consecutively added 4f electrons. For europium and ytterbium two maxima are obvious. In these cases semi and fully occupied 4f configurations shield especially the outermost 6s electrons somewhat better than the others due to the symmetric charge distribution. The improved shielding yields a relative expansion of both outer shells, and thus larger atomic radii.

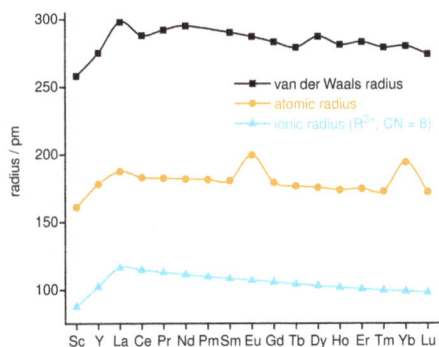

Figure 3.1: Van der Waals (black), atomic (orange) and ionic radii (blue) of the rare-earth atoms, the ionic radii represent those of eightfold coordinated trivalent ions.

In the case of the *ionic radii* of the trivalent ions, a smoothly decreasing trend is observed well in accordance with the above mentioned increase of the nuclear charge. Since the 6s states are empty here, the differing shielding of the 4f states has no impact. From a chemical point of view, it should be noted that trivalent yttrium ions are approx-

https://doi.org/10.1515/9783110680829-003

imately as large as holmium and erbium, and accordingly behave similarly like these regarding structures and chemistry. Interestingly, a similar trend is observed for the *van der Waals radii*, which have been determined by interatomic distances of compounds disregarding oxidation states and coordination numbers [22]. In the case of rare-earth elements, predominantly trivalent ions R^{3+} are present and these accordingly dominate the data analysis leading to the plotted behavior. You may find an overview on the ionic radii of R^{3+}, but also on the less often found R^{2+} and R^{4+}, in Appendix A.

3.2 Densities, Melting and Boiling Points

The densities of the rare-earth elements evolve similarly like the melting and boiling points, plotted together in Figure 3.2. As one may expect from the atomic radii, minima of the densities are achieved where atomic radii are maximal, i. e., the case for europium and ytterbium. The increase of densities from lanthanum to lutetium is mainly due to the lanthanide contraction as discussed in Chapter 2.1.

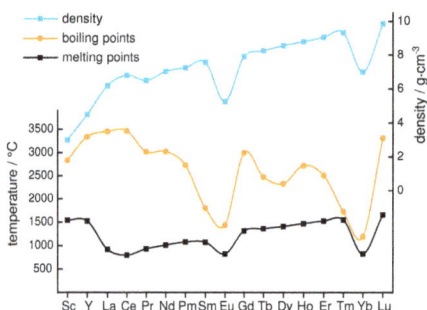

Figure 3.2: Melting points (black), boiling points (orange) and densities (blue) of the rare-earth atoms.

The melting points should be roughly a function of the bond strength within the element. According to Figure 3.2, the melting points smoothly increase with atomic mass and decreasing atomic radii within the row of lanthanide elements. Scandium and yttrium lack 4f electrons, and apparently feature a stronger interatomic bonding leading to similar melting points like lutetium, which in turn has a filled 4f shell and profits from the full lanthanide contraction. For europium and ytterbium, two minima are obvious in agreement with their larger atomic radii—due to a semi and a fully occupied 4f shell—and lower densities leading to slightly weaker interatomic interactions, and thus lower melting points. A more or less similar trend is found for the boiling points of the elements.

3.3 Crystal Structures of the Elements

The great advantage of crystal structures of metals throughout the Periodic Table is their simplicity as they derive from simple close packings of spherical atoms with only a few exceptions. In close packed layers—similar to oranges packed in a box—each atom is surrounded by six others leading to a hexagonal pattern. Adjacent close packed layers B follow in a way that its atoms reside above the voids of the first layer A. If the atoms of the adjacent layers show an eclipsed arrangement—layer sequence BAB—layer A is considered hexagonal (h). If they show a staggered arrangement according to a layer sequence BAC, layer A is considered cubic (c). Thus, two main typical metal structures are obtained: the cubic close packing only containing c layers comprising a layer sequence of ...ABC... (c) and the hexagonal close packing exclusively containing h layers with a general layer sequence of ...AB... (h). Moreover, primitively packed tetragonal layers are known yielding the slightly less close packed body-centered cubic arrangement of atoms (bcc).

The rare-earth metals show an easy solubility of hydrogen of up to approximately 21 at-%, which will be the subject of Chapter 6.1 on hydrides; this fact led to slightly differing structure determinations in early studies on the crystal structures of the elements as hydrogen had been overseen occasionally [27, 28]. This is a very important point that researchers on rare-earth elements should have in mind during syntheses, and especially during physical property measurements. Traces of hydrogen contaminations may yield astonishingly sensational results of the "elements" or their compounds.

Around room temperature the second half of the elements starting from gadolinium adopts a simple hexagonal close packing shown as structure **c** in Figure 3.3; the only exception here is ytterbium comprising a cubic close arrangement of atoms, depicted as **b**. In the first-half, different stacking orders of close packed layers are found: lanthanum, praseodymium, neodymium and promethium show a layer sequence ...ABAC... (hc), a so-called double hexagonal close packing displayed as structure **a** in Figure 3.3. Topologically, closely related to this arrangement is the unique crystal structure of samarium, in which the layer sequence ...ABABCBCAC... (hhc) is found, yielding the rhombohedral structure **d**. Finally, for cerium a cubic close packing (structure **b**) and for europium the only rare-earth element adopting a body centered cubic structure was reported at room temperature (structure **e**) [30]. The d shell elements scandium and yttrium adopt a simple hexagonal close packing h alike gadolinium.

In some cases, there are also low-, and in many cases, high-temperature phases as depicted in Figure 3.4. The transitions of terbium and dysprosium to orthorhombically distorted structures at low temperatures coincide with ferromagnetic ordering at the same temperatures; so these phase transitions are apparently magnetically driven ones as roughly discussed in Chapter 9.4. Cerium shows a transition from a cubic close packing c around −150 °C to the double hexagonal close packing hc and back around 50 °C; above 725 °C, a further transition to the body centered cubic packing (bcc) occurs. Since this latter packing features a lower density than close packings, it seems reasonable that

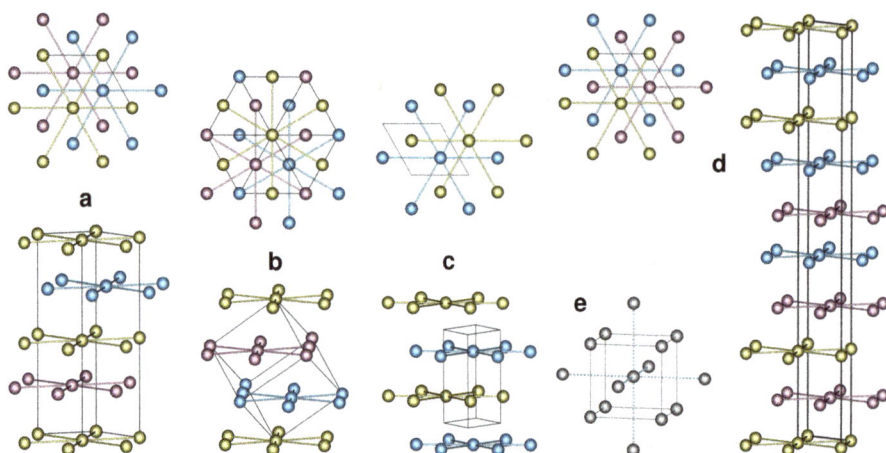

Figure 3.3: Crystal structures of the rare-earth elements; hc (**a**), c (**b**), h (**c**), body centered cubic (**e**) and hhc (Sm-type, **d**); for **a–d**, the layers are indicated by differently colored atoms and additionally shown from top; data: [23–26].

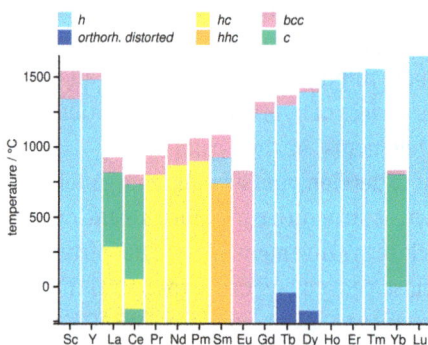

Figure 3.4: Crystal structures of the rare-earth elements at ambient pressure up to the melting points; data: [29–32].

most of the rare-earth elements adopt this structure type just below the respective melting point. Within the lanthanides, europium shows the lowest density, and throughout the temperature range, only the *bcc* type structure. The behavior of cerium is somewhat puzzling and behind these simple data more exciting details are hidden. Therefore, the properties of cerium metal will be subject of a short story in the following section. Here, I omit the also reported high-pressure phases in detail as these are beyond the scope of this book.

Watching out for a general trend between lanthanum and lutetium, we omit europium and ytterbium who always behave differently. As can be seen from Figure 3.4, the structures follow roughly a trend from *c* structures via *hc* and *hhc* (Sm-type) to *h*, apparently an increasing contribution of *h* layers. This apparent behavior is of interest because it helps to understand why these different crystal structures are realized. As mentioned earlier, an increasing nuclear charge leads to a contraction of the orbitals (Chapter 2.1). Accordingly, with growing nuclear charge the 4f states contract relatively stronger with respect to the 5d and 6s states. Calculations of the band-structure of gadolinium confirm an expected dominating role of broad 5d and 6s bands around the Fermi level, where the structural music plays; but these 5d states are apparently strongly influenced by the local 4f states [33]. This influence coincides with the energetic overlap between the exchanging states, and thus a roughly decreasing influence of 4f is to be expected going from left to right in the lanthanide series.

As mentioned above, looking at the crystal structures from left to right a relatively increasing contribution of h layers is observed. In the case of lanthanum, cerium and ytterbium the same can be deduced from the structures with decreasing temperature. Beyond the cubic close packing structures, the body-centered cubic ones are detected, apparently yielding an even lower exchange with 4f states as therein slightly larger interatomic distances are realized. Considering europium, only showing the body-centered cubic structure, this meets with the lowest density (Figure 3.2); moreover, calculations show a formal oxidation state transition within europium metal from a divalent ($Eu^{2+} \cdot 2\,e$) to a trivalent one ($Eu^{3+} \cdot 3\,e$) with increasing pressure—so, at high pressures a behavior comparable to the other rare-earth metals adopting even a hexagonal close packing can be expected, which is indeed the case [34, 25]. The phrasing $Eu^{2+} \cdot 2\,e$ might be new for many readers, so I add a few words here. In lithium metal, atomic cores Li^+ comprising a noble gas configuration host highly mobile *itinerant* electrons. Hence, lithium metal could also be described as $Li^+ \cdot e$. As the 4f orbitals are screened very well against neighboring atoms, they also form distinct bands with weak interactions with other valence orbitals like 5d or 6s within the metal. Therefore, the 4f configuration contributes rather to the core of the rare-earth atoms than to the itinerant electrons. Thus, in rare-earth elements the 4f configuration determines the number of itinerant electrons alike in $Eu^{2+} \cdot 2\,e$ with a core configuration $[Xe]4f^7$ and $Eu^{3+} \cdot 3\,e$ with a configuration of the core of $[Xe]4f^6$. We will see the consequences of such different core configurations in the following section.

3.4 On Cerium—a Short Story

It is impossible to portray all the interesting phases and fascinating properties coming along for the rare-earth elements in a single textbook. In this chapter, I will describe some of the features of cerium metal and you may then infer the diversity found throughout the other elements. Cerium is at least one of the most versatile metals of the Periodic Table if not the most versatile one. On a first glance, the schematic phase diagram depicted in Figure 3.5 looks quite simple but the story behind is puzzling and exciting at the same time. It certainly also illustrates some very basic characteristics of the behavior of rare-earth elements.

The shown phase diagram comprises seven solid phases and—this is unique among metals—a critical point within the solid regime! Further data on these phases are listed

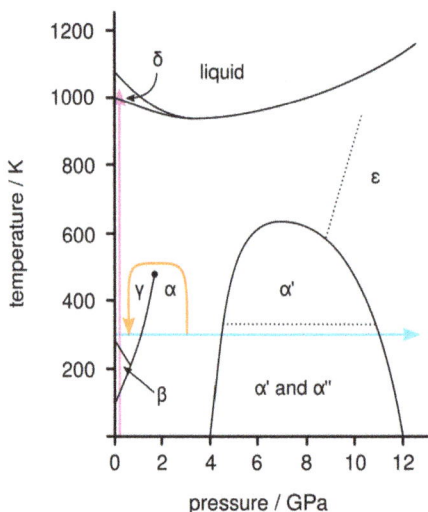

Figure 3.5: Schematic temperature-pressure phase diagram of cerium metal based on data in [31, 35–37]; the arrows relate to the discussion in the text.

in Table 3.1. Generally, it is often challenging to synthesize the pure phases because of strong hysteresis effects with delayed phase transitions. This holds especially for the martensitic transitions,[24] such as that of the γ into the β-phase, where layers of the structure are sheared with respect to the adjacent ones. Thus, tensions in the material are induced inhibiting further transition. Hence, it took several decades to achieve the knowledge we have today about cerium's phase diagram—and there is still discussion on some points like the dotted lines in Figure 3.5.

With increasing pressure following the horizontal blue arrow at room temperature after the γ-phase (c packing) follows the α-phase, also c packing. Further pressure increase yields the α'- (distorted h) and α''-phases (distorted c); finally, around 12 GPa the tetragonal body-centered ε-phase is obtained. In Figure 3.6, the structures of this sequence are depicted. As expected, the density of the phases rises with increasing pressure.

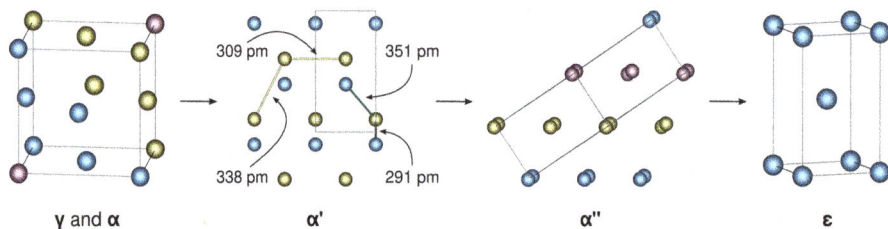

Figure 3.6: Structures observed along the blue arrow in Figure 3.5; all atoms shown are cerium. The colors indicate different layers of close packings; data: [38–40].

---

24 diffusionless, athermic transition which normally shows significant hysteresis.

Table 3.1: Solid phases of cerium metal up to 12 GPa and 1000 °C; c cubic close, h hexagonal close and bcc body centered cubic packing; given are also the densities and suggested 4f occupations at respective temperatures or at the critical point; data: [31, 35–37].

phase	structure	density / g/cm^3	$4f^n$
α	c	8.29 (r. t.)	0.33 (116 K), 0.74 (critical point)
α'	orthorhombically distorted h (α-U type)	9.55 (r. t.)	0
α''	monoclinically distorted c	9.81 (r. t.)	0
β	hhc	6.75 (r. t.)	0.96
γ	c	6.77 (r. t.)	0.94 (298 K), 0.74 (critical point)
δ	bcc	6.65 (1000 K)	0.94
ε	tetragonally distorted bcc	11.29 (r. t.)	0

Following the vertical red arrow with increasing temperature at ambient pressure first the α-, then the β-, the γ- and finally above 1000 K the δ-phase (bcc) with the lowest density of all phases are observed (Table 3.1). The body-centered cubic structure (bcc) is a comparably open structure with relatively large voids, and thus a lower volume efficiency compared with close packed ones as an increasing temperature fosters the average atomic movement. They occur frequently as high temperature phases of outer transition metals. With growing occupation of d states, the bcc packing becomes increasingly stable and is hence also found as high temperature phases from lanthanum through dysprosium, ytterbium, scandium and ytterbium.

The observations along the yellow arrow in Figure 3.5 are puzzling on the first glance as the structure remains a cubic close packing (c) and only the density declines until the turning point around 600 K. At the critical point, both α- and γ-phase are indistinguishable. But then within the γ-phase regime, the density declines significantly less and the density difference relative to the α-phase grows to approximately one-fifth approaching room temperature. Thus, the α- and γ-phase adopt the same structure but exhibit strikingly different densities. Calculations show that the energies of relatively localized 4f and delocalized 5d states are rather similar. Consequently, small shifts of relative stabilities lead to dramatically different electronic properties. For instance, upon pressure a contraction of a compound is promoted and indeed the density increases from the γ- to α-phase. Such a contraction easily occurs if the single f electron of the more or less $4f^1$ system in the γ-phase moves into the 5d states and becomes delocalized. The same effect occurs upon heating of the γ-phase when the f electron density is smoothly excited thermally into the 5d band from room temperature ($4f^{0.94}$) toward the critical point where $4f^{0.74}$ is achieved. Picking up the discussion at the end of the last section, this could also be formulated as $Ce^{3.26+} \cdot 3.26e^-$.

On the other side, the α-phase starts from $4f^{0.33}$ at low temperature and smoothly approaches the same value like the γ-phase with increasing temperature. Here, the thermal expansion apparently stabilizes the 4f states with respect to the 5d band. Nevertheless, this specific γ- to α-phase transition is still the subject of intense debates regarding the physical basics and reasons behind this behavior as some experimental results contradict this explanation [37, 41].

On cerium, a *Kondo effect*[25] behavior was reported. This effect describes the scattering of itinerant electrons in a metal by localized magnetic impurities. This may also be understood as a partial hybridization of the localized 4f states with itinerant electrons. Then with decreasing temperature an anomalous increase of the resistivity is observed, which collapses upon magnetic ordering. The magnetic impurities here are cerium atoms with localized f electrons. This Kondo behavior is one of the alternative explanations for the strange γ- to α-phase transition.

Interestingly, the β-phase shows two antiferromagnetic ordering temperatures at 12.5 and 13.7 K. This can be understood by the crystal structure of the *hhc* packing where cerium atoms in the *h* layers order earlier than those of the *c* layer. Below the Néel temperature, T_N = 14.4 K, the γ-phase orders antiferromagnetically. The on first glance small difference in the layer sequence yields a different magnetic ordering behavior. According to the phase diagram shown in Figure 3.5, at this temperature, the α-phase is the thermodynamically stable phase. But both the γ- as well as the β-phase, are metastable since the transition to the α-phase is impeded by tensions in the metal due to the volume contraction of more than 15 %.

At pressures between 2 and 4 GPa, even superconductivity with T_c ≈ 2 K was observed. According to Figure 3.5, either the α'- or the α''-phase should be responsible for this effect. Careful experiments and calculations suggest that indeed the orthorhombic α'-phase is the phase of choice as only here a sufficiently strong electron-phonon coupling was identified [42]. In this context, it is worth to mention that superconductivity up to a considerable T_c ≈ 115 K was predicted and experimentally confirmed for a so-called cerium superhydride, i. e., CeH_{10}, at a pressure of 95 GPa—but this is another story [43]. So, in my opinion, cerium is indeed an exciting metal, which nicely demonstrates that behind simple data there can be an intricate network of phase relationships and their properties puzzling physicists to date.

3.5 Ionization Behavior and Chemical Hardness

In Figure 3.7, the ionization energies of the rare-earth elements are displayed. The complete table of data may be consulted in Appendix B. Also, in this diagram, we find the expected trends similar to those discussed in the previous chapters. Almost constant figures for the first and second ionization energies, which remove both electrons from the outermost ns shell, are obvious. The third electron, predominantly present in the $(n-1)d$ states, is subject to a different shielding by core electrons. While 3d orbitals—not comprising any radial node—face a quite strong effective nuclear charge, 4d and 5d

---

25 *Jun Kondo*, Japanese theoretical physicist (*1930 †2022).

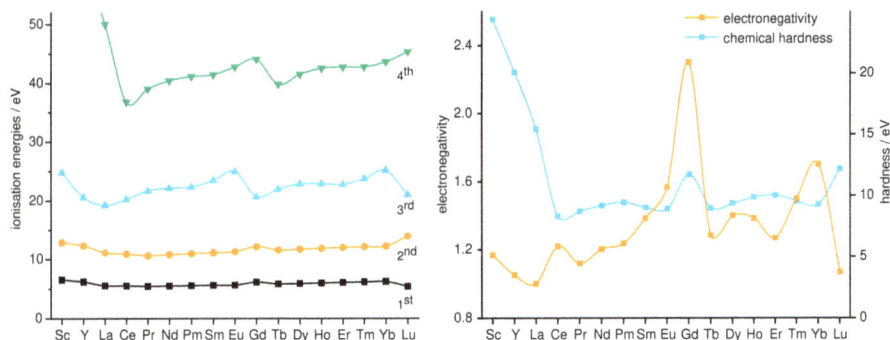

Figure 3.7: The first, second, third and fourth ionization energies of the rare-earth elements; electronega-tivities of the elements (orange) and chemical hardness of the R^{3+} (blue, given in eV); data: [20, 29].

orbitals—comprising one and two radial nodes—are better shielded. Accordingly, the ionization energies decrease along the row scandium, yttrium and lanthanum. Within the lanthanide series, the shielding declines with increasing number of 4f electrons crest two maxima for Eu^{2+} and Yb^{2+} where semifilled and fully occupied 4f states are achieved (Table 2.1). The same reason is responsible for the relatively low third ionization energies found for gadolinium and lutetium, respectively. In the case of the fourth ionization energy, the relatively higher stability of empty (Ce^{4+}) and semifilled 4f states (Tb^{4+}) is reflected in the found minima.

To estimate, understand and explain chemical behavior, the concept of *hard and soft acids and bases* (HSAB concept) has been proven helpful in many cases. The chemical hardness may be understood as the charge density of a chemical moiety and it increases with charge and decreasing size. It essentially uses the moiety's polarizability to estimate whether two particles behave similarly, and if these prefer either covalent or electrostatic interactions. If they feature a high polarizability they are soft, and they are hard if the polarizability is low. Pearson[26] proposed that the chemical or *absolute* hardness η can be quantified by the HOMO-LUMO gap[27] according to

$$\eta = \frac{I - A}{2} \tag{3.1}$$

using the ionization energy I and electron affinity A. In the case of monoatomic trivalent ions of the rare-earth series, the fourth and third ionization energies may be used in good approximation for I and A, respectively [44, 45]. The more diffuse an orbital is the larger the difference between ionization energy (I) and electron affinity (A) would be for the same electron

---

26 *Ralph G. Pearson*, American theoretical chemist (*1919 †2022).

27 energy between the highest occupied molecular orbital (HOMO) and the lowest unoccupied molecular orbital (LUMO).

$$M \xrightarrow{\;I\;} M^+ + e^- \xrightarrow{\;A\;} M \qquad\qquad (3.2)$$

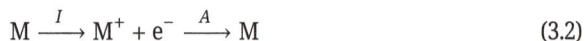

Considering the Ln^{3+} ions, this difference can be neglected in good approximation. Employing this approach, the hardness values displayed in Figure 3.7 (right, blue squares) and listed in Appendix B were calculated. Considering the increasing ionic radius from scandium via yttrium to lanthanum, a decreasing hardness is reasonable. Within the Ln^{3+} series, almost constant figures indicate very similar behavior, only Gd^{3+} and Lu^{3+} are slightly harder, presumably due to the symmetric charge distribution typical for semi and fully filled orbitals. Such minor gradual differences make a separation of a mixture of such ions very difficult and, therefore, sophisticated techniques had to be developed to get the pure elements from natural ores, which will be discussed in Chapter 5. It also confirms the great challenges the discoverer of rare-earth elements were facing (Chapter 1.2).

Chemical bonding properties may be estimated in a straightforward way by looking at the electronegativities of the involved elements (Figure 3.7). Unfortunately, neither Pauling nor Allen provided them for the complete series of rare-earth elements, and thus only estimated values were listed in textbooks [46, 19]. Here, I shall use the recently published values listed in Table 2.1 [20]. Since electronegativity can be directly connected to the ionization energies of the considered chemical species as done by Allen, it is not further surprising that the shape of the evolving orange curve in Figure 3.7 resembles features of the ionization energies' development.

On average, the electronegativities of the rare-earth elements are lower than those of the outer transition metals, the d-block elements. The 4f states are localized significantly closer to the nuclei, and accordingly show higher electron binding energies yielding higher electronegativities for the lanthanide elements compared to the remaining rare-earth elements without 4f valence electrons. Among scandium, yttrium, lanthanum and lutetium, scandium has the highest electronegativity as there 3d states contribute, lacking radial nodes. The lanthanide series can then be halved; between cerium and gadolinium as well as from terbium to ytterbium the electronegativities increase smoothly. The elements of the second half are somewhat more electronegative due to the higher nuclear charges and smaller radii. Gadolinium shows an exceptionally high electronegativity due to its highly symmetric charge distribution with a half-filled, and thus contracted 4f shell yielding relatively high electronic binding energies. This holds also for ytterbium with its fully occupied 4f shell. Apart from that, the elements frequently adopting the divalent state, i. e., samarium, europium, thulium and ytterbium are comparable to the platinum elements regarding their electronegativities.

3.6 Oxidation States—Colors in an Aqueous Solution

According to the previously scrutinized screening behavior of core electrons, one expects that the outermost two 6s electrons will be ionized quite easily. Moreover, this also holds for the third one, normally localized in the relatively diffuse 5d states. Thus, the trivalent state is the most common oxidation state of all rare-earth elements.

Nevertheless, there are a few exceptions from the trivalent state in the cases where half or fully occupied 4f states are present. These are of specific stability due to the highly symmetric charge distribution even better shielding the next electron. In the case of a half-filled shell, this next electron features opposite spin and is screened even better. Accordingly, Eu^{2+} and Tb^{4+} may be obtained under suited conditions. A similar situation is found for trivalent cerium where only the last remaining 4f electron has to be ionized to achieve the xenon noble gas configuration; further, ytterbium can achieve a fully occupied 4f shell in its divalent state. Thus, Yb^{2+} and Ce^{4+} are also quite stable. Additionally, the nuclear charge is lower and the screening thereof is better in the case of Ce^{3+} compared with praseodymium and terbium, so the fourth ionization energy is higher in the latter cases. The general trend, that 4f electrons are rarely ionized, is contrary to the early actinoids where the 5f states are rather easily further ionized due to the presence of the first radial node causing a clearly higher diffuseness as discussed in Chapter 2.1. The relative stability of different oxidation states in aqueous media is reflected in the reduction potentials of the respective elements. Figure 3.8 displays the *Frost diagram*[28] of the rare-earth elements including the alkaline-earth metal calcium for comparison. All reduction potentials are listed in Table 3.3.

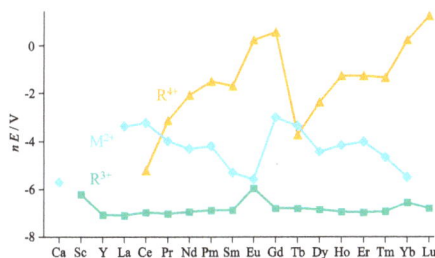

Figure 3.8: The Frost diagram[28] of the rare-earth elements and calcium; the pure elements are at 0 V.

As discussed thoroughly in Chapter 8, most of the optical transitions occur between 4f terms and are accordingly more or less forbidden. Consequently, the colors of such ions in a solution given in Table 3.2 remain pale. Only a few ions show intense colors due to 5d-4f transitions like the Ln^{2+} or due to charge-transfer transitions like Ce^{4+}.

For the trivalent ions of holmium and neodymium, more than one color is given, which depends from the illumination source as discussed in Chapter 8.4.6 and sometimes from the chemical surrounding. Occasionally, Eu^{3+} already under daylight shows red emission, which mixes with the otherwise colorless impression to a pinkish hue.

---

28 Frost diagrams show the relative stabilities of oxidation states based on the free enthalpy ΔG, the Faraday constant F, and the reduction potentials E^0 multiplied with the number of transferred electrons n: $\frac{\Delta G}{F} = n \cdot E^0$.

Table 3.2: Colors of the chemically most relevant lanthanide ions in aqueous solution and within solids with a large band-gap under daylight.

name	Ln^{2+}	Ln^{3+}	Ln^{4+}
lanthanum		colorless	
cerium		colorless	orange
praseodymium		yellow/green	
neodymium		violet, blue	
promethium		pink, violet	
samarium	orange-yellow/brown-red	pale yellow	
europium	colorless/yellow	colorless, pinkish	
gadolinium		colorless	
terbium		colorless	
dysprosium		pale yellow	
holmium		yellow, pink	
erbium		pink	
thulium	red	pale green	
ytterbium	yellow/green	colorless	
lutetium		colorless	

3.7 General Reactivity

The chemical behavior of the most reactive rare-earth element europium is akin to that of the alkaline-earth metals calcium, strontium and barium. This holds already for the stability of the hydrides. The rare-earth elements react readily with hydrogen to form hydrides, slowly at room temperature, readily above 400 °C (see also Chapter 6.1). These are among the most stable hydrides of all transition metals. For the careful investigation of metals' properties it is therefore essential to exclude hydrogen contaminations. The general reactivity declines roughly in the row europium, cerium, lanthanum, neodymium, praseodymium and then samarium through lutetium. In this section, the reactivity against the halogenes, oxygen and air is described; for the reactivity against other elements and for the formation of alloys, [47] is suggested.

Also, against halogenes the reactivity of the rare-earth elements is only somewhat lower compared with the alkaline-earth elements by reacting smoothly at room temperature but vigorously with increasing temperature. The most stable binary compounds among the halides are the trifluorides. Especially calcium is clearly more attractive for fluoride, which is nicely reflected in the fact that the remarkably stable *fluorite*, CaF_2, forms upon reaction of rare-earth fluorides with calcium metal. This provides an easy metallothermic process to obtain rare-earth metals; only samarium, europium and ytterbium cannot be achieved this way as these are too volatile to separate surplus calcium by distillation under vacuum. The fluorides are so stable due to their higher chemical hardness compared with oxide that the fluorides may be even obtained from the rare-earth oxides by reacting these with hydrogen fluoride above 600 °C. Moreover, the fluorides show a limited solubility in aqueous solutions for the same reason. More details like the synthesis of the pure binary halides will be a topic of Chapter 6.2.

The high stability of the sesquioxides R_2O_3, the *rare earths*, reflects the high affinity of rare-earth elements against air. Figure 1.1 displays photographs of common rare-earth oxides. Especially in humid air, the most reactive rare-earth element europium quickly develops a film on its surface consisting of yellow $Eu(OH)_2 \cdot H_2O$. Besides the special role of europium, the reactivity against air generally increases with humidity, temperature and the lighter the rare-earth element is. Ytterbium and yttrium can be handled in dry air without significant oxidation, for instance. Further details on oxides can be found in Chapter 6.3. The reactivity regarding humidity increases further with pure water, mineral acids and bases. The formed compounds on the surface flake-off and allow for further oxidation.

3.8 Solution Chemistry

3.8.1 Reduction Potentials

The reduction potentials of the rare-earth elements are listed in Table 3.3. If a potential is negative, the left side of the reaction is more stable and vice versa. And certainly, the more negative, the more stable the left side is. On a first glance, it is obvious that the trivalent state is the most stable throughout the whole series. For some not even a second reduction potential has been reported. Certainly, the noble gas configuration of xenon, realized in La^{3+} and Ce^{4+}, is of specific stability. The same holds for a half-filled $4f^7$ shell, realized for Eu^{2+}, Gd^{3+} and Tb^{4+}, and a fully occupied $4f^{14}$ configuration like in Yb^{2+} and Lu^{3+}. With increasing distance from these specifically stable configurations, the respective ions become less stable. Considering the divalent ions, this trend can be followed by decreasing reduction potentials along the series Eu^{2+}, Yb^{2+}, Sm^{2+} and Tm^{2+}, followed by Pm^{2+}, Nd^{2+} and Pr^{2+}. Thus, the most stable divalent ions are those of europium, ytterbium, samarium and thulium. To assess whether these ions were stable in degassed aqueous solutions, the potentials

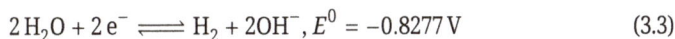

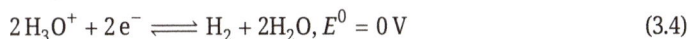

$$2\,H_2O + 2\,e^- \rightleftharpoons H_2 + 2OH^-, E^0 = -0.8277\,V \tag{3.3}$$

$$2\,H_3O^+ + 2\,e^- \rightleftharpoons H_2 + 2H_2O, E^0 = 0\,V \tag{3.4}$$

are relevant. Comparing these with the potentials $R^{3+} + e^- \rightleftharpoons R^{2+}$ in Table 3.3 shows that only divalent europium is stable in neutral or alkaline water; already in acidic solutions it will be readily oxidized to the trivalent state. Nevertheless, many of the aforementioned divalent ions can be stabilized in solid compounds, many of which are even stable in humid air.

Table 3.3: Standard reduction potentials E^0 of relevant rare-earth ions and the alkaline-earth elements (for comparison) given in volts; the precision of all values relates roughly to the number of significant figures, those marked with * were calculated from the remaining data of the same element; data: [29, 48].

R	$R^{2+} + 2e^- \rightleftharpoons R$	$R^{3+} + 3e^- \rightleftharpoons R$	$R^{3+} + e^- \rightleftharpoons R^{2+}$	$R^{4+} + e^- \rightleftharpoons R^{3+}$
Ba	−2.912			
Sr	−2.899			
Ca	−2.868			
Mg	−2.372			
Sc		−2.077		
Y		−2.372		
La	−1.70*	−2.379	−3.74	
Ce	−1.62*	−2.336	−3.76	1.76
Pr	−2.0	−2.353	−3.03	3.9
Nd	−2.1	−2.323	−2.62	4.9
Pm	−2.2	−2.30	−2.67	5.4
Sm	−2.68	−2.304	−1.57	5.2
Eu	−2.812	−1.991	−0.36	6.2
Gd	−1.51*	−2.279	−3.82	7.4
Tb	−1.69*	−2.28	−3.47	3.1
Dy	−2.2	−2.295	−2.42	4.5
Ho	−2.1	−2.33	−2.80	5.7
Er	−2.0	−2.331	−2.96	5.7
Tm	−2.4	−2.319	−2.27	5.6
Yb	−2.76	−2.19	−1.05	6.8
Lu		−2.28		8.1

Europium and ytterbium dissolve like alkaline and alkaline-earth metals in liquid ammonia forming a blue electride solution according to

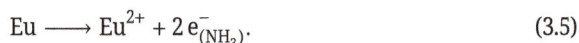

$$\text{Eu} \longrightarrow \text{Eu}^{2+} + 2\,e^-_{(\text{NH}_3)}. \tag{3.5}$$

A good review on the solution chemistry of europium is given in [49]. Looking now on the other side, out of the tetravalent ions, only Ce^{4+} can be handled in an aqueous solution. The respective relevant potentials

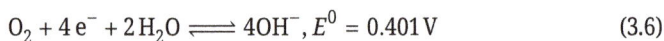

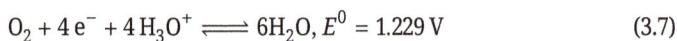

$$O_2 + 4\,e^- + 2\,H_2O \rightleftharpoons 4\,OH^-, E^0 = 0.401\,\text{V} \tag{3.6}$$
$$O_2 + 4\,e^- + 4\,H_3O^+ \rightleftharpoons 6\,H_2O, E^0 = 1.229\,\text{V} \tag{3.7}$$

suggest an instability of tetravalent cerium because of the larger reduction potential even in an acidic solution. But luckily, a high kinetic overpotential of oxygen prevents direct reduction of Ce^{4+} and development of oxygen. Therefore, the *cerimetry* can be conducted under ambient atmosphere; because of at least partial precipitation of hydroxides, or the dioxide, an acidic solution is compulsory. Cerimetry is an analytical method, a type of redox titration, where a sulfuric, deep orange solution of Ce^{4+} upon reduction to Ce^{3+} turns colorless. Generally, it can be employed instead of all quantitative determinations where manganometry is used since the volumetric cerium solution is clearly more stable. In many cases, the color change may be hard to see, and then a respective redox indicator like ferroin is recommended.

3.8.2 Acid-Base Chemistry and Simple Complexes

According the discussion in Chapter 3.5, all trivalent rare-earth ions R^{3+} are hard acids. Their hardness increases along the whole series with decreasing ionic radius. The hardness is emphasized by the closed $5s^2p^6$ shell because the potentially more polarizable 4f states are invisible to any solvent as they are hidden behind that shell (see Figure 2.4). With the declining size of the ions, also the basicity decreases from $La(OH)_3$ through $Lu(OH)_3$. Comparing Lu^{3+} and Sc^{3+} with Al^{3+}, the ionic radii decrease for a common octahedral surrounding from 100 via 89 to 68 pm [50]. While $Al(OH)_3$ is a prime example for amphoterism, this behavior is unknown in rare-earth solution chemistry. But in the solid state quite recently hexahydroxides $Sr_3[R(OH)_6]_2$ were characterized for the smallest ions starting from Sc^{3+} and continuing with yttrium, holmium through lutetium. These compounds were synthesized in an extremely basic hydroflux [51]. Up to then, only a few solid alkaline scandates like $K[ScO_2]$ or $Cs[ScO_2]$ had been described. Here, octahedral ScO_6 moieties were reported quite some time ago [52]. This underlines the by far less tendency of amphoterism for any rare-earth ion compared with Al^{3+}. Trivalent scandium can also be found in a tetrahedral surrounding like in the quite particular suboxoscandate $Cs_9(ScO_4)$ which looks metallic golden [53]. This is presumably due to the presence of itinerant electrons according to $(Cs^+)_9(ScO_4)^{5-} \cdot 4\,e^-$.

Thus, the tendency to form covalent bonds to ligands, such as water molecules, is low. Hence, the interactions are electrostatic, allowing for a broad range of coordination numbers. This certainly also holds for the solid state, but also in a solution, systematic variations were detected. Below a pH of approximately 5, the rare-earth elements form pure aqua complexes and above the bound water molecules are increasingly deprotonated [49]. The coordination number should increase with growing size of the cation, and this is indeed the case. Considering the *aqua complexes*, experiments and molecular dynamic calculations confirm that the average coordination number changes from nine to eight along the series La^{3+} through Lu^{3+} as schematically depicted in Figure 3.9. The larger ions prefer a ninefold coordination $[R(H_2O)_9]^{3+}$ with the water molecules most probably forming a tricapped trigonal prismatic coordination polyhedron. The smallest ions prefer an eightfold coordination $[R(H_2O)_8]^{3+}$ with a square antiprismatic surrounding. However, between both a broad transition regime was ascertained where predominantly a fluctuating surrounding between bi and tricapped trigonal prismatic coordination was found. A central finding is that the exchange frequency is dominant in this region where the third capping water molecule hops on and off the trigonal prism as is

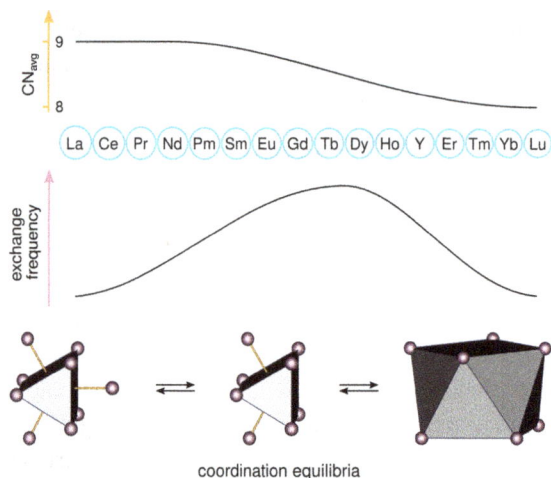

Figure 3.9: Schematic development of the average coordination number in aqueous solution, the evolvement of the ligand exchange frequency and the determined coordination polyhedra according to [54, 55].

obvious from Figure 3.9. This exchange frequency reaches its maximum for different solvents between gadolinium and holmium. Thus, here the greatest dynamic and according reactivity is expected. As soon as the square antiprismatic surrounding is achieved, the ninth solvent molecule does not find its place that easy anymore. Although there is only a minor difference between the square antiprismatic and bicapped trigonal prism, the somewhat larger vacant coordination site in the latter matters.

Historically, and for a broader comprehensive understanding of the complex chemistry of the rare-earth elements, the nitrato complexes are of great relevance. These demonstrate the relevance of available space and the dominance of electrostatic interactions. Moreover, they laid the foundation for the development of the separation chemistry of rare-earth elements, and they were employed in the course of fractional crystallization of double salts with magnesium, ammonium or alkaline metals. As previously stated, in an aqueous solution there is almost no chemical difference between La^{3+} and Lu^{3+}. But by complexation, chemically more diverse moieties with different solubilities and so forth are obtained—and that is what separation chemistry is all about as you will read in Chapter 5. Therefore, the nitrato complexes were among the first thoroughly characterized ones.

For the largest trivalent rare-earth ions, a twelvefold coordination building an icosahedron is found where each nitrato group acts as a bidentate ligand. Hence, the cerium atom is coordinated by six nitrato groups, which are arranged octahedrally as depicted in Figure 3.10. With the decreasing size of the R^{3+}, less space is available, and so only five nitrato ligands coordinate Pr^{3+} ions assisted by two water molecules. Thus, the icosahedral surrounding is more or less maintained; the nitrato ligands find themselves in the more spacious equatorial positions. But chemically it makes a difference whether six

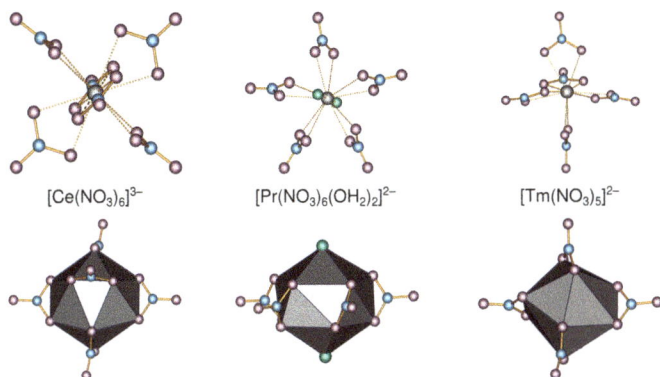

Figure 3.10: Structures of nitrato complexes, oxygen atoms of coordinating water molecules green, nitrato oxygen atoms red, nitrogen blue and rare-earth atoms grey; data: [56–59].

nitrato ligands like in $[Ce(NO_3)_6]^{3-}$ are seen by a solvent or five nitrato and two fairly acidic water ligands as in $[Pr(NO_3)_5(OH_2)_2]^{2-}$. For the smallest R^{3+} ions, the coordination is only tenfold by five bidentate nitrato ligands, e. g., in $[Tm(NO_3)_5]^{2-}$ (Figure 3.10). Here, the nitrato ligands are arranged in a trigonal bipyramidal fashion. And also these complexes look chemically different for a solvent compared with both types described previously.

It is really astonishing that the complexes of trivalent and tetravalent cerium are isostructural. Ce^{4+} is substantially smaller than Ce^{3+} and comparable to the size of Lu^{3+}. Therefore, tetravalent cerium features a significantly higher charge density than the latter. On the other hand, considering the electrostatic coordination strengths of an ion i,

$$s_i = \frac{z_i}{k_i}, \tag{3.8}$$

introduced in Chapter 6.5.1, both Ce^{4+} and Lu^{3+} exhibit similar electrostatic coordination strengths of $s_{Ce} \approx 0.33$ and $s_{Lu} = 0.30$ in their respective twelve or tenfold surrounding.

In the same context, the complexes with the hexadentate ligand ethylene-diamine-tetraacetate ($EDTA^{4-}$) are of interest. In Figure 5.1, the complex constants are given. *Chelate complexes* are of special interest since their formation is not only kinetically, but also entropically promoted as according to

$$[R(OH_2)_8]^{3+} + EDTA^{4-} \longrightarrow [R(EDTA)(OH_2)_2]^- + 6\,H_2O \tag{3.9}$$

from two educt particles seven product particles are obtained. This reaction holds for the smaller rare-earth elements, which form the most stable complexes with EDTA as they apparently fit best into the chelate cavity displayed in Figure 3.11 (left). Here, typically two water molecules complete the eightfold coordination while for larger rare-earth elements like neodymium a third water molecule enters the coordination sphere.

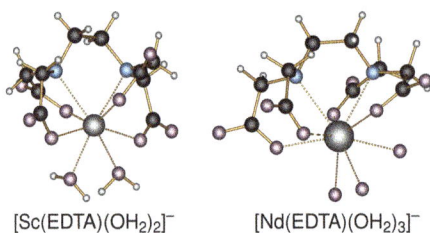

Figure 3.11: Structures of EDTA complexes; oxygen red, nitrogen blue, carbon dark grey and the lanthanide elements grey; data: [60, 61].

[Sc(EDTA)(OH₂)₂]⁻   [Nd(EDTA)(OH₂)₃]⁻

Finally, I will roughly discuss a complex out of the broad field of acetylacetonato complexes. The typically bidentate ligands comprise a central 1,3-dioxo-group capable of coordinating metal atoms with a conjugated double bond system after deprotonation so that both terminal carbon-oxygen bonds are of equal length. The ligand under consideration here is called 6,6,7,7,8,8,8-heptafluoro-2,2-dimethyl-3,5-octanedion, or shorter Hfod. Three fod⁻ ligands coordinate here trivalent europium ions yielding a sixfold coordination in $Eu(fod)_3$. Normally, two more Lewis base ligands complete the eightfold coordination sphere as depicted in Figure 3.12.

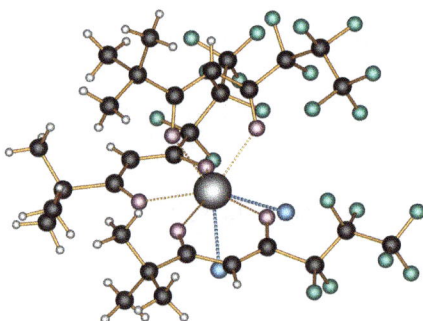

Figure 3.12: Structure of $Eu(fod)_3L_2$ with two free coordination sites L to be coordinated (blue); oxygen red, fluorine green, carbon dark grey and the lanthanide atom grey; data: [62].

$Eu(fod)_3$ is strongly paramagnetic due to the Eu^{3+} ions with an electronic configuration of $4f^6$. Moreover, $Eu(fod)_3$ is a chiral complex and soluble in nonpolar organic solvents. As $Eu(fod)_3$ is attractive for Lewis bases, molecules featuring ether groups or similar basic atoms may occupy the free coordination sites. This is especially interesting if the coordinating molecules are also chiral. If two different chiral molecules interact— one of which is enantiopure, the other not—, then diastereomers form. Such diastereomers are distinguishable regarding their physical properties like the chemical shifts in n. m. r. spectra. Employing this complex, the enantiomeric excess of the coordinating molecule can be quantified. Moreover, the paramagnetic contribution of europium to the local field influences the chemical shifts of atoms in their neighborhood significantly. Thus, complicated n. m. r. spectra can be analyzed easier. Such complexes are called *n. m. r. shift reagents* [63].

Furthermore, Eu(fod)$_3$ catalyses stereoselective cyclization reactions like aldol and Diels–Alder where the formation of the thermodynamically favored *exo* product can be hindered by the complex formation to yield the kinetically controlled *endo* product. Here, the bulky substituents end up on the side with the bulky bridge. For a deeper understanding, the respective organic literature should be consulted.

4 Natural Resources

4.1 General Aspects

The overall abundance of rare-earth elements follows a characteristic zigzag pattern with the elements with odd atomic numbers being less abundant than those with even ones as shown in Figure 4.1. This reflects the relative nuclear stability as this is enhanced by even numbers of protons and neutrons, which is incompatible with an odd atomic number. Both the classical *liquid drop model* by Weizsäcker[29] as well as the quantum mechanical *nuclear shell model* may serve to understand the lower stability of elements with an odd atomic number further and especially the absence of stable isotopes of promethium [64, 65]. The essence is that nuclei comprising an even number of protons and neutrons (g = *gerade*, German) generally are more stable than an those with odd numbers (u = *ungerade*, German) like the carbon isotope $^{12}_{6}C$, a $^{g+g}_{g}E$ nucleus of an element E. Moreover, an even number of protons yields a stronger stabilization than an even number of neutrons. Generally, $^{u+g}_{g}E$ nuclei are more stable than $^{g+u}_{u}E$ nuclei. Any isotope of a given number of nucleons can only be considered *stable* if no other isotope with the same number of nucleons but different ratio of protons-to-neutrons exists, which is more stable. For promethium with no. 61, the most stable isotopes are those with an even number of neutrons like $^{145}_{61}Pm$—but $^{145}_{60}Nd$ is slightly more stable due to its even proton count. Accordingly, the natural abundance of promethium is almost zero—it can be found in low concentration in *uraninite* as radioactive decay product of uranium [66].

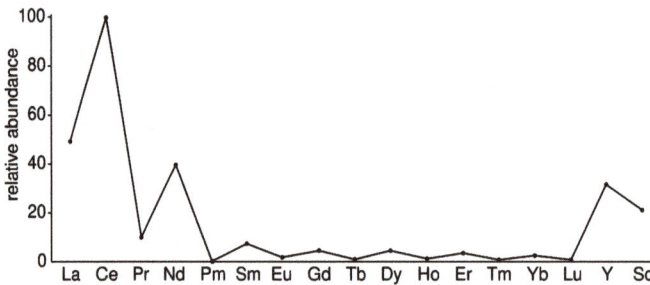

Figure 4.1: Average relative abundance of rare-earth elements in the upper crust; data from [67].

If you are interested in reading more about the formation of the rare-earth elements themselves and how they arrived on earth, please refer to [68].

---

29 *Carl Friedrich von Weizsäcker*, German physicist and philosopher (*1912 †2007).

https://doi.org/10.1515/9783110680829-004

As outlined in the previous chapter, the chemical properties of the rare-earth elements develop smoothly over the whole series with increasing atomic number. Not actually surprisingly, this played an important role during formation of ores containing rare-earth elements. Thus, usually all elements are present simultaneously with different ratios, though. Typical rare-earth contents of minerals were given in Table 1.1. None are found as pure elements on earth. Since all of them are oxophilic, striving for a chemically harder oxidic coordination, they are predominantly dispersed in phosphates, silicates or other oxidic minerals. Regarding silicates, the trivalent ions favor those comprising less-condensed silicate anions with decreasing size [68]. A possible explanation delivers the increasing hardness with decreasing ionic radius of the rare-earth ions and the decreasing hardness of silicate anions with increasing condensation degree. Moreover, generally in silica-rich rocks tend to be found smaller rare-earth ions compared with silica-poor rocks where the larger ones tend to be favored. And again europium adopts a diva-like role as it is—depending on the oxygen content of the surrounding atmosphere—somewhat torn between the common trivalent and the divalent oxidation state. Moreover, its physical properties are different due to its special position in the periodic table. Europium is accordingly enriched in minerals like *feldspar* avoided by its colleagues, and in contrast depleted in the more common ones, thus reducing the yield of one of the most important elements from common ores further.

All those minerals hosting preferentially the larger rare-earth cations also concentrate the naturally abundant actinoid ions of thorium and uranium; thus, these minerals usually show significant radioactivity, which is a certain problem especially during exploitation of the desired elements. The world's mine production in 2020 and 2021 and the reserves of rare-earth oxide equivalents are given in Table 4.1. It may be worth

Table 4.1: The world's mine production in 2020 and 2021 and reserves of rare-earth oxide equivalents are given in thousand tons (Data: [3], n. a. = not available).

country	2020	2021	reserves
China	140	168	44,000
United States	39	43	1,800
Burma	31	26	n. a.
Australia	21	22	4,000
Thailand	4	8	n. a.
India	3	3	6,900
Madagascar	3	3	n. a.
Russia	3	3	21,000
Vietnam	<1	<1	22,000
Brazil	<1	<1	21,000
other	<1	<1	4,290
total	240	280	120,000

mentioning here that the radioactivity of thorium was discovered by the great Marie Curie.[30]

4.2 Ores and Minerals

To become a commercially interesting ore, the rare-earth elements have to be concentrated. Generally, this may occur already by nature via weathering of less stable minerals around very stable ones, such as *bastnäsite* or *monazite*. Subsequent washing separates it further due to its higher density from other stable minerals of lower density; such ores are found as sediments or in clays. In an earlier stage, the elements may be concentrated by phase equilibria. This happens if minerals favored by rare-earth elements crystallize due to higher melting points earlier than other competing ones. For instance, highly melting perovskites crystallize first, if sufficient amounts of titanium are present. Eventually, ores are formed, which contain several rare-earth minerals as minor admixture to other minerals. Such ores are exploited for rare-earth production, which are topic of this section.

Concordantly with the trending ionic radii, minerals of the rare-earth elements are roughly classified as Ce-group, Y-group and complex ones. Cerium and yttrium are the most abundant members of the respective groups; scandium is usually contained in the latter. Although this suggests a clear separation, this classification just indicates the main contributors to the respective mineral with the remaining rare-earth elements being also present as minor contribution. The most important class of rare-earth minerals for exploitation are carbonate fluorides like *bastnäsite* and phosphates such as *monazite*.

We will shed some light on a selection out of the well over 100 rare-earth minerals in this paragraph sorted chemically and starting with phosphates followed by silicates, carbonates, fluorides and oxides. The contents of radioactive admixtures are listed in [80]. The respective crystal structures and coordinations of the rare-earth ions are depicted in Figure 4.2. With decreasing ionic radius nature bestows three different phosphate phases on earth. The most important phosphate mineral is *monazite*, i. e.,$R(PO_4)$ of the Ce-group, containing predominantly larger rare-earth elements including up to 20 wght-% of thorium dioxide and 16 wght-% uranium dioxide. Therein the rare-earth ions are coordinated by nine oxygen atoms providing a convenient coordination environment for ions of similar size of cerium; it crystallizes monoclinically and is found associated with *xenotime*, namely $R(PO_4)$ of the Y-group, a tetragonal phosphate mineral. Here, the rare-earth ions are coordinated only by eight surrounding oxygen atoms, thus suited for slightly smaller ions—containing only up to 5 wght-% uranium dioxide. Accordingly, *xenotime* contains predominantly yttrium and the smaller rare-earth ions. In *apatite*,

---

30 *Maria Skłodowska-Curie*, Polish-French physicist, Nobel Prizes in Physics 1903 and Chemistry 1911 (*1867 †1934).

namely $Ca_5(PO_4)_3X$ (X = F, Cl, OH, CO_3), the rare-earth ions partially replace seven- and nine-fold coordinated Ca^{2+} ions[31] only slightly favoring the larger ones as the space of the two sites differs. All phosphates may be found in pegmatites or granite-related rocks and, of course, in weathered rocks as part of sands. *Apatites* are also present in marine sediments.

By far, the largest group of minerals in the earth's crust are silicates, and naturally there are also relevant silicates hosting rare-earth ions. In *allanite*, namely $CaR(Al_2,Fe)[Si_2O_7][SiO_4]O(OH)$, distinct disilicate and orthosilicate moieties coordinate a series of different cations; here, almost exclusively larger rare-earth ions and actinoid ions up to 3 wght-% of thorium dioxide were found. This is consistent with the high coordination number of eleven of the rare-earth site. *Titanite*, a comparably frequent nesosilicate of the typical composition $CaTi(SiO_4)X$ (X = O, F, OH), solely contains orthosilicate anions. They normally prefer to host the larger rare-earth ions in contrast to *zircon*, $Zr(SiO_4)$, another nesosilicate in which the rare-earth ions and between 0.1 and 0.8 wght-% Th^{4+} ions find an eight-fold coordination of Zr^{4+}. This coordination fosters the presence of smaller rare-earth ions—if found in a silicate-rich granite-related environment. In pegmatites larger rare-earth ions dominate generally, and *titanite* and *allanite* are found in pegmatites or granite-related rocks. The noncondensed SiO_4 tetrahedra in *cerite*, $(R,Ca)_9(SiO_4)_3(SiO_3(OH))_4(OH,F)_3$, form closely packed layers and host preferably the larger rare-earth ions. Further rare-earth silicates are described later in Chapter 6.5.2.

Due to the low content of radioactive ions of up to 0.3 wght-% of thorium dioxide and 0.1 wght-% uranium dioxide carbonate fluorides are the mainly employed minerals for the production of rare-earth elements. *Bastnäsite*, $R(CO_3)X$ (X = F, OH), is found associated with carbonatite and magmatic alkalic rocks containing relatively low amounts of silicon dioxide. Closely related to *bastnäsite* are the *synchysites*, which can be understood by the formal addition of alkaline-earth carbonate according to $MR(CO_3)_2F$ (M = Ca, Ba). In both of the discussed carbonate fluorides, the cation sites are nine-fold coordinated. Since calcium and barium ions resemble the sizes of the larger rare-earth ions, these are favored here. CaF_2, *fluorite*, crystallizes late, and thus rather purely from magmatic melts; astonishingly, therein the smaller rare-earth ions are found. This is presumably because of the higher fluorine affinity of the harder rare-earth ions, which holds these in the melt longer than the softer ones.

Regarding oxides, as a prominent example, *samarskite* is mentioned here. It is comprised of a simple hexagonal closely packed arrangement of oxygen atoms (red and blue in Figure 4.2) with the cations occupying zig-zag chains of octahedral voids in each layer leading to the composition $(R,U,Th,Sn,Fe,Ti)(Nb,Ta)O_4$. It may contain considerable amounts of 20 wght-% uranium and 5 wght-% thorium, but essentially also the whole rare-earth series (Table 1.1). Finally, also the famous *perovskites* may feature rare-earth

---

31 $r_{i,CN=7} \approx 106\,pm$, $r_{i,CN=9} \approx 118\,pm$ [50]

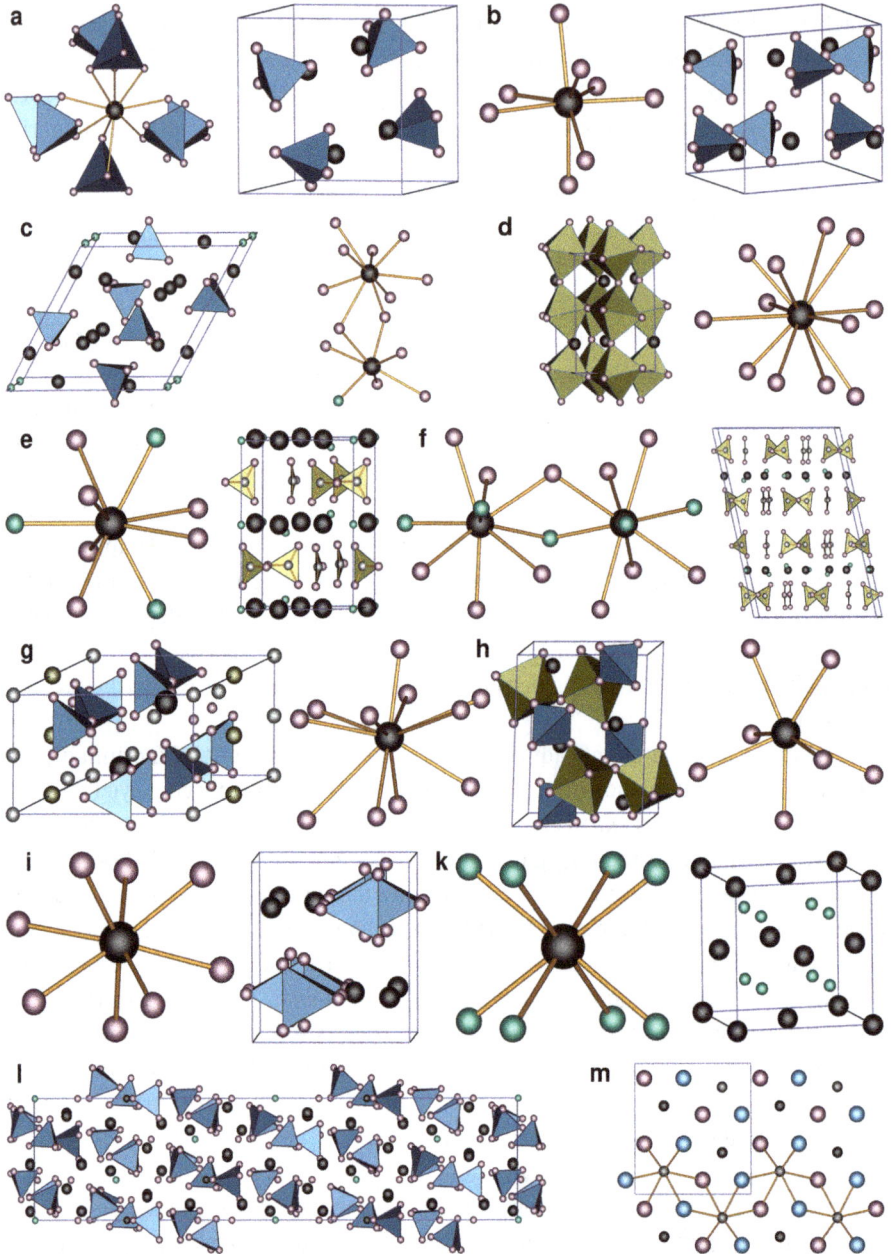

Figure 4.2: Crystal structures of important rare-earth minerals and local surroundings of the sites populated by R^{3+}; (a) *monazite*, (b) *xenotime*, (c) *apatite*, (d) *perovskite*, (e) *bastnäsite*, (f) *synchysite*, (g) *allanite*, (h) *titanite*, (i) *zircon*, (k) *fluorite*, (l) *cerite*, (m) *samarskite*; phosphate and silicate tetrahedra blue, carbonate triangles and titanate octahedra yellow, oxygen red or blue, fluorine green and rare-earth site dark grey; data: [69–79].

ions. *Perovskite*, in its pure form found as $Ca[TiO_3]$ and more generally denoted as $A[BO_3]$, may host a series of elements replacing both cations. On the octahedrally coordinated B site, for instance, niobium, tantalum, iron or magnesium may be found; on the twelve-fold coordinated A site, rare-earth and radioactive actinoid ions and many others coping with large coordination numbers are situated. Since the coordination number is rather large, here the larger rare-earth elements are preferred.

On earth, the rare-earth deposits are broadly distributed as shown in Figure 4.3 confirming the simple fact that the rare-earth elements are not actually *rare*, but if you find them somewhere, they are a *rare* content only. For quite sometime, production sites were almost exclusively situated in China, but during recent years several new sites were and are being set up.

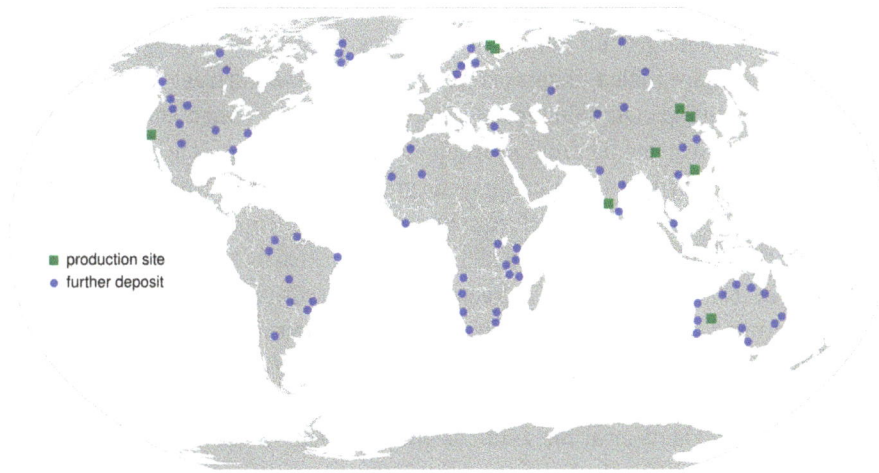

Figure 4.3: Overview on world deposits and main production sites for rare-earth elements; data from [81, 82].

5 Production

We saw in Figure 4.3 that almost in every region of the world rare-earth deposits are known. But the concentration of rare-earth minerals in the ores is rather low. Moreover, we learned from Chapter 3 that the chemical and physical properties develop smoothly over the whole series of elements. Accordingly, they are found to be associated with almost all others in relevant minerals as shown on some examples in Table 1.1; this leads to the main two problems of any rare-earth production process, namely to concentrate and to purify the elements.

5.1 Concentration of Rare-Earth Minerals

After surface or underground mining of hard rock deposits by drilling and blasting methods, the ores are milled to approximately 0.1 mm sized particles. This powder is then suspended in aqueous solutions of detergents or other suited chemicals allowing for separating hydrophilic from hydrophobic particles, which is achieved by froth flotation methods. Further techniques include gravity methods to separate minerals of differing density, magnetism or surface charge. Eventually, concentrates of 60 to 70 % rare-earth oxide content are achieved and separated primarily in *bastnäsite* and *monazite* fractions [83]. After this physical treatment, the *bastnäsite* concentrates are treated with concentrated hydrochloric or sulphuric acid to remove alkaline earth carbonates and other remains like SiO_2, $Zr(SiO_4)$ or TiO_2; subsequent calcination yields rare-earth oxide contents of approximately 90 % [84]. To remove phosphate from *monazite* concentrates, either alkaline or acidic solutions are employed. As the latter does not yield sufficiently pure products, the basic approach is the one discussed here. After dissolution of *monazite* in a hot concentrated sodium hydroxide solution, the rare-earth ions including thorium are precipitated as hydroxides; the phosphate remains in solution and is later separated as $Na_3(PO_4)$. Thorium is separated as ThO_2 by dissolving the previous product in hydrochloric acid.

Softer placer deposits are extracted using methods similar to those employed for hard rock deposits, typically without the need for drilling and blasting. However, complications may arise, especially when mining underwater. Since placers feature quite different compositions of minerals, the extraction methods have to be adapted accordingly to those mentioned above. The even softer deposits in clays allow to extract the adsorbed rare-earth ions *in situ* by leaching the deposit via drilled holes. Through these solutions like aqueous ammonium or alkaline sulphate penetrate the material to exchange more than 90 % of the rare-earth ions by other cations. Via drainage pipes, the aqueous rare-earth solution is collected; the rare-earth ions are precipitated as oxalates or carbonates from this solution, dried and decomposed to oxides containing almost exclusively rare-earths [82].

https://doi.org/10.1515/9783110680829-005

Unfortunately, rare-earth metals and oxides are being produced also in countries with insufficient environmental legislation and control. Accordingly, there the radioactivity of many ores is not appropriately addressed with a series of dirty consequences while there are bans on processing highly radioactive ores in Australia, Europe and China. Further, issues related more or less to any of the employed processes are the introduction of hazardous chemicals into the earth's crust, and the excessive water and solvent consumption—recurrent news report water pollution, poisoning of farmland and of the local population. For further reading, I recommend [83] and the citations therein.

5.2 Separation and Purification of the Elements

For applications where a separation of the elements is not urgent, the pyrophoric alloy *mischmetal* is produced. Mischmetal contains the elements in their natural ore-specific ratio as only the most relevant ones are separated from the initial solution of all rare-earth elements. A typical mischmetal comprises 50 % cerium, 25 % lanthanum, 18 % neodymium, 5 % praseodymium and mainly iron and magnesium among the remaining 2 %. The mixed oxides are reduced via the metallothermic route described in Chapter 5.3. Applications of mischmetal regarding everyday life are in flint ignition devices of lighters and in industrial applications as alloy additives in iron and steels. There mischmetal should react with impurities to improve the mechanical properties. In total, around one-fifth of the global rare-earth consumption per volume is employed as mischmetal [82].

Among the rare-earth elements, cerium and europium show a distinct redox behavior in an aqueous solution; cerium may be oxidized to the tetravalent state, brightly orange Ce^{4+} ions, and europium may be reduced to colorless or pale yellow Eu^{2+} ions (Table 3.2). Hence, these elements can be separated from the others in a quite elegant way by redox processes and subsequent precipitation. Since cerium usually forms the major portion, it is advantageous to remove it as early as possible from the feed. On the contrary, as europium is one of the really rare lanthanide elements, it should be further concentrated to facilitate efficient precipitation as a divalent ion later in the process.

Every production site employs its own separation process dependent from the specific ore used. So, Figure 5.1 displays a typical scheme based on a solvent-solvent separation. It does not cover scandium and promethium. The former is only rarely present in rare-earth ore; instead, it is obtained from scandium deposits described below in Chapter 5.2.3. Promethium as a radioactive element is not present in significant amounts. EDTA complexes (ethylen-diamin-tetraacetate) have been employed for many years, but certainly many other ligand systems are used nowadays. A concise discussion on formerly used complexes is given in Chapter 3.8.2. For example, the complex formation constants of the respective EDTA complexes are given in the figure to get an impression of the separation efficiency.

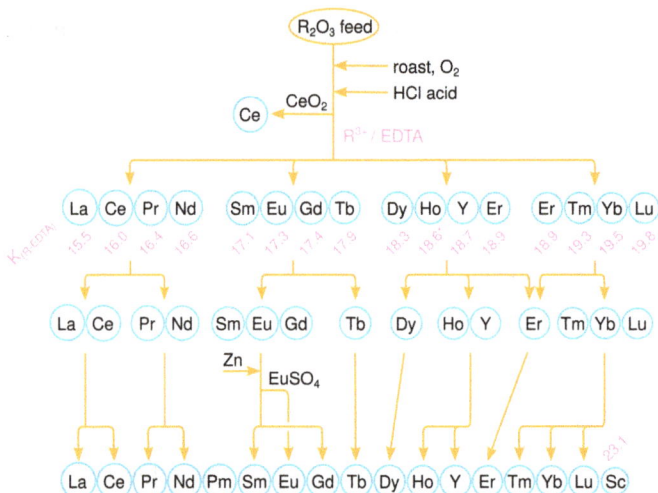

Figure 5.1: Scheme of relevant steps of a typical separation process starting from the concentrated rare-earth oxide feed; the stability parameters of the complexation of R^{3+} with EDTA (ethylen-diamin-tetraacetate) [85–87] are given in red, the asterisk indicates an estimated value therein; the circle diameters around the elements show the relative sizes of R^{3+}.

Initially, the abundant cerium is oxidized by roasting the purified and dried hydroxides of the previous step in air around 600 °C. It can then be separated as CeO_2, which is insoluble in aqueous diluted hydrochloric acid—in contrast to the other rare-earth oxides R_2O_3. In this context, the actinoid oxides ThO_2 and UO_2 can be partially removed as they behave chemically similar.

Praseodymium and terbium are partially oxidized during the initial roasting process. Hence, they are reduced back to the trivalent state prior to the separation of the remaining trivalent rare-earth ions. The following steps are not really sophisticated but are time, energy and solvent consuming because the R^{3+} ions behave so similar. Among these, quite inefficient separation techniques like fractional crystallization or precipitation were used in the past. Nowadays, ion exchange columns or solvent extraction are used. Since the distribution coefficients are quite similar for the R^{3+} due to alike chemical hardness (see Table 3.7 on p. 25) many subsequent cycles have to be conducted. This leads to the need for huge amounts of water and solvents, which are still contaminated by radioactive thorium or uranium traces afterwards. After the liquid-liquid extraction that is discussed below, europium is sufficiently concentrated and becomes reduced, e. g., by the addition of zinc metal, and separated as hardly soluble carbonate $Eu(CO_3)$ or sulfate $Eu(SO_4)$. This can be understood chemically. Eu^{2+} behaves similarly to Sr^{2+} due to the matching size and charge. Thus, like the respective strontium compounds, the carbonate and sulfate of divalent europium are hardly soluble in aqueous solutions.

5.2.1 Solvent-Solvent Extraction

Let us now consider the separation of R^{3+} via a solvent-solvent extraction. All these procedures suffer from low distribution coefficients leading to separation factors σ,

$$\sigma = \frac{c(R_A^{3+})}{c(R_B^{3+})} \tag{5.1}$$

using the concentrations of adjacent R^{3+} ions A and B just above 1. A first approach to enhance the distribution coefficients for the solvent extraction is the addition of chelating ligands, and a second to play skillfully with the acidity of the solutions. In the following paragraph, this approach is demonstrated on a process suggested and reported in Green Chemistry [88]. After that, another really innovative approach employing rare-earth-sensitive bacteria is described. Both might reduce some of the negative implications on the environment.

The feed employed for the herein reported solvent extraction process has the composition given in Table 5.1. Solvent extraction works via two immiscible solvents, typically water and organic kerosene as shown in Figure 5.2. The aqueous solution initially hosts the mixture of trivalent rare-earth ions. In the aqueous phase ligands are dissolved carrying hydrophobic tails and acidic heads, which are deprotonated in neutral water. In our considered example, the protonated ligand 2-Ethylhexyl-hydrogen-(2-ethylhexyl)phosphonate (Figure 5.2, right) is denoted as HL, the deprotonated form accordingly L^-. Such L^- anions occur as dimers with the protonated ligand $(LH \cdot L^-)$ and typically prefer the boundary layer with the charged and polar head dissolved in

Table 5.1: The rare-earth oxide content of a typical feed concentrate employed for solvent extraction compared to the terbium product after forty extraction cycles of the described process in an industrial setting given in wght-% [88].

oxide	feed	terbium product
La_2O_3	5.04	–
CeO_2	0.46	–
Pr_6O_{11}	1.29	–
Nd_2O_3	4.21	–
Sm_2O_3	1.21	<0.001
Eu_2O_3	0.42	<0.001
Gd_2O_3	42.06	0.008
Tb_4O_7	45.31	99.97
Dy_2O_3	<0.01	0.017
Ho_2O_3	–	0.002
Yb_2O_3	–	<0.001
Lu_2O_3	–	<0.001
Y_2O_3	<0.01	<0.001

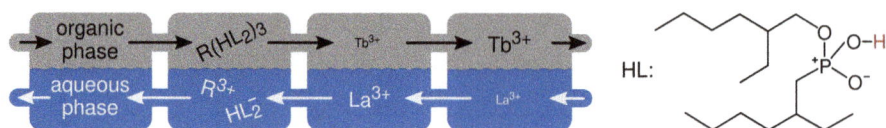

Figure 5.2: Scheme of a solvent extraction process with indication of phases, feed directions, relevant dissolved species, increasing enrichment of smaller ions like Tb^{3+} vs. larger ones like La^{3+} and structural formula of HL (acidic proton red).

the aqueous phase and the hydrophobic tail in the organic kerosene phase. Alternatively, they might form micelles in the aqueous phase. The positively charged R^{3+} then react in the aqueous phase with the $LH \cdot L^-$ dimers to give relatively huge neutral $R(LH \cdot L)_3$ complex molecules in which the rare-earth ions are octahedrally coordinated. These neutral complex molecules dissolve somewhat better in the organic phase; thus, the rare-earth ions are extracted from the aqueous phase. Certainly, equilibria depending on the acidity of the aqueous solution as well as the complex formation parameter K_R and the distribution coefficient of the neutral molecules between organic and aqueous phase are achieved:

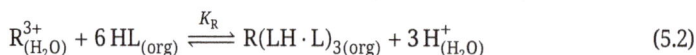

$$R^{3+}_{(H_2O)} + 6\,HL_{(org)} \underset{}{\overset{K_R}{\rightleftharpoons}} R(LH \cdot L)_{3(org)} + 3\,H^+_{(H_2O)} \tag{5.2}$$

The key figure for separation are the different stability constants K_R. In the case of the aforementioned ligand HL, smaller R^{3+} form more stable complexes than the larger cations as the six-fold coordination as well as the chemical hardness of L^- apparently better suits the smaller ions. This behavior can be found for many complex ligands employed for the rare-earth separation [85]. Consequently, the organic phase is enriched with smaller rare-earth ions, and the aqueous solution is enriched with larger R^{3+} ions. Further, the addition of ammonia to the aqueous solution somewhat buffers the basic solution, but also ammine complexes can be formed. Since the latter are fairly flexible regarding the coordination number, in contrast to the chelate ligand system $LH \cdot L^-$, an even better stabilization of the larger R^{3+} in the aqueous phase is achieved. After equilibration, both solutions are transferred to the next chamber according to Figure 5.2. The organic phase contacts with another aqueous solution $R^{3+}_{(H_2O)}$ depleted of smaller rare-earth ions, while the aqueous phase now equilibrates with another organic solution $R(HL_2)_{3(org)}$ enriched of smaller rare-earth ions. After equilibration, both original phases are further enriched in smaller R^{3+} in the organic phase and in larger rare-earth ions in the aqueous phase.

Because of the reverse feed of organic and aqueous phase, the separation advances quite fast (Figure 5.2). Although this treatment is quite efficient, depending on the desired purity of the rare-earth ions, normally a few tens up to a thousand of such cycles have to be conducted. After the desired separation degree is achieved, the complex molecules dissolved in the organic phase are destroyed by the addition of basic ammonia shifting the equilibrium in Equation (5.2) to the left side as then all HL of the treated fraction will be deprotonated and the R^{3+} will be precipitated as hydroxides. Eventually, oxides are obtained.

5.2.2 Further Separation Approaches

Many more processes for the separation of rare-earth elements have been proposed, such as to extract rare-earth elements by ionic liquids or by employing nonaqueous solvents like polar organic ones [89, 90]. Here, I will describe two other approaches: a biogenic one and some ideas for the recycling of rare-earth elements from devices.

A few years ago, a very interesting biogenic approach to separate rare-earth elements employing marine bacteria of the *roseobacter* family was reported [91]. As marine bacteria, they are tolerant against concentrated salt solutions. Remarkably, they also tolerate really acidic solutions and provide three surface sites of different acidy with approximate pK_a values of 2, 3.7 and 5.5. These presumably consist of phosphate groups. For the experiments, the bacteria were immobilized on an assay filter and fully protonated before being exposed to an aqueous solution containing equimolar amounts of rare-earth ions, which then are exchanged with the surface-bound protons. The rare-earth ions are subsequently desorbed with increasing pH; the smaller (and harder) the rare-earth ions are, the stronger they are bound to the bacterial sites yielding an astonishingly good separation efficiency.

The truly rare rare-earth elements should be recycled from wasted materials as their mining, separation and production is extremely time, money and energy consuming. Moreover, the respective applications already concentrate on certain suited element combinations. Therefore, the recycling should be done sorted by the applications, such as permanent magnets, accounting for about 40 %, phosphors (30 %) and nickel alloy batteries (10 %) [92]. Permanent magnets mainly contain neodymium, praseodymium and dysprosium. These elements have to be separated as their ratio varies with the concrete application. For instance, dysprosium enhances the stability against demagnetization, which in many cases, is urgently necessary while it is not in others. Phosphors, on the other hand, may be recycled as phosphor materials, though. The main problem here is contained mercury if recycled from compact fluorescent lamps, which requires a thermal treatment above 600 °C to get rid of the toxic metal. Moreover, upon the crushing of the lamps, the phosphors are contaminated with silicon dioxide and alumina. From such mixtures, the phosphor particles may be separated magnetically [93]. Unfortunately, the recycling of battery materials is considered seldom. Here, on the other hand, often mischmetal is employed, which is quite cheap, and together with nickel, such waste is added to steel.

5.2.3 Scandium

Scandium is not a rare, but a very scarce element. Looking at Figure 4.1, it is less abundant than cerium, lanthanum, neodymium and yttrium, but its distribution within the earth crust is finer. Sc^{3+} are the chemically hardest ions of the rare-earth elements and resemble the behavior of hard acids like zirconium, tantalum and especially aluminium in many cases—accordingly, one of its major applications is the hardening of aluminium alloys. The main scandium minerals are *thortveitite*, $Sc_2[Si_2O_7]$ (Chapter 6.5.2 and Figure 6.23) and *kolbeckite*, $Sc(PO_4) \cdot 2\,H_2O$ shown in Figure 5.3, which are mainly found in Norway and Madagascar. However, these deposits are not exploited to date. It is also contained in *zircon* where it usually replaces $Zr(SiO_4)$ with $Sc(PO_4)$, sometimes as its own mineral, *petrulite*. Scandium is occasionally found in metorites as *allendeite*, $Sc_4Zr_3O_{12}$, a scandium zirconate adopting the structure of Tb_7O_{12} (see Figure 6.17) [94]. Finally, there is also a very rare mineral containing tantalum named *heftetjernite*, $ScTaO_4$, crystallizing in the *wolframite* type and also displayed in Figure 5.3.

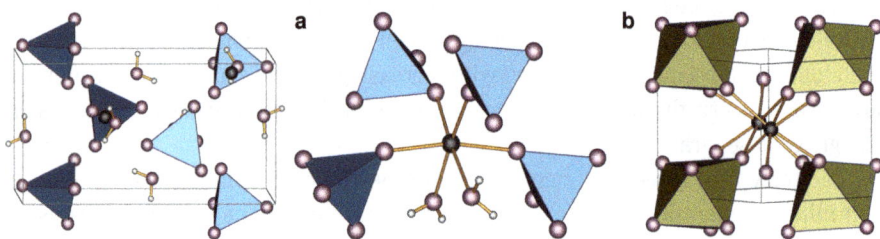

Figure 5.3: Crystal structures of selected scandium minerals and local surroundings of the sites populated by Sc^{3+} (dark grey spheres); (a) *kolbeckite*, (b) *heftetjernite*; phosphate tetrahedra blue, tantalate octahedra yellow, oxygen red and hydrogen white; data: [95, 96].

Scandium was historically produced in regions of the former Soviet Union, i.e., Russia, Kazakhstan and Ukraine. Nowadays, the main production sites are in China, the Philippines and Australia using any mineral or waste where it occurs as a minor contribution. Thus, it is normally recovered from various sources such as uranium leach liquors, titanium pigment production, tungsten refineries and certainly also from rare-earth element production residues. Since precipitation methods do not yield pure products, also here a solvent extraction resembling the aforementioned approach is the method of choice. For further reading regarding this subject, I refer to [97]; there are also further techniques being reviewed such as ion exchange or liquid membrane processes. From these sources, Sc_2O_3 is the starting material for the production of the element, which then is normally converted into the trifluoride ScF_3 likewise cerium. For the reduction to the pure element, the same metallothermic approach as for almost all other rare-earth elements is used which is the subject of the next section. Here, in case of scandium calcium metal is employed as reduction agent.

5.3 Obtaining the Elements

At the end of the industrial separation process, cerium oxide, europium carbonate or europium sulfate were obtained besides the oxides of the other rare-earth elements. To get the pure elements by a suited reduction an electrolysis of the halides would be desirable. A review of the reduction potentials in Table 3.3 (see Chapter 3.8) and considering the relevant potential in neutral water

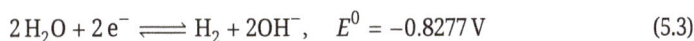

$$2\,H_2O + 2\,e^- \rightleftharpoons H_2 + 2\,OH^-, \quad E^0 = -0.8277\,V \tag{5.3}$$

tells us that it is impossible to get the reduction done from an aqueous solution. Looking at Figure 6.4 (Chapter 6.2), the chlorides seem interesting due to their comparably low melting points, and indeed the europium metal is obtained by electrolysis of $EuCl_3$ in an eutectic mixture with rock salt. Certainly, the chlorides have to be synthesized first, and this is also challenging to avoid formation of the halide oxides ROX as discussed in Chapter 6.2.1.

The alternative method is the destillation of europium metal after a *metallothermic reduction* by reacting the oxide with less noble metals such as lanthanum; with more reactive metals, the reactions proceed too vigorously in the case of europium. Pure cerium can be obtained by the reduction of its anhydrous fluoride CeF_3 with calcium metal; this latter method is a quite general method to obtain the remaining rare-earth metals,

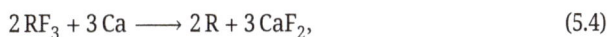

$$2\,RF_3 + 3\,Ca \longrightarrow 2\,R + 3\,CaF_2, \tag{5.4}$$

which yields the very stable CaF_2. Calcium fluoride melts around 1420 °C and can be separated either due to its differing melting point or its differing density from the respective rare-earth metal. The metals exhibit a silvery luster, but show different hardnesses. The softer ones like cerium, europium and ytterbium may be cut with a knife, the others with pincers. Scandium and lutetium are significantly harder [82]. For the reactivity of the pure elements, please refer to Chapter 3.7.

6 Basic Compound Classes

6.1 Hydrogen Compounds

Personally, I prefer to name them hydrogen compounds, although in most cases, it is well justified to call them hydrides since without question the hydrogen atoms will be negatively polarized due to the higher electronegativity of hydrogen compared with the rare-earth elements. But you should not expect to find salt-like hydrides containing H^- anions in the rare-earth hydrogen compounds and I certainly do not intend to suggest that in the cases where I call them hydrides.

It is generally quite important that the rare-earth elements consume gaseous hydrogen already at room temperature to a certain extent. This has to be considered for any property measurements of metallic samples as traces of hydrogen might change the results significantly. Due to its low electron density, hydrogen might be overseen in X-ray diffraction, so a careful treatment of all respective samples is strongly recommended. Under moderate conditions below 500 °C, the rare-earth elements react readily with hydrogen to form hydrides of different compositions. All react via an exothermic reaction,

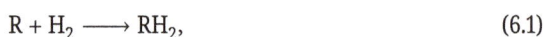

$$R + H_2 \longrightarrow RH_2, \tag{6.1}$$

to the dihydrides. Their enthalpies of formation resemble more these of the heavier alkaline-earth metals than those of the outer transition metals with hydrogen. Most of the rare-earth elements proceed further to RH_3. Typically, the resulting structures are intercalation compounds hosting the negatively polarized hydrogen atoms in the voids of close packings related to the metal's structure. Close packings provide two tetrahedral voids per atom, which are predominantly filled by hydrogen in the first instance, and one octahedral void per atom. The latter is increasingly occupied during hydrogenation toward RH_3. Moreover, the hydrogen compounds rarely form stoichiometric compounds but feature a significant phase width. This behavior is in accordance with a subsequent intercalation into the metal's crystal and electronic structure. Therefore, the hydrogen compounds start as metallic intercalates and smoothly approach semiconducting behavior [98].

Scandium, ytterbium and europium do not form trihydrides at all under ambient pressure. As stated previously, europium shows a tendency to the divalent state as then a half-filled $4f^7$ shell is achieved and especially then realized in compounds with reducing anions such as hydride H^- or iodide I^-. Compounds of Eu^{3+} with reducing anions are commonly unstable. Scandium on the other hand is the smallest of all rare-earth elements and crystallizes in a hexagonal close packing where the tetrahedral voids share common faces. Hence, in a hypothetical ScH_3 the hydride anions came too close. Only under high pressure, hydrides up to $YbH_{2.55}$ were reported, which adopt the LaH_3 structure.

https://doi.org/10.1515/9783110680829-006

The hydrogen compound with the highest hydrogen content, i. e., LaH_{10}, was obtained in a diamond anvil cell around a high pressure of 170 GPa heated to 2000 K. Therein, the lanthanum atoms adopt a face-centered cubic arrangement enclosed by a hydrogen clathrate featuring H_{32} cages around each lanthanum atom [99]. The authors describe this (under ambient pressure unstable) compound as a form of metallic hydrogen doped with lanthanum metal. For materials chemistry, it remains a curiosity despite its relevance for chemistry as a whole. Further details on the hydrogen compounds and their properties may be studied in [100] where most of the herein discussed data were taken.

6.1.1 Structures and Bonding of RH$_2$ and RH$_3$

The hydrogen compounds RH_2 of *scandium*, and from *lanthanum to neodymium* adopt the structure of *fluorite*, CaF_2. Therein the R atoms replace calcium and the hydrogen atoms fluorine. These hydrogen compounds form solid solutions with the respective RH_3 in which also the octahedral voids are occupied. Finally, for RH_3 a structure isotypic with the *Heusler phase* $AlCu_2Mn$ is obtained where aluminium atoms (blue) form a cubic close packing with copper atoms in all tetrahedral (yellow) and manganese atoms in all octahedral voids (white) as displayed in Figure 6.1 in the example of LaH_3 or Appendix D.

Important crystal structure types no. 1–fluorite, CaF₂: The crystal structure of the CaF₂ type is illustrated in Figure 6.1 (left) by omitting the white spheres. It consists of a cubic close packed arrangement of calcium atoms (blue) with all tetrahedral voids occupied by fluorine atoms (yellow); in each closely packed structure, two tetrahedral and one octahedral void per packing atom is found. In the case of a cubic close packing, also dubbed face centered cubic packing, adjacent voids only share common vertices. This structure type is adopted by various rare-earth element compounds. For the basics of this structure related to others, see also Appendix D.

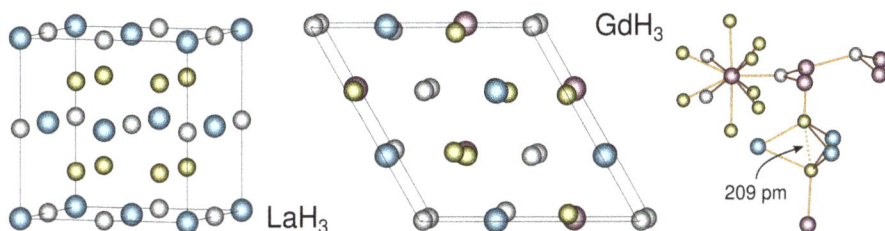

Figure 6.1: Crystal structures of RH₃; shown are the structure of LaH₃ and *fluorite* by omitting the white spheres (left), structure of and coordinations in GdH₃ (right), the hydrogen atoms are colored in yellow (situated in tetrahedral voids) and white (left: octahedral voids, right: triangular voids), the rare-earth atoms in blue and red; data: [101, 102].

A slightly different behavior is found for the elements from *samarium through lutetium and yttrium*, except europium and ytterbium. These also adopt the *fluorite*-type structure for the composition up to RH_2. For higher hydrogen contents, the structures switch to a hexagonal close arrangement of the cations consistent with a relative movement of adjacent cation layers—thus going from an ABCABC to an ABABAB sequence. In Figure 6.1 (right), the cation layers are indicated by blue and red spheres. Still both tetrahedral voids are occupied (yellow spheres), but now they share common faces. The third hydrogen atom (white) is hosted on two sites more or less within the cation layers and coordinated by three cations as shown in Figure 6.1 on the example of GdH_3. This resulting structure is that of the *fluocerite* or *tysonite* structure type (LaF_3). In the halide section, an alternative structure description based on a close packing of anions is discussed. Interestingly, there was a controversy about this very crystal structure since the small electron density of hydrogen inhibited a straightforward structure determination. A careful combination of neutron and X-ray diffraction revealed a threefold superstructure perpendicular to the c-axis and finally yielded the ordered satisfactory structure model discussed in the previous paragraph [102]. Here, the third hydrogen atom is distributed over two sites. Both are coordinated by three rare-earth cations, one slightly off the triangular plane, the other within the plane. This superstructure slightly reduces the average hydrogen-hydrogen as well as the R–H distances.

Considering the ionic radii of La^{3+} and Gd^{3+}, it is reasonable that the former are coordinated by fourteen in LaH_3 and the latter are coordinated by only eleven hydrogen atoms in the GdH_3 structure (Table 6.1). Although it sounds strange to me that apparently the larger La^{3+} ion prefers a structure, in which the third hydrogen atom occupies a comparably larger octahedral site, while in GdH_3 the third hydrogen atom only experiences a threefold coordination. Moreover, the smaller tetrahedral voids sharing common faces are found in the GdH_3 structure, where the hydrogen atoms come even closer to each other (209 pm vs. 243 pm). To get an idea about the electrostatic interactions within both structure types, MAPLE calculations were conducted (see the info box). Detailed results of the calculations are listed in Appendix F.1. Indeed for both, LaH_3 and GdH_3, the experimentally found crystal structure matches the electrostatically favored one with respect to the other structure type. I certainly adjusted the lattice parameters to get similar R–H

Table 6.1: Comparison of relevant parameters regarding the structures of LaH_3 and GdH_3, the ionic radii of coordination number (c. n.) 9 [50], minimal hydrogen distances, coordination number of the rare-earth ions, Madelung constant M, Madelung Part of Lattice Energies.

compound	$r_{ion}^{[9]}$	$d(H–H)_{min}$ / pm	c. n.(R^{3+})	M	MAPLE / kJ/mol
LaH_3	122	243	14	9.58	5484
(in GdH_3 structure)			11	8.84	5394
GdH_3	111	209	11	8.84	5870
(in LaH_3 structure)			14	9.58	5772

distances for the tetrahedral voids. The results suggest that indeed electrostatic reasons cause the choice of structure type here.

Madelung Part of Lattice Energy (MAPLE): The MAPLE concept was developed by Rudolf Hoppe.[32] Thereby the electrostatic interactions in ionic crystals are quantified and used to determine effective coordination numbers, effective charges and to check the reasonability of structure models [103–105]. The calculation also yields a good estimate of the electrostatic contribution to the lattice energy of the material under consideration as used in the examples LaH_3 and GdH_3. Further applications are the localization of protons, the determination of ordering of differently charged cations like Si^{4+} and Al^{3+} in cases where diffraction methods do not deliver reliable evidence.

The dihydrides of *europium and ytterbium* behave like the ionic alkaline-earth hydrides and are thus isolators, in contrast to all other dihydrides. Likewise the alkaline-earth hydrides, both realize the *cottunite* type structure ($PbCl_2$). Here, the hydrogen atoms form a distorted hexagonal close packing. In Figure 6.2, the two layers in EuH_2 are indicated by yellow and white spheres, respectively. The europium atoms (blue) are for several reasons too large for both, the tetrahedral as well as the octahedral voids. Accordingly, they formally expand the close layer locally and move into the triangular face shared by two adjacent octahedral voids; this is illustrated by the larger triangles spanned by yellow atoms and emphasized by broken lines (Figure 6.2, left). Thus, a larger ninefold-coordinated void is achieved where the metal atoms are surrounded by a three-capped trigonal prism. For the basics of the *cottunite* structure related to others, see also Appendix D.

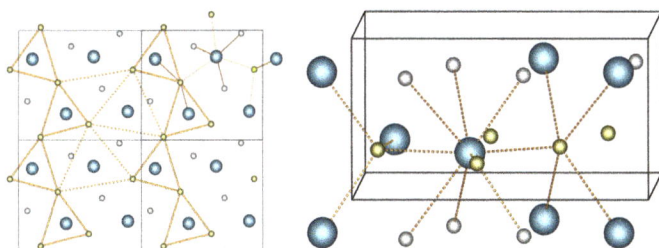

Figure 6.2: Crystal structure of EuH_2 along two perspective views, actually shown is the structure of the isotypic BaH_2; the distorted closely packed layers of hydrogen atoms in yellow and white, the metal atoms in blue; data from [106].

Due to the higher electronegativity, the hydrogen states are localized at significantly lower energies relative to the Fermi level and the rare-earth elements states. Hence, upon hydrogen uptake, electron density is transferred from the highest metals' states—typically 5d and 6s—into the hydrogen band yielding polar hydrides with fairly localized

---

32 *Rudolf Hoppe*, German chemist (*1922 †2014).

electrons therein. With the increasing transfer of electron density, the conductivity of the materials shrinks until a semiconducting behavior is observed. In agreement with this preliminary remark, the *chemical bonding* is generally described as metallic for all dihydrides except EuH_2 and YbH_2, which are predominantly ionic compounds. All existing trihydrides are semiconductors, and somewhere between RH_2 and RH_3 a metal-to-semiconductor transition occurs.

Magnetically there is almost no change with increasing hydrogen uptake for the elements cerium through samarium. Here, the 4f electrons remain at the rare-earth core. Between europium and thulium slightly decreasing magnetic moments were observed with increasing hydrogen content. In the case of gadolinium, the ferromagnetic ordering of the metal vanishes upon hydrogenation. The ytterbium hydrides show increasing paramagnetism with increasing hydrogen content while the metal is diamagnetic; apparently, some electron density is transferred from the fully occupied $4f^{14}$ states in ytterbium onto the hydrogen atoms. In contrast to europium metal, EuH_2 orders ferromagnetically below 25 K.

6.1.2 Hydrogen Storage—LaNi$_5$

Not only, but mainly climate change issues, foster the development of hydrogen storage materials. Over the last decades, many compounds have been investigated and novel materials are being developed, but there are a few really prominent evergreens in this field. Among these, $LaNi_5$ is the one which contains rare-earth elements as a crucial component. Nowadays, instead of pure lanthanum, usually *mischmetal* is employed as this features a very similar chemical behavior, provides similarly sized voids, and by the way, is significantly more affordable than the pure metals.

The late outer transition metals like nickel exhibit too small voids to allow for quick uptake and release of hydrogen. Moreover, the chemical bonding mainly stems from rather weak interactions with the very broad 4s band since there are only antibonding interactions with the 3d states [17]. The consequence here is that metals like nickel nicely work as catalysts where hydrogen is briefly bound to the metallic surface, and thus activated. As discussed in the previous chapter, rare-earth elements readily form stable metallic dihydrides RH_2 since these metals provide sufficiently large voids and bonding interactions. The latter are too strong for releasing hydrogen again easily, so the pure compounds RH_2 are not really suited as hydrogen storage materials. So, it seems promising to combine a metal with excellent hydrogen uptake behavior like lanthanum with a metal like nickel, which only weakly interacts with hydrogen. Moreover, these metals should be as cheap as possible—and there we are with lanthanum or mischmetal and nickel. The stable compound that forms is $LaNi_5$. In a typical approach, this hydrogen storage material reacts readily at room temperature under a smooth overpressure

according to

$$LaNi_5 + 3\,H_2 \xrightleftharpoons[80\,°C,\,2\,bar]{20\,°C,\,4\,bar} LaNi_5H_6 \qquad (6.2)$$

and yields $LaNi_5H_6$. The hydrogen can then be released again by slight heating. Apparently, the aforementioned combination yields a material, which provides sufficiently large voids, sufficiently strong bonding and a sufficient reversibility of hydrogen uptake and release. The stored hydrogen per cubic meter amounts to 88 kg, which is more than the 71 kg a cubic meter of liquid hydrogen weighs. On the other hand, the overall weight of the material itself is quite high with only 1.5 wght-% hydrogen storage capacity rendering it unattractive for applications as long as weight plays an important role.

The crystal structure of $LaNi_5$ adopts the $CaCu_5$-type structure and is shown in Figure 6.3. Here, a layer A comprises a kagomé[33] net of nickel atoms, a frequently occurring motif in metal-rich phases. Such a kagomé net may be understood as a deficient close packed layer where a quarter of all atoms was removed. Layer B comprises a close packed layer of lanthanum atoms with further nickel atoms in any triangular void. Following a layer sequence AB, the structure is obtained where the lanthanum atoms of layer B lie directly below and above the voids of the kagomé net of layer A. The lanthanum atoms are coordinated by eighteen nickel atoms while the nickel atoms are surrounded by four (Ni situated on layer A) or three (Ni situated on layer B) lanthanum atoms as well as eight or nine nickel atoms—yielding a total coordination number of twelve. The voids for the hydrogen atoms indicated as white spheres in Figure 6.3 are

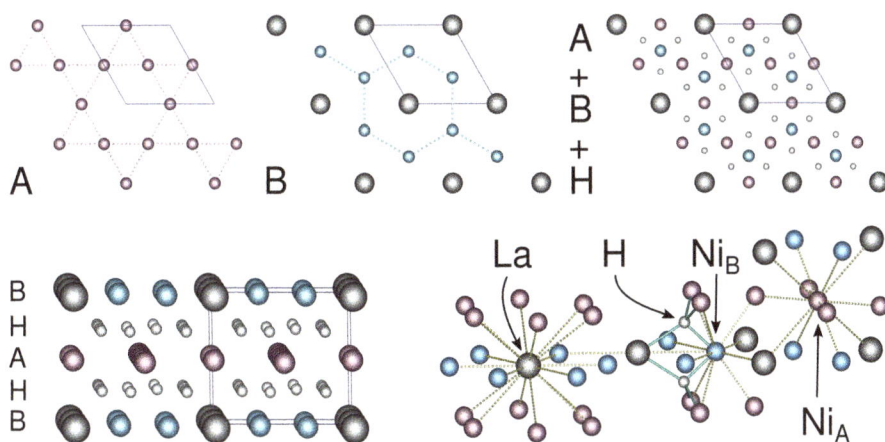

Figure 6.3: Crystal structure of $LaNi_5$; shown are layers A and B, the full structure including hydrogen atoms from two directions (A+B+H) and the surroundings of all atoms; data: [107].

---

33 *Kagomé* is the designation of a typical Japanese pattern of woven bamboo fabric.

found between both layers. There the hydrogen atoms are situated in distorted tetra-hedral voids formed by a lanthanum and three nickel atoms, essentially one void per metal atom.

6.2 Binary Halides and Halide Oxides

Amid the numerous halides of the rare-earth elements, the most common ones are the trivalent, tetravalent and divalent ones. Aside of these, also various somewhat reduced halides were characterized. In this book, I will certainly focus on the most common ones and just mention some more general remarks on the remaining ones as relevant prop-erties of the latter are also found among the former. Moreover, the pure halides show interesting structure systematics, which I will address in this chapter. Figure 6.4 gives an overview about the polymorphism and the melting points of the known divalent, trivalent and tetravalent halides. For *further reading* [108–111] are recommended.

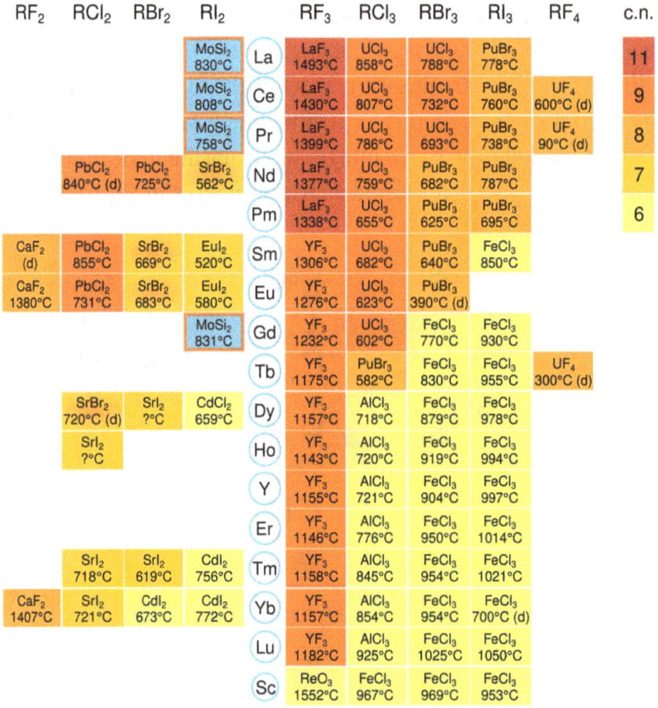

RF_2	RCl_2	RBr_2	RI_2		RF_3	RCl_3	RBr_3	RI_3	RF_4	c.n.
			MoSi₂ 830°C	La	LaF₃ 1493°C	UCl₃ 858°C	UCl₃ 788°C	PuBr₃ 778°C		11
			MoSi₂ 808°C	Ce	LaF₃ 1430°C	UCl₃ 807°C	UCl₃ 732°C	PuBr₃ 760°C	UF₄ 600°C (d)	9
			MoSi₂ 758°C	Pr	LaF₃ 1399°C	UCl₃ 786°C	UCl₃ 693°C	PuBr₃ 738°C	UF₄ 90°C (d)	8
	PbCl₂ 840°C (d)	PbCl₂ 725°C	SrBr₂ 562°C	Nd	LaF₃ 1377°C	UCl₃ 759°C	PuBr₃ 682°C	PuBr₃ 787°C		7
				Pm	LaF₃ 1338°C	UCl₃ 655°C	PuBr₃ 625°C	PuBr₃ 695°C		6
CaF₂ (d)	PbCl₂ 855°C	SrBr₂ 669°C	EuI₂ 520°C	Sm	YF₃ 1306°C	UCl₃ 682°C	PuBr₃ 640°C	FeCl₃ 850°C		
CaF₂ 1380°C	PbCl₂ 731°C	SrBr₂ 683°C	EuI₂ 580°C	Eu	YF₃ 1276°C	UCl₃ 623°C	PuBr₃ 390°C (d)			
			MoSi₂ 831°C	Gd	YF₃ 1232°C	UCl₃ 602°C	FeCl₃ 770°C	FeCl₃ 930°C		
				Tb	YF₃ 1175°C	PuBr₃ 582°C	FeCl₃ 830°C	FeCl₃ 955°C	UF₄ 300°C (d)	
	SrBr₂ 720°C (d)	SrI₂ ?°C	CdCl₂ 659°C	Dy	YF₃ 1157°C	AlCl₃ 718°C	FeCl₃ 879°C	FeCl₃ 978°C		
		SrI₂ ?°C		Ho	YF₃ 1143°C	AlCl₃ 720°C	FeCl₃ 919°C	FeCl₃ 994°C		
				Y	YF₃ 1155°C	AlCl₃ 721°C	FeCl₃ 904°C	FeCl₃ 997°C		
				Er	YF₃ 1146°C	AlCl₃ 776°C	FeCl₃ 950°C	FeCl₃ 1014°C		
	SrI₂ 718°C	SrI₂ 619°C	CdI₂ 756°C	Tm	YF₃ 1158°C	AlCl₃ 845°C	FeCl₃ 954°C	FeCl₃ 1021°C		
CaF₂ 1407°C	SrI₂ 721°C	CdI₂ 673°C	CdI₂ 772°C	Yb	YF₃ 1157°C	AlCl₃ 854°C	FeCl₃ 954°C	FeCl₃ 700°C (d)		
				Lu	YF₃ 1182°C	AlCl₃ 925°C	FeCl₃ 1025°C	FeCl₃ 1050°C		
				Sc	ReO₃ 1552°C	FeCl₃ 967°C	FeCl₃ 969°C	FeCl₃ 953°C		

Figure 6.4: Overview on binary rare-earth halides, given are the respective structure type where the c. n. is encoded by color and the melting point or decomposition temperature (d); the circles around the element symbols resemble the relative size of the ion; data: [29, 112–114].

6.2.1 Trihalides RX$_3$

Overview and Syntheses

Considering Figure 6.4, almost all possible combinations of rare-earth elements and fluorine, chlorine, bromine and iodine have been reported. As frequently mentioned before, again europium and ytterbium behave like divas because both may be reduced quite easily to the divalent state; hence, all trihalides of both with reducing halides are fairly unstable. While EuI$_3$ has not been reported at all, EuBr$_3$ decomposes [115] according to

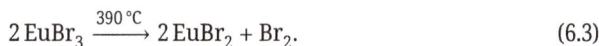

$$2\,EuBr_3 \xrightarrow{390\,°C} 2\,EuBr_2 + Br_2. \tag{6.3}$$

The melting point of EuBr$_3$ given in some places is in fact that of the hexahydrate [R(H$_2$O)$_6$]X$_3$ [113]. For YbBr$_3$ at least a melting point could be determined, but YbI$_3$ analogously decomposes around 700 °C. This nicely shows the slightly higher relative stability of europium's divalent state.

Generally, the hygroscopicity of the halides increases from the fluorides to the iodides following the increasing softness of the anions. The smallest anion in this series is the fluoride yielding the shortest R–Hal interatomic distances and the highest coordination numbers. Accordingly, the lattice energy is maximized here, and thus the fluorides show the highest melting points ranging between 1143 °C for HoF$_3$ and 1552 °C for ScF$_3$. The chlorides' melting points were reported between 582 °C for TbCl$_3$ and 967 °C for ScCl$_3$.

According to the HSAB concept, R^{3+} ions are relatively hard cations, and thus prefer interaction with hard bases such as fluoride or oxide. Because of this high oxygen affinity, the synthesis of anhydrous halides is challenging with exception of the *fluorides*. These can be obtained with good yields starting from the aqueous chlorides via

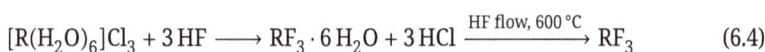

$$[R(H_2O)_6]Cl_3 + 3\,HF \longrightarrow RF_3 \cdot 6\,H_2O + 3\,HCl \xrightarrow{HF\ flow,\ 600\,°C} RF_3 \tag{6.4}$$

by reaction with hydrogen fluoride and subsequent dehydration. Alternatively, they may be synthesized via the following ammonium chloride route employing (NH$_4$)HF$_2$. Here, the reaction of R$_2$O$_3$ with NH$_4$HF$_2$ at temperatures around 100 °C yields hydrated ammonium hexafluorolanthanidate (NH$_4$)$_3$RF$_6$, which can be smoothly decomposed to give pure RF$_3$. The *chlorides and bromides* are most easily obtained via the ammonium halide route from the oxides, and later improved and developed further by Meyer[34] [116, 111]. A drawback for this approach is the high oxygen affinity of the rare-earth ions preventing quantitative reaction and leaving significant amounts of oxygen in the final products; this especially holds for the softer halides of chloride, bromide and iodide.

---

34 *Gerd Meyer*, German chemist (*1949).

In a typical sequence, the sesquioxides react with ammonium halide NH_4X and the respective hydrogen halide acid HX according to

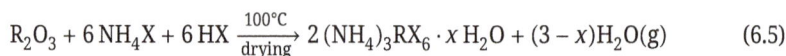

$$R_2O_3 + 6\,NH_4X + 6\,HX \xrightarrow[\text{drying}]{100°C} 2\,(NH_4)_3RX_6 \cdot x\,H_2O + (3-x)H_2O(g) \qquad (6.5)$$

Drying results in fair yields in hydrates of the triammonium-hexahalido-salt. This is further reacted under nitrogen atmosphere

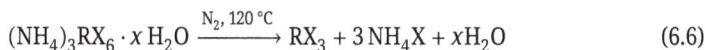

$$(NH_4)_3RX_6 \cdot x\,H_2O \xrightarrow{N_2,\ 120\ °C} RX_3 + 3\,NH_4X + x\,H_2O \qquad (6.6)$$

to give the trihalide. Unfortunately, this careful dehydration still yields significant amounts of halide oxides like ROX as side products. These can be separated by sublimation under high vacuum and at temperatures of up to 950 °C in tantalum vessels since any contamination with oxygen has to be avoided. A similar, even quicker approach via the reaction of the chloride hydrates of the most basic rare-earth elements with sulfuryl chloride, featuring a boiling point of 69 °C, under reflux [112] via

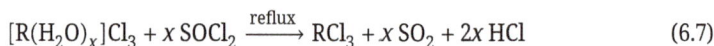

$$[R(H_2O)_x]Cl_3 + x\,SOCl_2 \xrightarrow{\text{reflux}} RCl_3 + x\,SO_2 + 2x\,HCl \qquad (6.7)$$

also yields contamination with chloride oxides.

The most straightforward and efficient approach for rare-earth *iodides* is the reaction of the elements according to

$$2\,R + 3\,I_2 \xrightarrow{300\ °C,\ \text{tantalum crucible}} RI_3 \qquad (6.8)$$

because of predominant formation of ROI by the aforementioned routes. For further reading regarding the syntheses of the trihalides, the already cited literature is recommended. The trihalides typically show the colors of the R^{3+} ions as mentioned earlier (Chapter 3.6).

Structures and Structure Systematics

With the decreasing size of the cations less anions can coordinate, and certainly the larger the anions become the smaller the coordination number. Since almost all trivalent halides are known and all of these compounds are ionic, this trend can be seen nicely in Figure 6.4 starting from a coordination number (c. n.) of eleven in LaF_3, decreasing via nine for the YF_3 and UCl_3 structure types to c. n. eight in the $PuBr_3$ structure, and finally dropping to c. n. six for representatives like ScI_3 adopting the $FeCl_3$, the related $AlCl_3$ or the ReO_3 structure types. In this section, these structure types will be briefly described and compared. The crystal structures discussed here are those reported at room temperature and ambient pressure. For instance, for SmF_3 at room temperature the YF_3 structure is found, but at elevated temperatures the LaF_3 structure type was

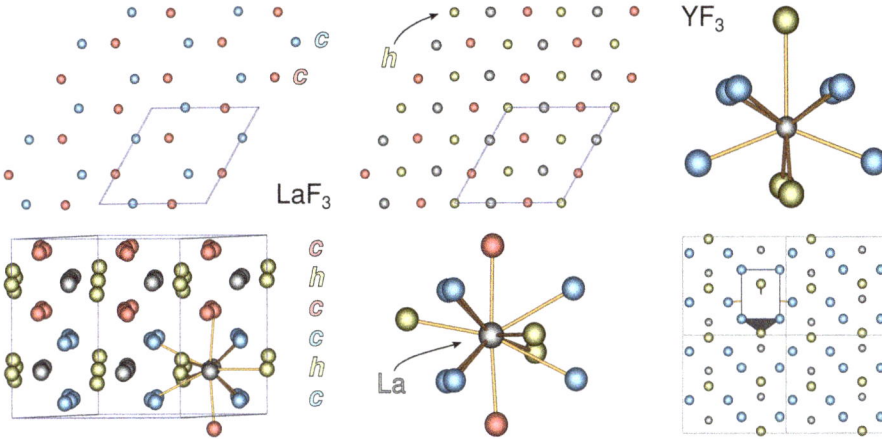

Figure 6.5: Crystal structure of LaF$_3$ (left); shown are the c (blue, red) and h (yellow) layers of F atoms, the full structure including La atoms and their surroundings; in YF$_3$ (right), the twin planes run through the yellow F atoms; data: [117, 118].

reported [119]. Or for ScF$_3$ at ambient pressure, the ReO$_3$ structure type is described but at high pressures the YF$_3$ type structure is achieved; these details will not be treated here. For further structure relations especially at higher temperature and pressure, please refer to the literature referenced in the introduction to this chapter.

LaF$_3$ consists of closely packed fluoride layers with a layer sequence *hcc* indicated in yellow, red and blue in Figure 6.5 (left). All layers are expanded so much that the La^{3+} fit into the triangular voids of the yellow distorted layer. Thus, lanthanum is coordinated by three fluorine atoms in the same layer (h layer), by six of the blue and by further two of the red layer (c layers). In total, eleven fluorine atoms coordinate every lanthanum atom forming a so-called Edshammar polyhedron [122]. An alternative structure description based on close packed lanthanum layers is given in Chapter 6.1.1.

YF$_3$ adopts the anti*cementite* structure, i. e., Fe$_3$C, with fluorine on the iron and yttrium on the carbon sites. This crystal structure can be understood by a concept named *chemical twinning* [123, p. 121]; we will make use of this approach in this chapter several times. Here, relevant structure motifs such as fragments of close packed layers are formally twinned by virtual mirror planes like in the YF$_3$ structure. Apparently, the octahedral voids in a close packed structure of fluorine atoms were too small to host the large yttrium ions properly. So, fragments of a hexagonal close packing of fluorine layers, indicated in blue in Figure 6.5, are twinned via virtual mirror planes running through the yellow fluorine atoms. By this 3,3-twinning trigonal prismatic voids are generated instead of octahedral voids with contacts to further three fluorine atoms resulting in a tricapped trigonal prismatic coordination around the yttrium atoms.

In ScF$_3$, i. e., the *ReO$_3$ type*, all scandium atoms are surrounded octahedrally by fluorine atoms. These octahedra share common vertices along all directions resulting in

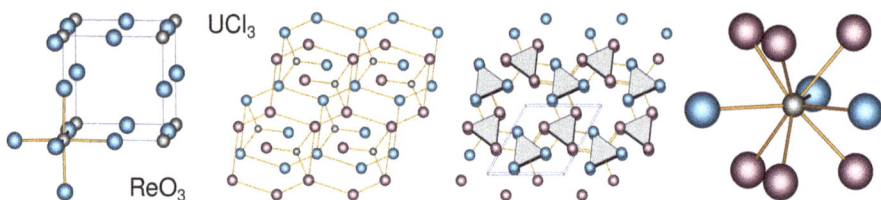

Figure 6.6: Crystal structures of the ReO$_3$ type on the example of ScF$_3$ (left) and the UCl$_3$ type on the example of GdCl$_3$ (right); metal atoms grey, fluorine atoms blue as well as the distorted layers of chlorine atoms (blue, red) in GdCl$_3$; data: [120, 121].

a structure closely related to *perovskite* lacking only the alkaline-earth atoms therein (Figure 6.6).

The uranium atoms in the *UCl$_3$ structure type* choose a similar approach like the lead atoms in PbCl$_2$—realized in EuH$_2$ (Chapter 6.1.1). Here, the chlorine atoms form a hexagonal close arrangement; the layers are indicated with blue and red spheres in Figure 6.6. To host the uranium atoms, both layers are distorted to fit the uranium atoms in selected triangular voids. Thus, two adjacent octahedral voids fuse via their shared face and lead to a ninefold-coordination sphere around the uranium atoms, a tricapped trigonal prism. In this structure type, the anion-anion distances are potentiallly smaller compared with the structure type of YF$_3$. But here the anions are larger, and thus experience less electrostatic repulsion.

In the *PuBr$_3$ structure type*, essentially the same layers indicated in blue and red like in UCl$_3$ are found. Also here, they form a hexagonal close packing. But these layers are somewhat more distorted than there, thus reducing the coordination number of the cations from nine to eight—a bicapped trigonal prism (Figure 6.7). Due to this distortion,

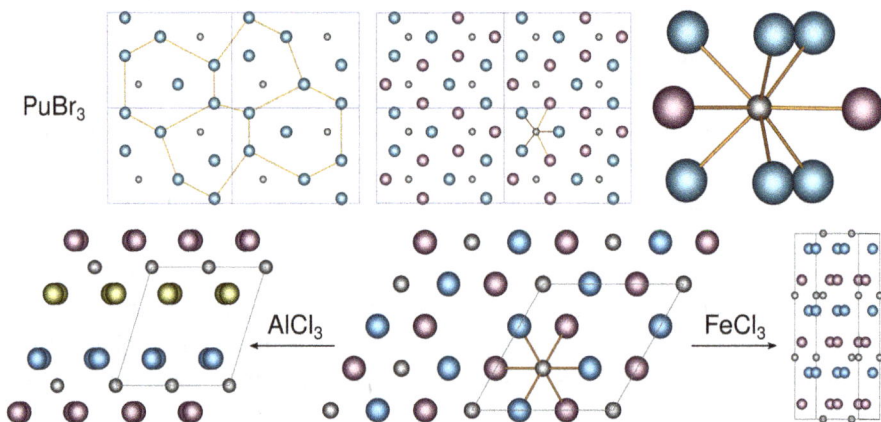

Figure 6.7: Crystal structures of PuBr$_3$ on the example of NdBr$_3$ (top), AlCl$_3$ and FeCl$_3$ on the example of ScCl$_3$ (bottom); shown are the metal atoms (grey) and the halide atoms colored according to the layers (blue, red and yellow); data: [124–126].

in $PuBr_3$ the empty channels are clearly smaller than in UCl_3. The difference between the hexagonal close packing with the layer sequence AB and the cubic close packing with layer sequence ABC is that in the former the atoms experience in average slightly smaller distances. There is no difference in the first two coordination spheres where twelve and six atoms are seen; but in the third—the next but one layer of atoms further—the atoms of the next A layer show smaller distances than those of the respective C layer in cubic packings. Therefore, more ionic compounds prefer the cubic close, more covalent compounds the hexagonal close packed structures. Considering the halides, the tendency to covalent and dispersive interactions markedly increases from fluoride to iodide.

So, the *AlCl₃ structure type* where a cubic close packing of chlorine atoms hosts aluminum atoms in two-thirds of every second layers' octahedral voids is indeed preferred by the harder halides than the *FeCl₃ structure type* adopted by the softer halides. Here, the chlorine atoms form a hexagonal close packed structure with the same occupation of octahedral voids than in $AlCl_3$. Both structures are depicted in Figure 6.7.

6.2.2 Tetrafluorides RF_4

Only a few tetrafluorides have been reported so far. Already chloride obviously provides sufficiently reducing conditions to endure besides the most stable tetravalent rare-earth ion Ce^{4+}. CeF_4 can be synthesized employing pure fluorine via

$$2\,CeF_3 + F_2 \xrightarrow{\text{r.t., HF (l)}} 2\,CeF_4 \tag{6.9}$$

from the trifluoride. The two other fluorides require harsh conditions and via u. v. light activated fluorine according to

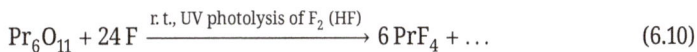

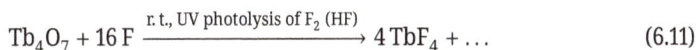

$$Pr_6O_{11} + 24\,F \xrightarrow{\text{r.t., UV photolysis of } F_2 \text{ (HF)}} 6\,PrF_4 + \ldots \tag{6.10}$$

$$Tb_4O_7 + 16\,F \xrightarrow{\text{r.t., UV photolysis of } F_2 \text{ (HF)}} 4\,TbF_4 + \ldots \tag{6.11}$$

starting from the already mixed valent oxides of both [127]. All three tetrafluorides suffer from rather easy decomposition under the evaporation of fluorine gas

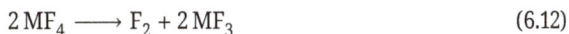

$$2\,MF_4 \longrightarrow F_2 + 2\,MF_3 \tag{6.12}$$

reflecting the decreasing stability with the decomposition temperatures of 600 °C (CeF_4), 300 °C (TbF_4) and 90 °C (PrF_4). All three compounds crystallize isotypically with UF_4, where the metal atoms are coordinated by eight fluorine atoms forming a square antiprism. These antiprisms build a hexagonal close packing (blue and red in Figure 6.8) with further (yellow) antiprisms in voids of the packing [128].

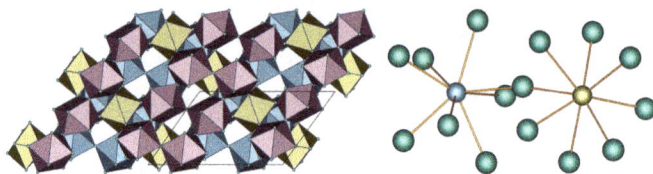

Figure 6.8: Crystal structure of CeF_4; shown are h packed CeF_8 antiprisms (blue and red) with further antiprisms in voids (yellow) as well as the surroundings of the two crystallographically distinct metal atoms (blue and yellow); data: [129].

6.2.3 Reduced Halides

Like the hydrogen compounds, the halogen compounds also feature a broad variety of compositions between the full oxidation states. In this chapter on binary reduced halides, my focus lies on the simple divalent compounds RX_2. At the end of this chapter, properties of selected examples of reduced halides will be mentioned. Especially in this topic, the famous chemists Bärnighausen,[35] Simon,[36] Corbett[37] and Meyer[34] contributed groundbreaking work.

Overview and Syntheses

Figure 6.4 illustrates impressively the dominance of the oxidation state +III in rare-earth chemistry; almost all combinations of rare-earth elements with halogens have been reported over the last century. Considering the divalent compounds, only a few elements form truly ionic halides, and certainly those where a reduction from trivalent to the divalent state is conceivable— such as europium, ytterbium and samarium (Table 3.3, p. 30). For these, even fluorides can be synthesized, which is remarkable as fluorides usually stabilize higher oxidation states. Iodides however stabilize lower oxidation states due to their reduction potential. Accordingly, a majority of the rare-earth elements at least forms the diiodides RI_2 with the exception of holmium, where only $HoCl_2$ was reported to date [130–132].

Generally, the reduced halides show considerable phase widths; so normally long reaction times are necessary to achieve homogeneous products. A typical synthetic approach is the comproportionation of the respective rare-earth elements and its trivalent halide in sealed tantalum containers enclosed in glass ampoules according to endothermic reactions

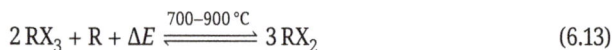

$$2\,RX_3 + R + \Delta E \xrightleftharpoons{700-900\,°C} 3\,RX_2 \qquad (6.13)$$

---

35 *Hartmut Bärnighausen*, German chemist and crystallographer (*1933).

36 *Arndt Simon*, German chemist (*1940).

37 *John Corbett*, American chemist (*1926 †2013).

for which the reaction enthalpy almost solely depends from the different lattice energies, e. g.,

$$\Delta E \approx 2U(\text{RF}_3) - 3U(\text{RF}_2) \tag{6.14}$$

with the lattice energy U [133–136]. Since the fluorides are by far the smallest ions allowing for the largest coordination numbers, the lattice energy differences are comparably large there. This is reflected in the reaction enthalpies, and thus confirms the aforementioned better stabilization of higher oxidation states. Further, the synthesis of reduced rare-earth halides can succeed via a metallothermic reduction with alkaline metals according to

$$\text{LaI}_3 + 2\,\text{Na} \xrightarrow{550\ ^\circ\text{C}} \text{LaI} + 2\,\text{NaI} \tag{6.15}$$

at comparably lower temperatures as long as no ternary phases form [137–139]. The alkaline halides may be removed by treatment with solvents like anhydrous diglyme[38] as the rare-earth compounds are sensitive to humidity. Another synthetic approach is the reduction of the trihalides with hydrogen—only successful for europium, ytterbium and samarium—or a direct electrochemical reduction [140]. Detailed syntheses are described in [112, 141].

Structures and Bonding

Most of the dihalides are ionic compounds except those of the rare-earth elements which either occur only as trivalent ions or even tetravalent ions—lanthanum, cerium, praseodymium and gadolinium. While the former can safely be described as salts $R^{2+}(X^-)_2$, the latter are metallic according to $R^{3+}(X^-)_2 \cdot e^-$ where the rare-earth atoms are present as trivalent ions with the surplus electron being itinerant in the conduction band. The latter feature the structure of $MoSi_2$ while the former adopt typical structures of ionic compounds discussed previously.

The only known fluorides (SmF_2, EuF_2 and YbF_2) crystallize isotypically with *fluorite* (Figure 6.1). The dichlorides show a nice trend of decreasing coordination numbers from the larger to the smaller rare-earth elements starting from the *cottunite* type structure (Figure 6.2) with a ninefold coordination. Eightfold coordination is then realized in NdI_2, $SmBr_2$, $DyCl_2$ and $EuBr_2$ by adopting the *SrBr$_2$* structure type. Figure 6.9 displays motifs of the crystal structure of $EuBr_2$. Here, the europium atoms form a distorted cubic close packing. In the tetrahedral voids, the bromine atoms are situated. You might now expect a structure similar to that of *fluorite* and a cubic surrounding of the cations. But here, the distortion of the close packing yields a square antiprismatic surrounding of the first europium atom and to a quite similar coordination of the second for which a

---

38 diglyme = diethylene glycol dimethyl ether.

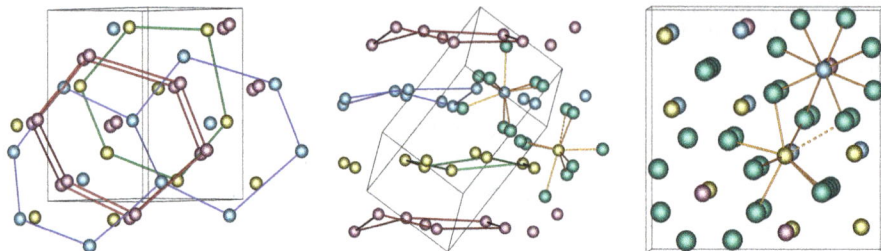

Figure 6.9: Crystal structure of EuBr$_2$; shown are the three layers of the distorted cubic close packing of europium atoms (blue, yellow and red) in two different orientations including the local surroundings with (green) bromine atoms as well as the unit cell; data: [142].

7+1 coordination is found with a single bromine atom further away (Eu–Br: 309–328 and 352 pm). MAPLE calculations (see the info box in Chapter 6.1.1 and Appendix F.2) confirm a significant electrostatic contribution of the eighth bromine atom to the coordination.

In the monoclinic *EuI$_2$ structure type*, the coordination number is further reduced to seven. This structure may be understood by using the chemical twinning approach discussed earlier in this chapter. A 3,3-twinning was discussed there for YF$_3$; here, a 2,2-twinning of a hexagonal close packing of bromine atoms can be deduced from Figure 6.10. On the mirror planes, also trigonal prismatic voids are generated. In contrast to YF$_3$, here only one further halide ion is in reach to complete the sevenfold coordination around europium.

A slightly less denser structure than that of EuI$_2$ is the orthorhombic *SrI$_2$ structure type*, which is related to that of EuI$_2$. Simply spoken, the mirror plane containing the yellow atoms in Figure 6.10 becomes broader due to a relative tilting of the coordination polyhedra, which apparently is the reason for the doubling of the unit cell here. The coordination polyhedra are very similar. YbBr$_2$ and the remaining ionic diiodides

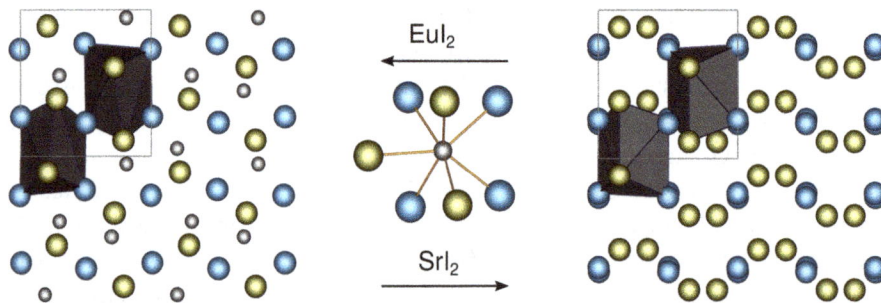

Figure 6.10: Crystal structures of EuI$_2$ (left) and SrI$_2$ on the example of YbCl$_2$ (right); illustration of the chemical twinning where the iodine atoms spanning the twin planes are shown as yellow and the others solely belonging to the h layers as blue spheres, in between the coordination polyhedron of EuI$_2$ is depicted, which is very similar to that in SrI$_2$; data: [130, 143].

of dysprosium, thulium and ytterbium adopt the structures of $CdCl_2$ and CdI_2 discussed in the info box. The rare-earth atoms are therein only coordinated octahedrally according to their smaller ionic radii.

Important crystal structure types no. 2–CdCl₂ and CdI₂: The crystal structures of $CdCl_2$ and CdI_2 are illustrated in Figure 6.11. $CdCl_2$ consists of a cubic close packed arrangement of chlorine atoms (red, blue and green layers) with all octahedral voids of every second layer occupied by cadmium atoms (grey); in each close packed structure, two tetrahedral and one octahedral void per packing atom is found. In the case of CdI_2, the same occupation scheme of the octahedral voids is found, but this structure is based on a hexagonal close packing of the iodide atoms. These structure types are adopted by various rare-earth element compounds. For the basics of this structure related to others, see also Appendix D.

The metallic diiodides RI_2 (R = La, Ce, Pr, Gd) crystallize in the *MoSi₂ structure type* shown in Figure 6.11. Here, the metal atoms are coordinated by ten iodide atoms in the form of a bicapped cube. The whole structure can be understood as it is built up by CsCl units shifted with respect to their neighbors. In CsCl, a primitive cubic packing of chlorine hosts caesium atoms in the cubic voids.

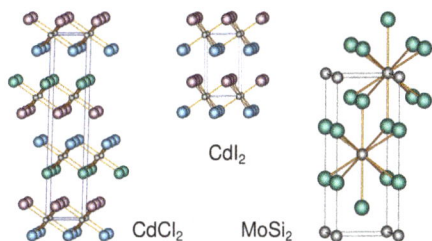

Figure 6.11: Crystal structures of $CdCl_2$ (left), CdI_2 on the example of TmI_2 (middle) and $MoSi_2$ (right); metal atoms in grey, halide atoms in green, red and blue; data: [144–146].

Selected Further Examples and Their Properties

On mixed valent europium halides intervalence charge transfer transitions and electron hopping were scrutinized. This will be addressed in Chapter 8.2.2. Metallic GdI_2 shows ferromagnetic ordering (T_C = 290 K) and a giant negative magnetoresistance (GMR)[39] of approximately 70 % at room temperature and 7 Tesla [147]. Thanks to the successful metallothermic reduction of LaI_3 with sodium and the separation of NaI phase pure samples of LaI could be thoroughly investigated. Electronic structure calculations confirmed the metallic behavior according to $La^{3+}(I^-) \cdot 2e^-$. Surprisingly, for a material of the NiAs structure type (Appendix D), almost no covalent metal-metal interactions along the *c* axis but within the *ab* plane were identified. Physical property measurements confirmed the metallic behavior despite a quite low density of states at the Fermi level [138].

---

39 under the influence of a magnetic field, the resistance decreases.

6.2.4 Halide Oxides

Overview and Syntheses

Also, for halide oxides—or oxyhalides—a manifold of compositions was reported, such as $Gd_4O_3F_6$, Sm_4OI_6, $Lu_3O_2F_5$ or Eu_3O_4Br. In this chapter, I will focus on the basic and most important ROX compounds. Their synthesis frequently occurs involuntarily if, e. g., an anhydrous halide was intended to be made from an oxygen containing source. Compared with the heavier halides, oxide is always the harder ion, and thus preferred according to

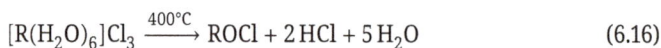

$$[R(H_2O)_6]Cl_3 \xrightarrow{400°C} ROCl + 2\,HCl + 5\,H_2O \tag{6.16}$$

forming the respective halide oxide in an open reaction vessel [112, p. 1085]. Further heating in air yields at first halide oxides with higher oxygen content and finally the sesquioxides. Another very reliable so-called *ammonium halide route* runs via partial formation of an ammonium halogeno complex

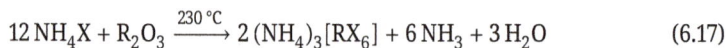

$$12\,NH_4X + R_2O_3 \xrightarrow{230\,°C} 2\,(NH_4)_3[RX_6] + 6\,NH_3 + 3\,H_2O \tag{6.17}$$

as discussed in the course of the synthesis of the trivalent halides. To target the halide oxides, usually a suited ratio of ammonium halide to sesquioxide corresponding to

$$2\,NH_4X + R_2O_3 \xrightarrow{230\,°C} 2\,ROX + 2\,NH_3 + H_2O \tag{6.18}$$

is employed, which requires a careful stoichiometric control of the starting materials of this latter reaction to obtain the desired halide oxide [148, 149].

Contrarily, fluoride and oxide show a comparable chemical hardness. Moreover, the fluorides show a certain volatility. Therefore, the ROF require syntheses in closed vessels, such as platinum tubes sealed by squeezing both ends, at quite high temperatures corresponding to

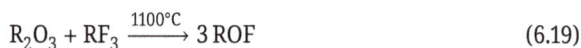

$$R_2O_3 + RF_3 \xrightarrow{1100°C} 3\,ROF \tag{6.19}$$

by reaction of the sesquioxide with the trifluoride [112, p. 256]. The discussion addressed all rare-earth elements except the scandium and cerium compounds. The above reactions yield only the dioxide CeO_2 in open systems due to its tendency to form oxidation state IV. Further, the higher oxophilia of scandium prevents success via these routes. Hence, scandium and cerium halide oxides have to be synthesized like the ROF in closed vessels under the exclusion of oxygen or in an alkaline metal halide flux [150, 151].

Structure Chemistry

Unfortunately, not all possible combinations of structure data are available as yet, but considering usual trends the information of reliable data bases and several original publications deliver a quite complete picture so that the missing members can be extrapolated [152–157]. Regarding the crystal structures, the cerium and scandium compounds play their own role; therefore, they will be discussed briefly at the end of this section. The general rule for the remaining halide oxides is that the coordination number decreases from the largest La^{3+} to the smallest Lu^{3+} ions (Figure 6.14). Since apparently the size of the ions plays a crucial role, the assumption may be made that the yttrium compounds are usually isotypic with the erbium compounds. Due to some reported polymorphy, e. g., in [155], I will only focus on the structures at ambient temperature and pressure.

The fluorides LaOF through ErOF adopt the crystal structure of NdOF, those from TmOF through LuOF adopt the *baddeleyite* structure type of ZrO_2. In the *NdOF structure type* according to Figure 6.12, double layers of oxygen and fluorine atoms form a cubic close packing c. The neodymium atoms occupy all octahedral voids between adjacent layers of oxygen and fluorine while the voids between two neighboring layers of the same anions remain empty. This enables a shrinking of the interlayer distance here and allows for the coordination of the neodymium atoms in the slightly compressed octahedral voids by two further atoms of the next, but one layer. Thus, an eightfold coordination is achieved where the two further atoms cap opposite triangular faces. Regarding structure systematics, this structure is strongly related to that of $CdCl_2$ (Appendix D).

The *baddeleyite structure type* is a distorted variant of the *fluorite* structure (Figure 6.1). Here, the lanthanide atoms are coordinated by four oxygen and three halogen atoms as depicted in Figure 6.12.

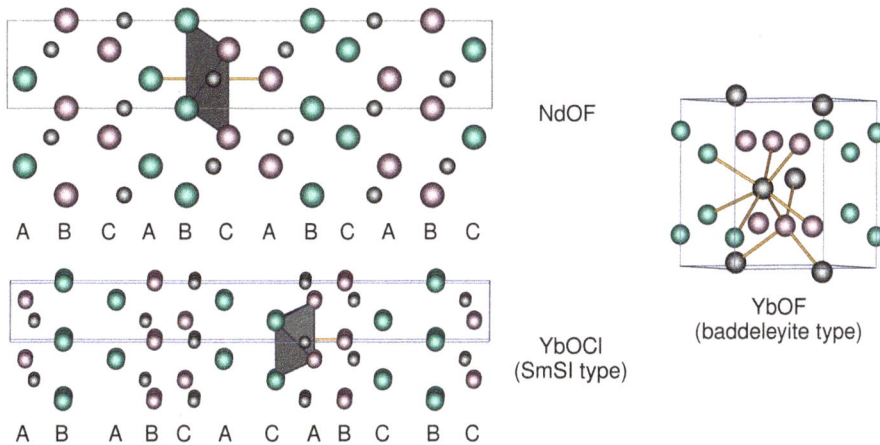

Figure 6.12: Crystal structures of NdOF, YbOCl (SmSI type) and YbOF (*baddeleyite* type); metal atoms grey, the halide and the oxygen atoms green and red; for NdOF and YbOCl, the halide layers are indicated by A, B and C; data: [156, 158, 159].

The chlorides LaOCl through HoOCl and YOCl crystallize in the PbFCl structure (*mat-lockite*) with oxygen replacing fluorine and the rare-earth elements replacing lead. The remaining ErOCl, TmOCl, YbOCl and LuOCl adopt the SmSI structure. The *NdOCl structure* (PbFCl type) is displayed in Figure 6.13 and is derived from *rutile* in the same manner as $PbCl_2$ from the α-PbO_2 structure. For further details, see Appendix D. In both structures, the hexagonal close packing of anions achieves ninefold-coordinated voids by formal fusion of two adjacent octahedral voids. Mixed oxygen-chlorine layers (red and green) form the *h* packing in NdOCl where neodymium is coordinated by five chlorine and four oxygen atoms.

The *SmSI structure type* is closely related to the previously elucidated NdOF type. Here, the sequence of double layers chalcogenide and halide switches from *c* to *hhcc* (Figure 6.12). Remember that a layer, which has the same adjacent layers is named *h* and a layer with two different neighboring layers, is named *c*. The occupation of octahedral voids of every second layer remains the same, but the change of the layer sequence yields a sevenfold coordination as one of the two capping atoms above the opposite triangular planes vanishes.

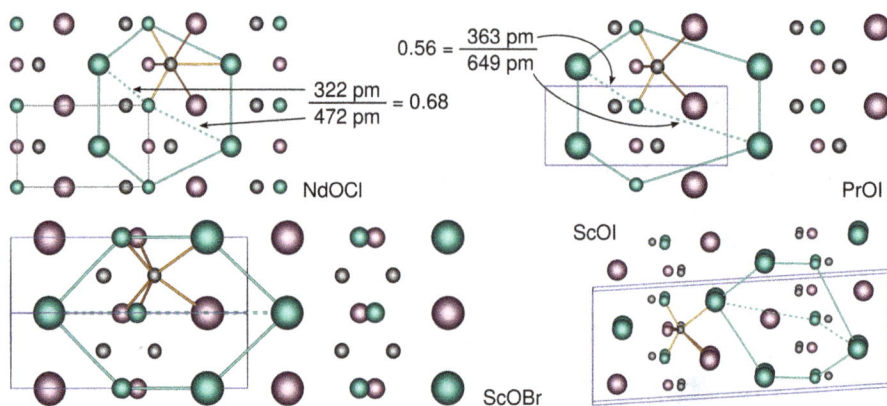

Figure 6.13: Crystal structures of NdOCl, PrOI, ScOBr and ScOI (top left to bottom right); metal atoms in grey, the larger halide and the smaller oxygen atoms are in green and red according to the layers in a hexagonal close packing; the indicated distances illustrate the different distortion; data: [160–163].

For LaOBr through NdOBr the PbFCl structure with ninefold coordination of the rare-earth atoms is found. As expected, from SmOBr through LuOBr and for all iodide oxides, the *PrOI structure type* is adopted, which provides a lower coordination number. This structure is a stretched variant of the PbFCl type so that one out of three capping atoms gets out of reach, and an eightfold coordination is achieved (Figure 6.13). This stretched distortion is quantified by the ratio of the indicated analogous distances within the green *h* layers of NdOCl and PrOI.

ScOF behaves like LuOF. For ScOBr, experimentally the FeOCl structure type was found, which is also the most stable according to calculations; the same holds for ScOCl

[155, 164]. The *ScOBr structure* is shown in Figure 6.13 and consists of a hexagonal close packed structure of oxygen and chlorine atoms similar but not identical to that of PbFCl. The *h* layers are indicated in red and green. The grey scandium atoms reside in half of the octahedral voids so that chains of edge sharing octahedra result. A distorted variant of ScOBr based on a packing of mixed *h* layers of oxygen and iodine is found in *ScOI* where the scandium atoms enter the layers as discussed above for lead atoms in PbFCl and PbCl$_2$. Thus, for the scandium atoms a sevenfold coordination is achieved (Figure 6.13). Regarding the cerium compounds, it was only reported that CeOF crystallizes in the *fluorite* type and CeOCl analogously to LaOCl in the PbFCl type. This can be extrapolated also for CeOBr, and for CeOI; thus, the PrOI structure can be assumed.

Among the ROX several structures with a layered ordering of the anions of different sizes and charges are observed, such as the NdOF, PbFCl and PrOI structure types. Bärnighausen[35] noted that throughout the ROX series with PbFCl structure the close distances R–O are shorter and the close R–X distances are longer than the sum of ionic radii; moreover, the X–X distances are smaller than the sum of their ionic radii, and the anions overlap. Therefore, he inferred interactions between the halogen atoms [153]. These trends also hold for the averaged distances within the coordination of the R^{3+}. Employing the revised ionic radii of 1976 [50], this also applies for the other structure types mentioned above as depicted in Figure 6.14 where also the interatomic X–X distances are shown. For each structure type, the respective ionic radii specific for the differing coordination numbers were used. The same behavior as observed by Bärnighausen is found for the similar structure type of PrOI, although to a lesser extent. This seems reasonable since the PrOI structure type is a stretched variant of the PbFCl type. In both cases, the overlap increases going from the harder to the softer anion—in absolute figures and relatively. Considering the NdOF type, there is no indication for an overlap of the hard fluorine atoms as the F–F distances are in all cases larger than the sum of the van der Waals radii. Also here, the doubly charged oxide ions are somewhat closer to the R^{3+} than expected, the singly charged fluoride ions somewhat further away. One might assume electrostatic reasons for this trend.

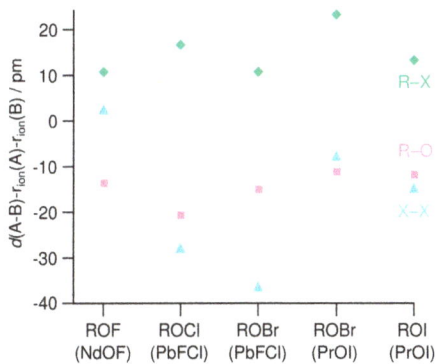

Figure 6.14: Differences between the average interatomic distances A–B (ions A and B) and the sum of the respective ionic radii, the structure types are given in parentheses.

6.3 Oxides

The oxides reflect the different stabilities of the oxidation states of the rare-earth elements. Accordingly, monoxides, sesquioxides R_2O_3 and dioxides RO_2 have been identified as the most stable ones discussed here. For elements like praseodymium and terbium, intermediate oxides between the latter two are found as a stable form indicating a broad phase width. The sesquioxides especially feature very high melting points between 2200 and 2500 °C, and CeO_2 melts around 2480 °C. Because of the higher lattice energy, these are significantly higher compared with the trifluorides ranging from roughly 1100 through 1550 °C. The boiling points of the Ln_2O_3 increase continuously from about 3620 °C for La_2O_3 to 3980 °C for Lu_2O_3 passing an intermediate maximum of 4070 °C for the ytterbium compound. *Further details* on the oxides and their properties, which are not treated in this chapter, may be studied in [165–167].

6.3.1 The Sesquioxides R_2O_3 and Mixed III/IV-Oxides

The sesquioxide—the rare earth—is for most of the rare-earth elements the most stable oxide. Thus, these are obtained during the industrial separation process as a first clean product from the hydroxides. Also, the sesquioxides form by simple reaction of the pure metals with air—except those of cerium, praseodymium and terbium; these can be obtained under a hydrogen atmosphere from the first oxidation product. Most easily, those oxides are obtained by smooth decomposition around 1000 °C of the respective hydroxides, carbonates, nitrates or oxalates as precursors, which are readily available [166]. As discussed in the previous chapter, thermal decomposition of the halide hydrates yields primarily halide oxides, but also finally the oxides. For the sesquioxides not less than five crystalline polymorphic forms, enumerated C to A, H and X, have been identified so far for ambient pressure and between room temperature and the respective melting points. An overview on the phase diagram for the lanthanide sesquioxides is given in Figure 6.15.

For the large lanthanide ions, the trigonal *A type structure* shown in Figure 6.16 is dominating. Here, two approaches to describe the crystal structure are used in literature. The first considers the lanthanide atoms as forming an almost undistorted hexagonal close packing of two layers indicated in blue and green. Red oxygen atoms occupy alternating layers of octahedral and tetrahedral voids. Thereby half of the octahedral and tetrahedral voids are occupied (Appendix D). Thus, the lanthanide atoms feature a sevenfold coordination, essentially a defect cube-like surrounding with one oxygen atom missing. The second, more classic approach, considers a packing of the anions. A distorted primitive cubic packing of oxygen atoms hosts the lanthanide atoms in two-thirds of all cubic voids, the latter being off-center to achieve a reduced coordination number of seven. This surrounding may be described as a monocapped octahedron. The first description may be more comprehensible; the latter one better illustrates the

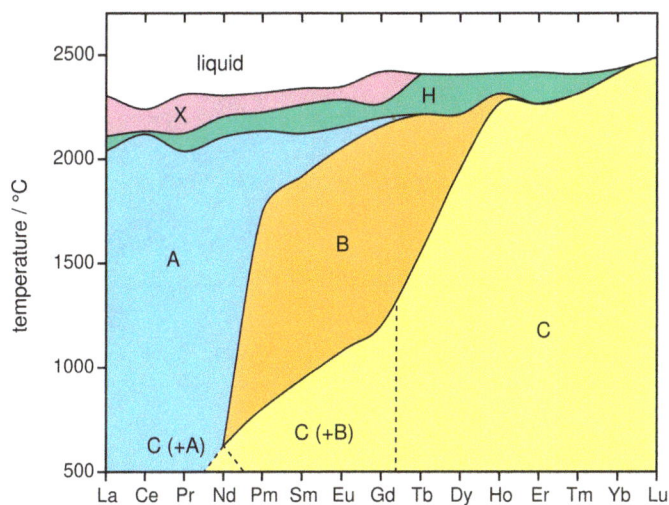

Figure 6.15: Phase diagram of the polymorphs A, B, C, H and X of lanthanide sesquioxides.

relationship to the thermodynamically neighbored phases as these are derived from the *fluorite* structure as shown in Figure 6.1.

The hexagonal *H type structure* is the high-temperature polymorph of the A type. Therein the oxygen atoms are disordered over all tetrahedral and octahedral voids of the A type structure. At even a higher temperature, the *X polymorph* was reported to feature a body centered cubic arrangement of the rare-earth ions where the oxygen atoms are disordered over all octahedral voids, which in this case, are slightly compressed octahedra. These voids are half-occupied since six voids and two rare-earth atoms are present per unit cell.

In the monoclinic *B-type structure* of Sm_2O_3, depicted in Figure 6.16, the oxygen atoms form a distorted hexagonal close packing with two-thirds of the octahedral voids being filled with samarium atoms [172]. The corrugated *h* layers show a tiny relative shift so the octahedral voids in this packing approach a prismatic appearance. While one out of the three different samarium atoms is indeed octahedrally coordinated, the other two experience a sevenfold, single-capped trigonal prismatic coordination sphere caused by the distortion of the close packed layers. These coordination numbers were also confirmed by MAPLE calculations (see the info in Chapter 6.1.1 and Appendix F.3).

The *C-type structure*, also known as the *bixbyite* type, crystallizes cubically and derives from the well-known *fluorite* type structure by a $2 \times 2 \times 2$ superstructure yielding a basic composition $R_{32}O_{64}$. To achieve the R:O = 2:3 ratio, ordered vacancies □ forming rows of empty tetrahedral voids along the unit cell's space diagonals are present according to $R_{32}O_{48}\square_{16}$. These vacancies reduce the coordination number of eight in *fluorite* to six causing a slight distortion of the whole structure with respect to the aristotype. Figure 6.16 displays a virtual face centered unit cell not identical with the real one. This representation simplifies the derivation from the *fluorite* type; the voids are indicated by different colored anions.

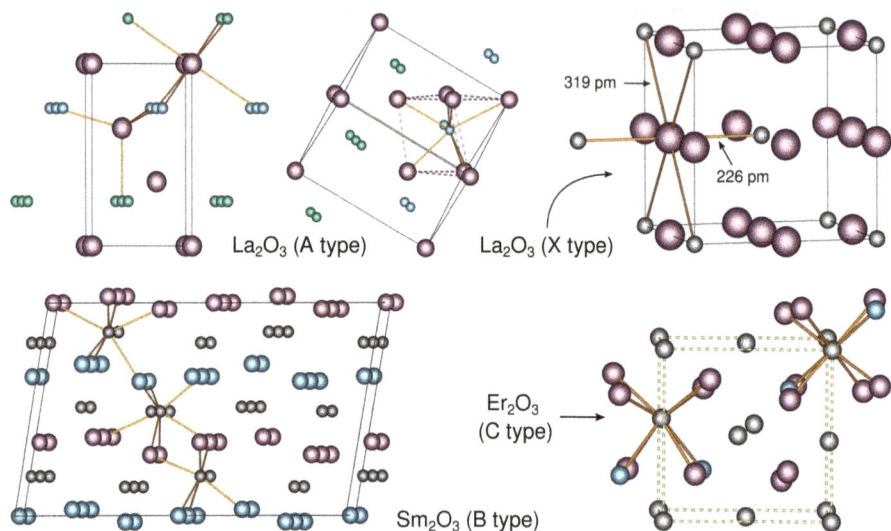

Figure 6.16: Structure types A (*h* layers of La in blue and green, oxygen atoms red), X (La grey, O red), B (*h* layers of oxygen red and blue, Sm grey) and C (dotted yellow lines indicate a unit cell analogous to that of *fluorite*, vacant positions in blue, oxygen atoms red); data: [168–171].

The decreasing coordination numbers from A via B to the C-type suggest that the decreasing ionic radii are the main origin of the phase stability reflected in Figure 6.15. In this context, the not yet mentioned oxides Sc_2O_3 and Y_2O_3 fit very well; both adopt the C-type structure, which apparently correlates with their respective ionic radii comparable with that of erbium (Y) or the smallest of all rare-earth elements (Sc) and the according preference for a smaller coordination number.

6.3.2 Syntheses and Structures of the Dioxides and Intermediate Oxides

The dioxides RO_2 have been described for all rare-earth elements, which tend to form tetravalent ions; these are cerium, praseodymium and terbium. For these three, numerous phases between R_2O_3 and RO_2 are mentioned and discussed; out of this series the most important are bright yellow CeO_{2-x}, brown $PrO_{1.8\pm x}$ ("Pr_6O_{11}") and deeply brown $TbO_{1.7\pm x}$ ("Tb_7O_{12}" or "Tb_4O_7") since they are usually obtained by simple oxidation of the pure metals in air. They are subject to considerable phase widths indicated by x. With increasing deviation from the composition RO_2, the oxides show intense coloring. These colors are due to charge-transfer transitions, which are the subject of Chapter 8.2. Charge-transfer transitions occur between oxide anions and Ln^{4+} as LMCT or as intervalence charge-transfer transitions (IVCT) between Ln^{4+} and Ln^{3+}. To get pure dioxides, the syntheses have to be conducted under a pure oxygen atmosphere for CeO_2 at ambient pressure, and in the case of PrO_2 and TbO_2 under high oxygen pressures.

Numerous crystal structure analyses illustrate the demand for more knowledge in these phase diagrams, to find more distinct phases—several of which have been claimed to exist. All of these apparently represent flat thermodynamic minima reflected in the broad phase widths. The structures derive from the *fluorite* type adopted by all dioxides; this can either be described as a face centered cubic close packing of cations with almost all tetrahedral voids filled with oxygen or—the anion packing related approach—as a primitive cubic packing of oxygen with the metal atoms in every second cubic void. Here, I prefer the former description. Anyway, the tetravalent ions thus are seven or eightfold coordinated, which is somewhat surprising regarding their size that decreases in the row Ce^{4+}, Pr^{4+}, Tb^{4+} and nicely reflecting their thermal stability or difficulty to be synthesized at all.

Pr_6O_{11} and Tb_7O_{12}

Both structures, albeit crystallizing in different crystal systems with very different unit cells, derive from the *fluorite* type structure discussed earlier in this chapter (Figure 6.1 and Appendix D). Therefore, in Figure 6.17 the *fluorite* type unit cells are illustrated by thin orange lines. The black spheres represent ordered defect sites within the packing of red oxygen atoms. Further, the actual unit cells of both structures are also shown.

For monoclinic Pr_6O_{11}, three *fluorite* type unit cells based on oxygen and praseodymium (blue) are shown. Without defects, these three unit cells represent $3 \times 4 = 12$ praseodymium and $3 \times 8 = 24$ oxygen atoms. Since there are two oxygen defects (black spheres), the overall composition $Pr_{12}O_{22}$ or Pr_6O_{11} is achieved. A similar game may be played with rhombohedral Tb_7O_{12} where at least seven *fluorite* type unit cells are

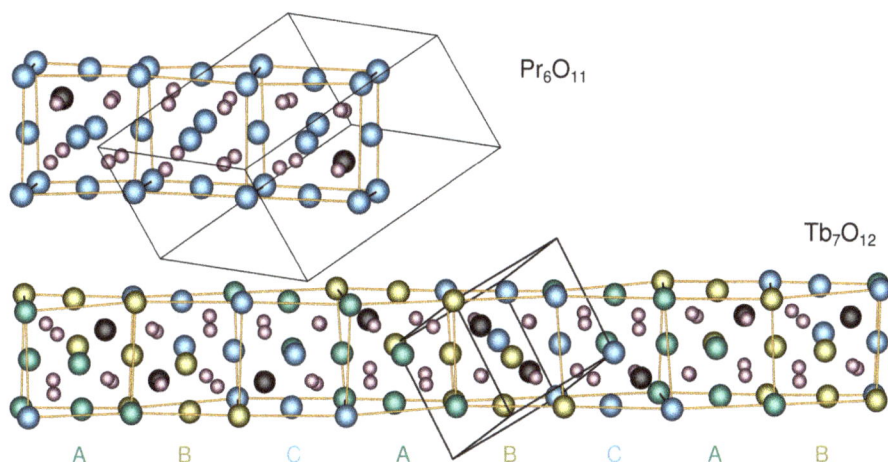

Figure 6.17: Structures of Pr_6O_{11} and Tb_7O_{12}. The oxygen atoms are shown as small red spheres. The oxygen defects as large black spheres. The rare-earth atoms as larger blue, green and red spheres; for Tb_7O_{12}, the different colors illustrate the cubic close packing of the R^{3+}; data: [173, 174].

needed. Thus, $7 \times 4 = 28$ terbium and $7 \times 8 - 8 = 48$ oxygen atoms are displayed, which fits with four formula units of Tb_7O_{12}. Moreover, in Figure 6.17 the terbium layers of the slightly distorted cubic close packing are indicated in different colors.

6.3.3 A Closer Look On CeO_2

For several reasons, cerium dioxide is a highly interesting oxide. Some of the most promising applications will be roughly discussed; for further reading, [175] is recommended. The phase diagram of the system Ce_2O_3–CeO_2 is well known [176]. Interestingly, we will come across the two compositions discussed just before. If CeO_2 is heated to temperatures above 680 °C under reduced oxygen pressure, the α-phase CeO_{2-x} forms. This phase adopts the cubic *fluorite* type with randomly distributed oxygen defects. Its typical phase width is indicated by x ranging between 0 and 0.29. Prolonged tempering leads to an ordering of the oxygen defects. Important phases formed are the β-phase, i. e., $CeO_{1.83}$ = Ce_6O_{11}, a rhombohedral phase, i. e., $CeO_{1.71}$ = Ce_7O_{12}, both of which are isostructural to the intermediate oxides of praseodymium and terbium described above. A further phase is the triclinic δ-phase $Ce_{11}O_{20}$ ($CeO_{1.82}$). This last example illustrates how many ordered phases may be found in this regime below 680 °C.

Further reduction, e. g., by hydrogen, yields the σ and ϑ-phases between $x = 0.29$ and $x = 0.5$ [177]. The σ-phase is a padded variant of the C-type structure Ce_2O_{3+y}, the ϑ-phase forms at compositions closer to Ce_2O_3 and adopts the A-type structure of the sesquioxide. While the latter is stable against reoxidation with oxygen at room temperature, the former is not and yields Ce_7O_{12} again. Any reduction of CeO_2 leads to an increase of the inner surface, which makes such materials interesting for catalysis.

Oxygen Storage
An obvious capability of the phases between Ce_2O_3 and CeO_2 is the uptake and release of oxygen with increasing and decreasing oxygen partial pressure, respectively. A typical equilibrium

$$2\,Ce_2O_3 + O_2 \underset{}{\overset{400°C}{\rightleftharpoons}} 4\,CeO_2 \tag{6.20}$$

operates around 400 °C. The *oxygen storage capacity* becomes even better in the presence of smaller cations such as Zr^{4+}.[40] Since ZrO_2 forms a tetragonal phase strongly related to *fluorite* when doped with rare-earth ions, solid solutions are easily obtained in the full regime ZrO_2–CeO_2. The same holds for Ce_2O_3 where a doping with Zr^{4+} results

---

[40] relevant ionic radii in this context: $r^{[c.n.=8]}_{Zr^{4+}} = 84\,pm$, $r^{[8]}_{Ce^{4+}} = 97\,pm$, $r^{[8]}_{Ce^{3+}} = 114\,pm$, $r^{[7]}_{Zr^{4+}} = 78\,pm$, $r^{[7]}_{Ce^{3+}} = 107\,pm$, $r^{[6]}_{Ce^{3+}} = 101\,pm$ [50].

in a stabilization of the *fluorite* type with random distribution of oxygen defects [178]. Moreover, since Ce^{3+} ions are significantly larger than Ce^{4+}, a relevant amount of Zr^{4+} ions inhibits the necessary expansion of the lattice, and thus the reduction.

The *fluorite* type and its related tetragonal phase provide numerous empty voids, such as the octahedral voids and with increasing reduction of Ce^{4+} tetrahedral voids. These voids allow for a high mobility of oxygen, and ZrO_2 is famous for its excellent oxygen conductivity at elevated temperatures. In such materials, an ion which features fairly stable adjacent oxidation states enables oxygen storage. It can further act as an oxygen pressure buffer as it is oxidized under high oxygen partial pressures—leading to an uptake of oxygen into the bulk structure—and vice versa. Right at the middle between ZrO_2 and CeO_2 a mild reduction yields the quite stable *pyrochlore* phase $Ce_2Zr_2O_7$, which also may be reoxidized under mild conditions [179].

Besides these more or less structural effects promoting the oxygen storage capacity, electronic effects can also further increase the capacity. For instance, supported platinum metals on the surface apparently provide low lying unoccupied states within the band gap. Here, the electrons formed during the release of oxygen via

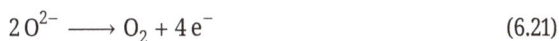

$$2\,O^{2-} \longrightarrow O_2 + 4\,e^-$$ (6.21)

can be consumed. Without supported platinum metals, the only option of the surplus electrons is to reduce Ce^{4+} to Ce^{3+}. This requires more energy as these states lie above the band gap as the lowest lying of the conduction band [180]. Consequently, the oxygen release occurs preferentially in the direct neighborhood of the supported metal. Thus, the deposition of respective metals fosters the entire process in both directions. Accordingly, the oxygen storage capacity is enhanced by several orders of magnitude and the kinetics are accelerated—and the material can be employed as an oxygen buffer in catalysis.

Catalysis

The interplay of such oxygen buffering and the phenomenon of comparable stability of adjacent oxidation states like Ce^{3+}/Ce^{4+} fulfills important preconditions for catalytic applications. So, among a series of other processes $Ce_{1-x}Zr_xO_2$ catalyzes the oxidation of SO_2 during the production of sulphuric acid, analogously to the common catalyst V_2O_4/V_2O_5 employed in the *contact process*.

Another very important application of supported solid solutions like $Ce_{0.76}Zr_{0.24}O_2$ happens in the so-called *three-way catalytic converters* to clean fossil car exhausts from toxic gases and pollutants. Here, three different reactions are triggered. One reaction is the oxidation of surplus fuel according to the schematic reaction

$$2\,C_8H_{18} + 25\,O_2 \longrightarrow 16\,CO_2 + 18\,H_2O,$$ (6.22)

and the second is the oxidation of toxic carbon monoxide via

$$2\,CO + O_2 \longrightarrow 2\,CO_2. \tag{6.23}$$

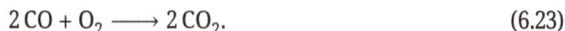

In this context, CO molecules are presumably caught by the supported platinum metal and subsequently oxidized by oxygen of the CeO_2, which is released more easily in the direct neighborhood as discussed above. The third reaction is the reduction of toxic nitrous gases such as

$$2\,NO + 2\,CO \longrightarrow N_2 + 2\,CO_2 \tag{6.24}$$

also catalyzed in the presence of platinum metals, presumably in a similar way. The challenge in this approach is that reactions 6.22 and 6.23 demand high oxygen to fuel ratios for an efficient conversion as in both oxygen is required while reaction 6.24 needs the reduction agent carbon monoxide, which is only present at low oxygen to fuel ratios. To maintain this ratio in the optimal regime, an excellent oxygen buffer is necessary—like $Ce_{0.76}Zr_{0.24}O$, which is by far better than the initially employed alumina. Furthermore, ceria catalyzes the oxidation of soot, that is normally poison for catalysts and which is significantly more relevant in exhausts from diesel engines [175].

6.3.4 The Monoxides and Eu_3O_4

After we shed some light on the oxides of rare-earth elements capable to form tetravalent ions, and thus mixed R(III/IV) or even pure R(IV) oxides, here we turn our attention to those being able to form comparably stable divalent ions R^{2+} like europium and ytterbium. The monoxides of the latter two are obtained as dark red powders either under strictly reducing conditions or reacting the pure metals with the sesquioxides under inert conditions. Both semiconductors crystallize in the *rock salt* structure type (for the structure details, see Figure 6.20 Appendix D). It should be noted that only EuO was obtained as a phase pure powder in larger quantities so far. A facile synthesis starts from excess lithium hydride and an europium halide oxide, with chloride as halide of choice. This yields via the reaction

$$2\,EuOCl + 2\,LiH \xrightarrow[\text{vacuum}]{600-800\,°C} 2\,EuO + 2\,LiCl(g) + H_2(g) \tag{6.25}$$

almost black europium monoxide EuO since the lithium halide sublimes [181]. This reaction also works with the bromide and iodide but not with the fluoride. Because of the higher stability of EuOCl against humidity, this educt is preferred. The magnetic properties are discussed in Chapter 9.5. Under high pressure conditions, the metallic monoxides from lanthanum to samarium can be synthesized; they are obtained as golden powders. Only in the case of SmO, the metal is approaching the divalent state while in the remaining ones magnetic measurements indicate trivalent lanthanide atoms according to $R^{3+}O^{2-} \cdot e^-$.

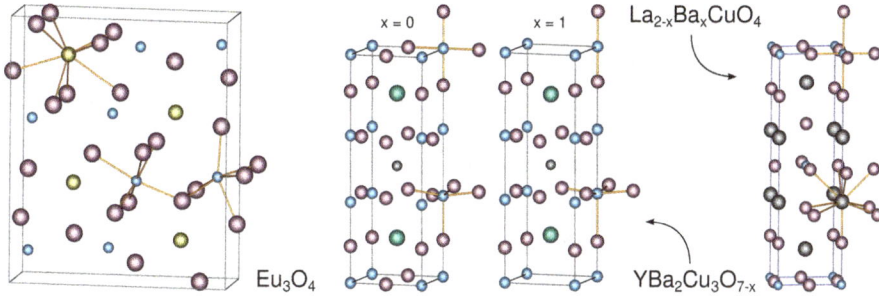

Figure 6.18: Structures of Eu_3O_4 (Eu^{2+} yellow and Eu^{3+} blue), $YBa_2Cu_3O_{7-x}$ (copper blue, Ba^{2+} green, Y^{3+} grey) and $La_{2-x}Ba_xCuO_4$ (copper blue, Ba^{2+} and La^{3+} grey), oxygen atoms red; data: [182–184].

Considering mixed valent oxides comprising Eu^{2+} and Eu^{3+}, one might think of a spinel analogue. And indeed in 1962, Bärnighausen[35] and Brauer[41] described a synthesis around 900 °C, reacting EuO with Eu_2O_3, and the orthorhombic structure of the dark red Eu_3O_4. It adopts the structure of $CaFe_2O_4$, where divalent europium is coordinated by a bicapped trigonal prism; both sites of trivalent europium are coordinated octahedrally as displayed in Figure 6.18 [185, 186, 181]. This structure can be understood by multiple chemical twinning not discussed here [123]. Its phase width is rather narrow and it forms solid solutions with either EuO or Eu_2O_3. Around 5 K, it shows antiferromagnetic ordering; the recorded effective magnetic moment of 5.3 μ_B agrees well with a ratio of 1:2 of ordered Eu^{2+} and Eu^{3+} ions in the structure [187].

6.3.5 The Superconducting Oxides $La_{2-x}Ba_xCuO_4$ And $YBa_2Cu_3O_{7-x}$

Because of their extraordinary superconducting properties, two further oxides are mentioned in this chapter—although they are not binary ones. In a superconducting state, an electrically conducting material shows no resistance anymore below a critical temperature T_c. Until 1986, the highest $T_c \approx 23$ K was known for Nb_3Ge. Thus, up to then, liquid helium or hydrogen were required for cooling to achieve superconductivity. In 1986, Bednorz[42] and Müller[43] discovered the first ceramic superconductors. The T_c of the first described compound within the system $La_{2-x}Ba_xCuO_4$ was already around 35 K for $x = 0.15$ [188]. Shortly thereafter, for $YBa_2Cu_3O_{7-x}$ critical temperatures up to 92 K were reported for $x = 0.07$ [189]. Now, also at a liquid nitrogen temperature of 77 K, superconductivity could be achieved. Until now, the application is inhibited as the ceramic compounds are brittle, and thus it is challenging to produce wires. Quite recently, the

---

41 *Georg Brauer*, German chemist (*1908 †2001).

42 *Georg Bednorz*, German mineralogist and physicist, Nobel Prize in Physics 1987 (*1950).

43 *Karl Alexander Müller*, Swiss physicist, Nobel Prize in Physics 1987 (*1927 †2023).

successful production of such a wire with a length of more than 300 km with a diameter of 4 mm was reported [190].

A typical synthesis of $YBa_2Cu_3O_{7-x}$ starts from the carbonates or nitrates of yttrium, barium and copper around 1000 °C. $La_{2-x}Ba_xCuO_4$ was obtained by coprecipitation of the aqueous solutions of the nitrates of lanthanum, barium and copper employing oxalic acid and heating the product to 900 °C [188].

The crystal structure of $YBa_2Cu_3O_{7-x}$ is derived from a threefold superstructure of the *perovskite* type. Here, ordered oxygen-vacancies are found; even more vacancies emerge increasing x from zero to one (Figure 6.18). These further vacancies alter the local surrounding of the copper atoms by oxygen. The copper atoms show either a square-planar or a linear coordination while the square-pyramidal environment of the other copper atoms remains intact. In $YBa_2Cu_3O_{7-x}$, the yttrium atoms experience a cubic coordination. In contrast to the yttrium compound in $La_{2-x}Ba_xCuO_4$, all copper atoms feature the same Jahn–Teller distorted elongated octahedral coordination. The lanthanum atoms share the site with barium atoms, and thus both are surrounded by nine oxygen atoms in a monocapped square antiprismatic fashion according to Figure 6.18. While in $YBa_2Cu_3O_{7-x}$, the electron concentration is driven by the oxygen content in $La_{2-x}Ba_xCuO_4$; this is controlled by the lanthanum-barium ratio.

6.4 Borides, Carbides, Nitrides and Sulfides

Selected further binary compounds of the rare-earth elements with main group elements will be considered in this section. In each group, there are numerous representatives, many with highly interesting properties. For closer looks on these, I recommend [191–198] for *further reading*.

6.4.1 Borides

Borides are usually obtained by direct reactions of the rare-earth elements or their oxides with boron at temperatures beyond 1800 °C. They show high melting points and a remarkable stability against acids and bases. Here, I will focus on two compounds, the binary RB_6 and the ternary boride $Nd_2Fe_{14}B$.

Among the RB_6, the europium and ytterbium compounds are isolators according to a configuration $R^{2+}(B_6)^{2-}$ while the others feature metallic behavior according to $La^{3+}(B_6)^{2-}\cdot e^-$. The latter are excellent thermal emitters of electrons or *hot cathodes* because of their high melting points and low work functions, especially the lanthanum and the cerium compounds. Moreover, the purple to violet LaB_6 becomes superconducting below 0.45 K [201]. Their crystal structure can be derived from the CsCl structure type, where a primitive cubic packing of chlorine hosts caesium atoms in the cubic voids. In RB_6 the caesium atoms and chlorine are formally substituted by R atoms and B_6 octa-

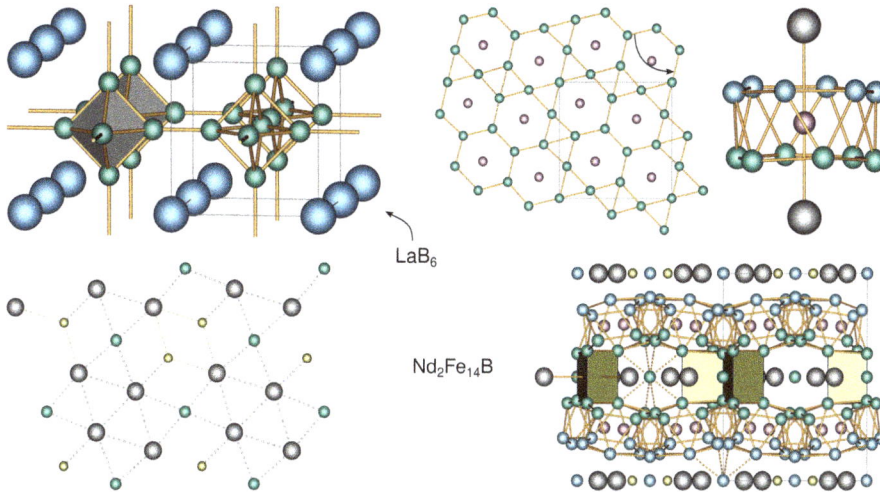

Figure 6.19: Structures of LaB_6 (top left, lanthanum blue and boron green) and $Nd_2Fe_{14}B$ (all other pictures, colors as discussed in the text); data: [199, 200].

hedra as depicted in Figure 6.19. These octahedra are covalently bound to further six octahedra, which form a 24-fold coordination of the rare-earth elements.

The second boride considered here is $Nd_2Fe_{14}B$. It is a prominent material employed, for instance, in wind turbines as permanent magnets. Its magnetic properties will be discussed in Chapter 9.6.4. The development of this material in the 1980s was motivated by quite expensive magnets comprised of samarium and cobalt. Nowadays, neodymium has been classified as a critical element like praseodymium and dysprosium due to their importance for the renewable energies sector. Nevertheless, the vast majority of wind turbines do not employ permanent magnets, but are rather geared toward turbine technology [202]. The hard magnetic phase $Nd_2Fe_{14}B$ is usually synthesized by rapidly melting the elements under inert atmosphere in induction furnaces; then the obtained flakes are milled under a hydrogen atmosphere. Here, brittle hydrogen containing $Nd_2Fe_{14}BH_x$ is formed, which can be ground very well to small particles. The preparation of the magnets under a strong magnetic field has to be conducted carefully as it has considerable impact on the magnetic properties of the final material. For instance, to get a permanent magnet with a maximal energy density, the individual grains are sintered with a non-magnetic phase while for a high remanence the contacts between the individual grains necessarily have to be as close as possible. Further details regarding synthetic aspects may be found in [203].

The crystal structure of $Nd_2Fe_{14}B$ depicted in Figure 6.19 is very complicated on a first glance, but very interesting on a second and quite well comprehensible on a third. For a better understanding, the relevant structure motifs are colored. Green and blue iron atoms build two kagomé[33] nets like those in $LaNi_5$ (Figure 6.3). In $Nd_2Fe_{14}B$, these layers are corrugated. Within this kagomé double layer further (red) iron atoms are situated, which are additionally coordinated by two grey neodymium atoms. Between the double layers planar layers containing the neodymium, further (green) iron and the (yellow) boron atoms are found, which are vaguely reminiscent of layer B in LaNi$_5$. In the full structure, the boron atoms occupy trigonal prismatic voids of iron atoms and connect both double layers. Half of the neodymium atoms are coordinated by sixteen iron, three neodymium and one boron atom, and the other half by sixteen iron, two neodymium and two boron atoms.

6.4.2 Carbides and Nitrides

The *nitrides* predominantly form simple binary and high melting compounds RN with *rock salt* structure; for the structure details, see the following info box and Figure 6.20. The same structure is found for the binary phosphides RP. The nitrides are formed around 1000 °C from the elements and develop ammonia upon hydrolysis due to the high oxygen affinity of the rare-earth elements. Therefore, during synthesis, oxygen has to be excluded carefully to avoid contamination. For some of them, in addition nitrogen deficient phases were reported. The nitrides RN were all reported to be semiconductors with band gaps between 0.6 and 1.6 eV. The magnetic properties of GdN will be the subject of Chapter 9.3.

> ℹ️ *Important crystal structure types no. 3–rock salt:* The rock salt structure type, chemically NaCl, consists of a cubic close packing of chloride ions; in all octahedral voids, sodium ions are situated (see also Chapter D). Accordingly, the chloride ions are also octahedrally coordinated by sodium cations and vice versa. This structure type is usually adopted by highly ionic compounds, and thus frequently found. In this book, GdN and the europium monochalcogenides adopt this structure.

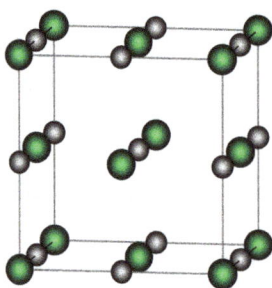

Figure 6.20: Representation of the crystal structure of rock salt, the sodium atoms are grey, the chlorine atoms green and the unit cell edges dark grey.

The Crystal Structure of Sm$_2$Co$_{17}$N$_3$

The crystal structure of Sm$_2$T$_{17}$N$_3$ (T = Fe, Co), depicted in Figure 6.21, is closely related to that of SmCo$_5$, and this is isotypic with that of LaNi$_5$ discussed in Chapter 6.1.2. Like there, two different layers can be identified. Layer A forms also a kagomé[33] net of nine cobalt atoms. Layer B is slightly distorted, and half of all samarium atoms (grey) are

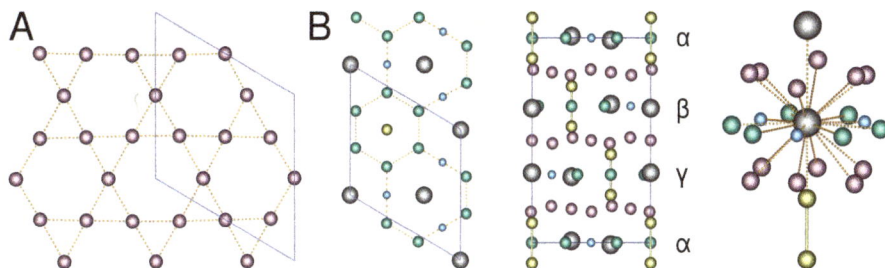

Figure 6.21: Crystal structure of $Sm_2Co_{17}N_3$; shown are layers A and B, the full structure showing the layer sequence of the B layer and the surrounding of the grey samarium atoms, data: [204].

substituted by cobalt dumbbells (yellow). Moreover, in layer B interstitial sites may be occupied by nitrogen (blue). Thus, the composition of this layer amounts to $Sm_2Co_8N_3$. In the unit cell, the layers A and B alternate. The B layers follow a sequence $\alpha\beta\gamma$, so do the A layers. The shown structure of the nitride is essentially the same as Sm_2Co_{17}, in the nitride the unit cell expands within the a-b plane. In $Sm_2Co_{17}N_3$, the samarium atoms are coordinated by nineteen cobalt and three nitrogen atoms. In the intermetallic phase Sm_2Co_{17}, it is accordingly only coordinated by nineteen cobalt atoms. For the sake of completeness, the coordination of the cobalt atoms in $Sm_2Co_{17}N_3$ should also at least be mentioned. The four chemically different cobalt atoms are coordinated by ten cobalt and two rare-earth atoms (Co1), by a nitrogen, ten cobalt and two rare-earth atoms (Co2), by a nitrogen, nine cobalt and three rare-earth atoms (Co3) and finally the dumbbell cobalt atoms are coordinated by thirteen cobalt atoms and a rare-earth atom (Co4).

The Carbides R_3C, R_2C_3 and R_4C_3

Regarding the *carbides*, many stoichiometries were reported. Besides the acetylides, intercalation compounds into graphene like EuC_6 or metal rich carbides were also reported, such as Sc_4C_3 or R_3C (R = Sm...Lu, Y). With increasing carbon content, the materials tend to very high melting points well above 2000 °C.

In the crystal structure of LaC_2, the lanthanum atoms build a tetragonally distorted body-centered cubic packing. The reason for this distortion—essentially an elongation along one axis—are the acetylide ions C_2^{2-}. These dumbbells occupy one-third of the pseudo-octahedral voids as shown in Figure 6.22. These voids were already discussed and shown in Figure 6.16 on the example of the X-type Nd_2O_3 where all of these voids are partially occupied. Here, the lanthanum atoms are coordinated tenfold by carbon; the structure type is that of CaC_2. The structure of Y_2C_3 adopts the structure type of Pu_2C_3 and also features C_2 dumbbells. As depicted in Figure 6.22, the structure of Y_2C_3 can be derived from an even stronger distorted body-centered cubic packing of yttrium atoms indicated by the broken blue lines. In such packings, eight atoms form lengthy *bisdisphenoid voids* colored yellow. These voids host the C_2^{2-} ions. The yttrium atoms are coordinated ninefold by carbon.

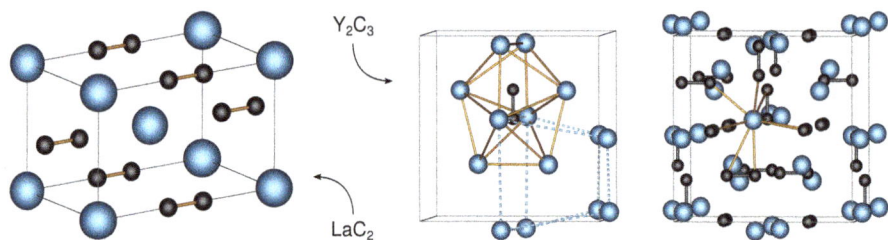

Figure 6.22: Structures of LaC_2 (left, lanthanum blue and carbon dark grey) and Y_2C_3 (center and right), yttrium blue and carbon dark grey; data: [205, 206].

A series of carbides containing C_2 dumbbells shows superconductivity. Interestingly, the transition temperature T_c seems related to the carbon-carbon bond length. The closer to 130 pm, the higher T_c; apparently, an optimum occupation of the antibonding σ^* states of the carbon-carbon bond plays a role here. The maximum is achieved for Y_2C_3 (T_c = 15 K, 129.8 pm), for La_2C_3 T_c = 13 K (129.4 pm), YC_2 T_c = 4.0 K (127.1 pm) and LaC_2 T_c = 1.6 K (127.5 pm) were reported [205].

6.4.3 Sulfides

Sulfide is clearly softer than oxide, which fits better for the hard rare-earth ions. Hence, the sulfides are easily subject to hydrolysis and the syntheses are achieved best reacting the elements, possibly by an assistant flux. Essentially, the sulfides adopt the stoichiometries already seen in the oxide section. Thus, for all rare-earth elements except europium the sulfides R_2S_3 are known. The structure systematics follow a similar trend from higher coordination numbers up to eight for lanthanum to lower coordination numbers of six for the lutetium compound. Moreover, sulfide is chemically more on the reducing side than the lighter chalcogenide oxide. For instance, paramagnetic CeS_2 does not contain tetravalent cerium like diamagnetic CeO_2 but trivalent ions and mixed valent sulfur according to $(Ce^{3+})_2(S^{2-})_2(S_2^{2-})$. This reducing side becomes obvious when considering the compositions RS and R_3S_4. The rare-earth elements europium, ytterbium and samarium show a certain tendency to the divalent state. Accordingly, these indeed form semiconducting monosulfides $R^{2+}(X^{2-})$ while the other elements form metallic monosulfides $R^{3+}(X^{2-}) \cdot e^-$. This can already be seen by the unit cell volumes of the structures isotypic with *rock salt* as they are considerably larger for Eu^{2+}, Yb^{2+} and Sm^{2+} compared with the trivalent ones. Semiconducting black SmS converts under slight pressure into metallic golden SmS containing the smaller Sm^{3+}; and by heating, it transforms back to the semiconducting form. Therefore, SmS is interesting for several applications such as data storage or infrared sensor [207]. Eu_3S_4 shows an interesting phase transition around $-70\,°C$, below of which an ordering of Eu^{2+} and Eu^{3+} occurs. At room temperature, the valences fluctuate and a structure closely related to that of $Y_2C_3 = Y_4(C_2)_3$

is found; here, europium replaces the C_2 dumbbells and sulfur the yttrium atoms iso-typic with the Th_3P_4 structure type [208]. Despite this report, there seem to still be some doubts about the existence of Eu_2S_3 [209, 210]. Further, also several polysulfides were reviewed [197].

6.5 Silicates And Selected Silicate-Analogous Compounds

Silicates are the most abundant class of minerals on earth and, therefore, also the silicates of the rare-earth elements are worth a closer look. Moreover, in the second part of this book, I chose several *silicate-analogous compounds* to act as examples for optical and magnetic properties. Silicate-analogous compounds in this sense contain tetrahedral building units centered by nonmetals like boron, silicon, phosphorus or sulfur and terminated by nitrogen, oxygen or fluorine atoms. After a short classification, the structures of ternary rare-earth silicates will be discussed before the examples of silicate-analogous compounds relevant to this book will be addressed. Among these compounds are also host structures of phosphors discussed later. These occasionally do not contain rare-earth elements but ions, which enable the doping with rare-earth ions. This will be indicated in the respective cases.

The main reason that silicates and silicate-analogous compounds are of interest also for this book is that they feature tetrahedra as basic building units. Tetrahedra do not have an inversion center, and thus structures become more probable lacking an inversion center or at least the sites on which the rare-earth ions reside have a good chance to be non-centrosymmetric. This is beneficial for the efficiency of optical transitions, which will be discussed in the second part. Silicates are classified according to the condensation degree of the tetrahedra and the resulting dimensionality of the anion. Classes of silicates and analogous materials are *nesosilicates*, i. e., silicates containing noncondensed SiO_4 tetrahedra, and *sorosilicates*, group silicates where tetrahedra form finite moieties. Further classes are chain silicates (*inosilicates*), *cyclosilicates*, *phyllosilicates* comprised of anionic layers and three-dimensional frameworks, i. e., *tectosilicates*.

6.5.1 Coordination Strength

Especially for the optical properties, the crystal chemistry of the local coordination of the rare-earth ions is relevant. The coordination strength of a ligand has two components—electrostatic and covalent interactions.

In order to roughly estimate the *electrostatic coordination strength* of the terminal atoms of such anions in this book, the *effective coordination charge* c_X of an ion X is introduced based on Pauling's rules for ionic crystal structures. Pauling[23] defined the electrostatic coordination strength of an ion i

$$s_i = \frac{z_i}{k_i} \qquad (6.26)$$

with its charge z_i and coordination number k_i [220]. For the charge of an ion j holds

$$z_j \approx - \sum_{i=1}^{k_j} s_i \qquad (6.27)$$

that its charge approximately equals the sum of coordination strengths of all surrounding oppositely charged ions. Therefore, the *effective coordination charge* c_X is the remaining charge on a terminal ion $X^{z_{theo}\,-}$ with its charge z_{theo} partially balanced by n neighboring ions

$$c_X = z_{theo} - \sum_{i=1}^{n} s_i \qquad (6.28)$$

and acting on a metal ion coordinated by this anion. In case of a silicate anion SiO_4^{4-} featuring a fourfold-coordinated Si^{4+} ion s_{Si} equals 1. For any terminal oxide ion then holds

$$c_{O^{2-}} = 2 - 1 = 1.$$

Besides this electrostatic influence, the *covalent coordination strength* plays an even larger role for ions like Ce^{3+} or Eu^{2+}. To estimate the covalency of a bond, Pauling also derived a formula based on the electronegativities of both atoms [46]. According to

$$p = 1 - e^{-\frac{1}{4}(\Delta\chi)^2} \qquad (6.29)$$

the polarity of a bond, p decreases, i. e., the covalency increases, with a decreasing difference of electronegativities $\Delta\chi$. For instance, the interaction between a R^{3+} ion and halides shows increasing covalency, i. e., *nephelauxetic effect*, going from fluorine to iodine. The coordination strengths will be discussed in this chapter briefly. For further details of the nephelauxetic effect, see Chapter 8.1.

6.5.2 Silicates

Chemically spoken, R^{3+} ions are quite hard and, therefore, prefer hard anions. Considering the previous considerations, the hardness of silicate anions decreases with an increasing condensation degree, because then the charge per volume ratio decreases. Thus, naturally occurring rare-earth elements have a strong preference for low-condensed neso and sorosilicates. The noncondensed anions will be enclosed in round parentheses, and condensed anions in square parentheses.

Considering first the *nesosilicates*, cerium and europium adopt their special roles as Ce^{4+} and Eu^{2+}. While in $Eu_2(SiO_4)$ divalent europium ions experience a nine and ten-fold coordination, tetravalent cerium ions get a square antiprismatic coordination in $Ce(SiO_4)$. $Ce(SiO_4)$ crystallizes in the *zircon* structure, which can be nicely related to that of *rutile* as shown in Figure 6.23. The 6+6-coordination of two adjacent octahedra in *rutile* evolves into a 8+4-coordination in the *zircon* type [123]. To a certain extent, natural *zircon* also hosts the other trivalent rare-earth elements. *Titanite*, another nesosilicate of the typical composition $CaTi(SiO_4)X$ (X = O, F, OH) also contains significant amounts of rare-earth elements; the smaller R^{3+} find a sevenfold coordination here (Figure 4.2). The trivalent ions form the nesosilicate oxide $R_2O(SiO_4)$ for the whole series R = La...Lu; on display in Figure 6.23 is the gadolinium compound. Here, the rare-earth ions are sur-

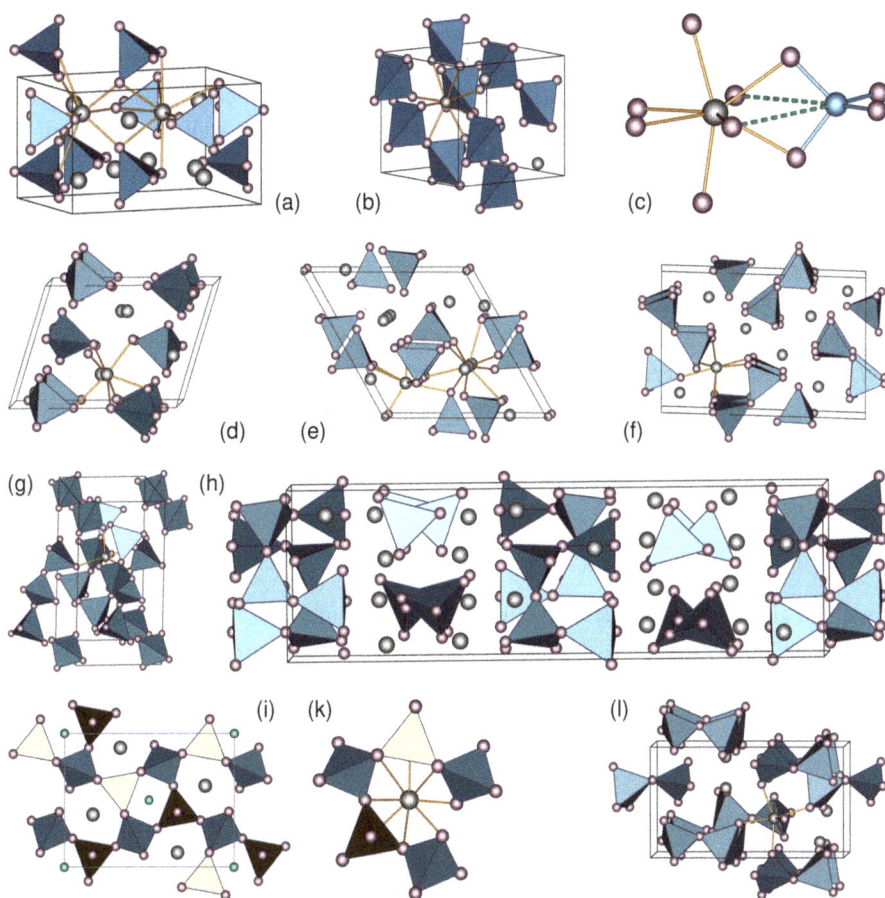

Figure 6.23: Structures of (a) $Eu_2(SiO_4)$, (b–c) $Ce(SiO_4)$ (unit cell and relationship to *rutile*), (d) $Gd_2O(SiO_4)$, (e) $R_{4.67}(SiO_4)_3O$, (f) $Pr_2[Si_2O_7]$, (g) $Ho_4[Si_3O_{10}](SiO_4)$, (h) $La_6[Si_4O_{13}](SiO_4)_2$, (i–k) *gadolinite* and (l) $Lu_2[Si_2O_7]$; oxygen atoms red, silicon centered tetrahedra blue and metal atoms grey; data: [211–219].

rounded by seven oxygen atoms. The larger rare-earth ions from lanthanum through gadolinium also adopt the famous *apatite* type structure. However, the fourteen negative charges of the anions in $R_{4.67}(SiO_4)_3O$ require a significantly reduced occupation of trivalent rare-earth ions or mixed occupation with other lower charged cations on the sites hosting the five cations to achieve charge balance. Another important silicate mineral is *cerite*, $(R,Ca)_9(SiO_4)_3(SiO_3(OH))_4(OH,F)_3$, where the silicate tetrahedra form close packed layers. This is shown in Figure 4.2.

The *sorosilicates* are also depicted in Figure 6.23. $R_2[Si_2O_7]$ for $R = Sc$, La…Lu is the most important scandium silicate and scandium mineral *thortveitite*. There the rare-earth ions are sixfold coordinated, and accordingly this structure is best suited for the smaller R^{3+}. Under high pressure, the coordination number rises to seven like in the example of $Lu_2[Si_2O_7]$, shown in Figure 6.23(l). The final two sorosilicates in this chapter comprise two different anions. Besides a noncondensed SiO_4 tetrahedron in $Ho_4[Si_3O_{10}](SiO_4)$ a condensed ensemble of three and in $La_6[Si_4O_{13}](SiO_4)_2$ an ensemble of four tetrahedra were reported. The chemically similar natural mineral *allanite*, namely $CaR(Al_2,Fe^{2+})[Si_2O_7][SiO_4]O(OH)$, features distinct disilicate and orthosilicate moieties and hosts preferentially larger rare-earth ions feeling well in an elevenfold coordination as depicted in Figure 4.2.

Gadolinite is a mineral comprised of silicate-analogous layers. In these layers, silicate tetrahedra alternate with boron or beryllium centered ones, so that it formally might be classified as a nesosilicate. Because of the similarity of boron and beryllium based networks, *gadolinite* is indeed a phyllosilicate-analogous compound. This coincides with a typical composition $(R,Ca)_2Fe_2[(B,Be)_2Si_2O_{10}]$ and a ratio of tetrahedral centers to oxygen of 4:10. As stated above, the higher condensed silicates are chemically softer and should therefore host the preferable larger rare-earth ions. This is indeed the case (Table 1.1).

6.5.3 Aluminates

Anions in aluminates comprise trivalent aluminium ions coordinated by oxide either tetrahedrally like in silicates or octahedrally. The effective coordination charge is generally higher in aluminates according to

$$c_{O-Al} = 2 - \frac{3}{4} = 1.25 \text{ (c. n. = 4)}, \quad c_{O-Al} = 2 - \frac{3}{6} = 1.5 \text{ (c. n. = 6)} \tag{6.30}$$

Moreover, also the covalency of the interactions between terminal oxygen atoms and metal atoms is stronger in aluminates as the aluminium atoms in the coordination sphere of oxygen reduces their electronegativity. Thus aluminates show a stronger coordination than silicates.

Yttrium aluminium garnet (YAG), $Y_3Al_2^{[o]}(Al^{[t]}O_4)_3$, is one of the most prominent host structures for phosphors. The structure displayed in Figure 6.24 is most easily com-

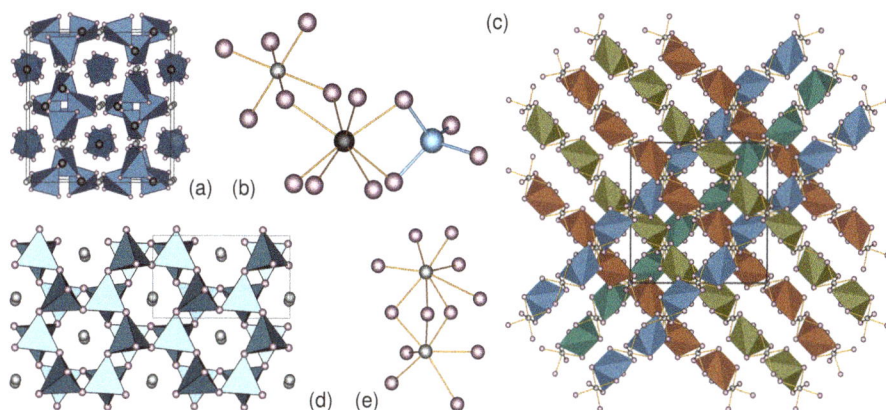

Figure 6.24: Structures of (a)–(c) $Y_3Al_2(AlO_4)_3$ (YAG, unit cell, surroundings and rod packing), (d–e) $Sr[Al_2O_4]$ (unit cell and surroundings); oxygen atoms red, aluminium and aluminium centered tetrahedra blue and metal atoms grey; data: [221, 222].

prehended by considering rods comprising alternating occupied octahedra and empty prisms. These rods align along the space diagonals of the unit cell. The resulting rod packing is shown in 6.24(c). In YAG, the aluminium atoms $Al^{[o]}$ are octahedrally, the atoms $Al^{[t]}$ are tetrahedrally coordinated and the yttrium atoms feature a slightly distorted cubic coordination. If YAG is doped with rare-earth ions, this solely occurs on the yttrium site. Since the AlO_4 tetrahedra are not condensed to others, YAG is a nesosilicate analogous aluminate.

The structure of the tectosilicate analogous strontium aluminate $Sr[Al_2O_4]$ can be understood as a stuffed *tridymite* type, a SiO_2 polymorph, where the aluminium atoms replace silicon. In $Sr[Al_2O_4]$, sechserrings of AlO_4 tetrahedra provide large cavities to host the strontium atoms, which are further connected perpendicular to the ring layer to build a three-dimensional framework. The Sr^{2+} atoms, which are replaced upon doping by Eu^{2+} and R^{3+}, are surrounded by eight and seven oxygen atoms forming irregular coordination polyhedra as depicted in Figure 6.24. Above 650 °C, a higher symmetric high-temperature polymorph is known.

$La[MgAl_{11}O_{19}]$ and $Ba[MgAl_{10}O_{17}]$ (BAM) are magnesium aluminates, which are derived from β-alumina. A typical synthesis is conducted according to

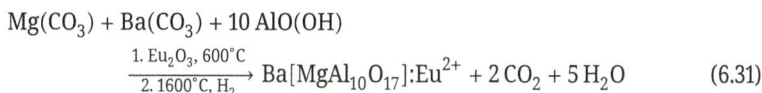

$$Mg(CO_3) + Ba(CO_3) + 10\ AlO(OH)$$

$$\xrightarrow[\text{2. 1600°C, } H_2]{\text{1. } Eu_2O_3,\ 600°C} Ba[MgAl_{10}O_{17}]{:}Eu^{2+} + 2\ CO_2 + 5\ H_2O \qquad (6.31)$$

and yields under reducing conditions blue luminescing $BAM{:}Eu^{2+}$. In both magnesium aluminates, close packed layers of oxygen form layer sequences as indicated in Figure 6.25. Magnesium and aluminium are surrounded tetrahedrally and octahedrally. The former are shown as closed polyhedra, the latter as light grey spheres. Indeed, both structures feature motifs of the spinel structure with the magnesium ions having a pref-

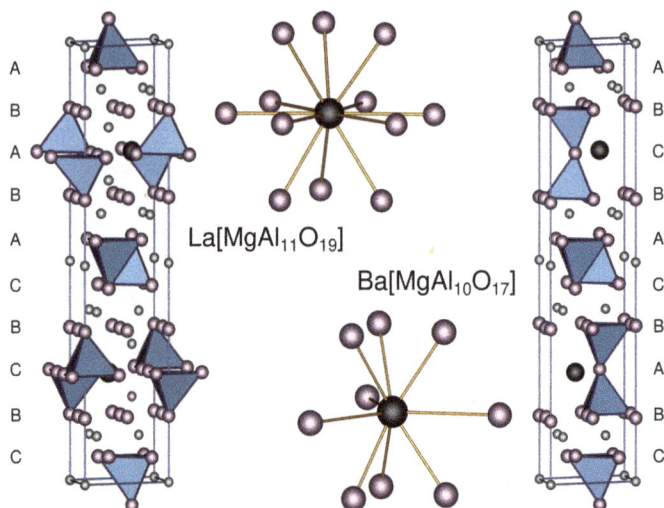

Figure 6.25: Structures of La[MgAl$_{11}$O$_{19}$] (left) and Ba[MgAl$_{10}$O$_{17}$] (right), data: [223, 224].

erence for the tetrahedral voids. In La[MgAl$_{11}$O$_{19}$], the rare-earth ions are coordinated by twelve, and the barium atoms in Ba[MgAl$_{10}$O$_{17}$] only by nine oxygen atoms. Both structures are important host structures for lanthanide ions, which are doped on the lanthanum and barium sites, respectively. The cerium compound doped with terbium, i. e., Ce[MgAl$_{11}$O$_{19}$]:Tb^{3+} is known under the acronym CAT.

6.5.4 Nitridosilicates, Nitridoaluminates and Oxonitridosilicates

Nitridosilicates are silicate analogous compounds based on SiN$_4$ tetrahedra (Figure 6.26). This compound class was mainly developed over the last decades by Schnick[44] and opened up for rare-earth based phosphors. Nitrogen tends much more than oxygen to bridge more than two tetrahedra in silicate analogous networks, and consequently higher condensation degrees and denser structures are possible. Considering the coordination strength nitrogen interacts more covalently than oxygen with metal atoms due to its lower electronegativity. Furthermore, the effective coordination charge is generally higher in nitridosilicates according to

$$c_{\text{N-Si}} = 3 - \frac{4}{4} = 2 \tag{6.32}$$

because of the higher charge of N^{3-}. Thus, nitridosilicates show a significantly stronger coordination than silicates and also aluminates. Further, in oxonitridosilicates mixed

---

44 *Wolfgang Schnick*, German chemist (*1957).

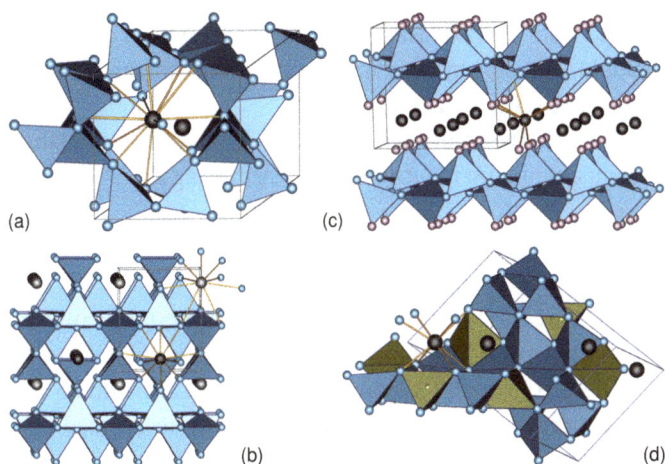

Figure 6.26: Structures of (a) $Ba[Si_7N_{10}]$, (b) $Sr_2[Si_5N_8]$, (c) $Eu[Si_2O_2N_2]$ and (d) $Sr[LiAl_3N_4]$; nitrogen atoms blue, oxygen red, silicon and aluminium centered tetrahedra blue, lithium centered tetrahedra yellow and metal atoms grey; data: [225–228].

coordination environments of O^{2-} and N^{3-} are found. This enables a certain tuning of the coordination strength.

The barium nitridosilicate $Ba[Si_7N_{10}]$ comprises a very dense three-dimensional, and thus tectosilicate-analogous anionic framework, which hosts the barium atoms in almost spherical spacious voids with an irregular thirteenfold coordination. This high coordination number yields comparably large coordination distances. If doped with smaller rare-earth ions like divalent europium ions, these presumably will experience an overall fairly weak coordination strength. $Sr_2[Si_5N_8]$ is also tectosilicate-analogous and provides two tenfold-coordinated sites for the strontium ions Sr^{2+}, which may be replaced with Eu^{2+} upon doping. One polyhedron is a four-capped trigonal antiprism, the other a four-capped trigonal prism of nitrogen atoms (Figure 6.26). Such nitridosilicates are typically synthesized under oxygen and humidity-free conditions starting from the so-called silicon diimide, a compound strongly related to SiO_2, which can be accessed via

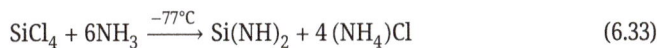

$$SiCl_4 + 6NH_3 \xrightarrow{-77°C} Si(NH)_2 + 4\,(NH_4)Cl \tag{6.33}$$

with subsequent evaporation of ammonium chloride and further reaction typically with the metals according to

$$5Si(NH)_2 + 2Sr \xrightarrow[Eu]{1500°C} Sr_2[Si_5N_8]:Eu^{2+} + 5\,H_2 + N_2 \tag{6.34}$$

The reaction conditions are so reducing that it is no challenge to obtain doping with divalent europium here. The oxonitridosilicate $Eu[Si_2O_2N_2]$ is a phyllosilicate-analogous compound comprising dense layers of condensed $SiON_3$ tetrahedra, which coordinate the europium atoms sevenfold with six oxygen atoms forming a prism and a further nitrogen atom as quite distant cap (Figure 6.26).

A further increase of the effective coordination charge is feasible by replacing tetravalent silicon by trivalent aluminium in nitridosilicates according to

$$c_{N\text{-}Si} = 3 - \frac{3}{4} = 2.25 \tag{6.35}$$

yielding nitridoaluminates. The example here, also employed as a host structure for phosphors, is the strontium lithium nitridoaluminate $Sr[LiAl_3N_4]$. The structure of the polyanion is obviously very dense (Figure 6.26). And considering that also lithium contributes LiN_4 tetrahedra, here the effective coordination charge is on average even higher. The coordination of strontium by AlN_4 and LiN_4 tetrahedra leads to an almost perfectly cubic environment.

6.5.5 Phosphates

Because of the higher charge of phosphorus phosphates almost exclusively form nesosilicate, inosilicate and cyclosilicate analogous compounds. The anions comprise pentavalent phosphorus ions coordinated tetrahedrally by oxide. The effective coordination charge of terminal oxygen atoms is generally lower than in silicates according to

$$c_{O\text{-}P} = 2 - \frac{5}{4} = 0.75 \tag{6.36}$$

Moreover, also the covalency of the interactions between terminal oxygen atoms and metal atoms is weaker in phosphates as the higher charged phosphorus atoms in the coordination sphere of oxygen increases their electronegativity. Thus, phosphates show a weaker coordination than silicates. All structure pictures of this section are depicted in Figure 6.27.

The famous mineral *monazite*, i. e., $Ce(PO_4)$, was already mentioned in Chapter 4. The coordination of the cerium atoms is ninefold and the polyhedron can be described as a tricapped trigonal prism where one of the caps actually caps a triangular face. It is apparently a nesosilicate analogous phosphate. In the inosilicate analogous $\alpha\text{-}Sr[PO_3]_2$, phosphate helices wriggle through a diamond-like arrangement of strontium atoms. This diamond-like arrangement is emphasized by a virtual yellow unit cell in Figure 6.27(d). In total, eight oxygen atoms forming a quite irregular polyhedron coordinate the strontium atoms. Such polyphosphates are typically synthesized via the reaction of the respective carbonate with concentrated phosphoric acid according to

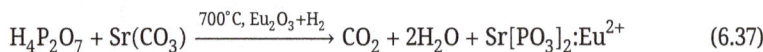

$$H_4P_2O_7 + Sr(CO_3) \xrightarrow{700°C,\ Eu_2O_3+H_2} CO_2 + 2H_2O + Sr[PO_3]_2{:}Eu^{2+} \tag{6.37}$$

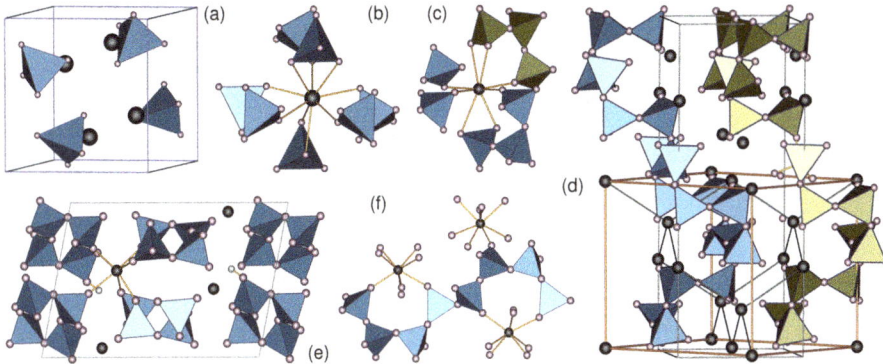

Figure 6.27: Structure motifs of (a–b) Ce(PO$_4$) (unit cell and coordination), (c–d) α-Sr[PO$_3$]$_2$ (unit cell and surrounding) (e) HoH[PO$_3$]$_4$ (f) Ce[PO$_3$]$_4$; oxygen atoms red, phosphorus centered tetrahedra blue or yellow and metal atoms dark grey; data: [69, 229–231].

where the doping with divalent europium on the strontium sites is achieved by adding the respective amount of any europium oxide and conducting the reaction under reducing conditions, like in the presence of hydrogen gas [229]. HoH[PO$_3$]$_4$ is a further inosilicate analogous compound where the holmium atoms are coordinated monocapped prismatically by oxygen. As mentioned in the introduction to this section, the electronegativity of the coordinating oxygen atoms in phosphates is reduced compared with silicates. Since a ternary silicate featuring Ce^{4+} is known, it is not surprising that this has also been reported for phosphates. The inosilicate-analogous Ce[PO$_3$]$_4$ contains zig-zag phosphate chains, which provide an eightfold coordination, best described as bicapped prismatic, for the tetravalent cerium atoms. Here, the charge balance is easily achieved while a proton was introduced by nature to compensate the anion's charge in HoH[PO$_3$]$_4$.

6.5.6 Borophosphates, Borosulfates and Fluorooxoborates

So far, I considered the classic, long known compound classes. In the final section, relevant examples of silicate-analogous compounds of three quite new compound classes will be discussed. The oldest of these are the *borophosphates*, which are solids in which phosphate tetrahedra are condensed with borate tetrahedra or borate triangles [232]. Typical borophosphates are built by a borate backbone and terminal phosphate groups. Accordingly, mainly the terminal oxygen atoms of the phosphate groups coordinate the metal atoms. Based on the previous considerations, these terminal oxygen atoms exhibit a coordination strength similar to that of phosphates. However, when compared to polyphosphates, borophosphates feature a slightly stronger coordination as in the second coordination sphere lower charged boron atoms follow. The structure pictures of this section are depicted in Figure 6.28.

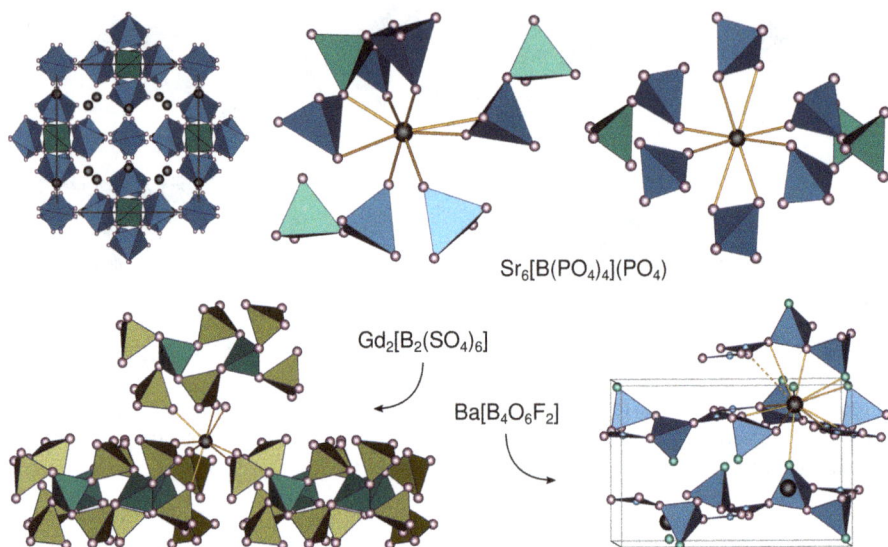

Figure 6.28: Structure motifs of $Sr_6[B(PO_4)_4](PO_4)$ (top row), $Gd_2[B_2(SO_4)_6]$ (bottom left) and $Ba[B_4O_6F_2]$ (bottom right); oxygen atoms red, fluorine green phosphorus centered tetrahedra blue, sulfur centered tetrahedra yellow, boron centered tetrahedra green and metal atoms dark grey; data: [233–235].

The strontium borophosphate phosphate $Sr_6[B(PO_4)_4](PO_4)$ is a quite typical representative. It contains $[B(PO_4)_4]$ moieties where a boron atom is surrounded tetrahedrally by four phosphate tetrahedra. Further noncondensed phosphate tetrahedra are present. Both anions coordinate the strontium atoms in an eightfold manner, resulting in the formation of a square antiprism and an irregular polyhedron. Also, here the strontium atoms may be replaced upon doping with Eu^{2+} ions.

From borophosphate to *borosulfates* is only a small formal step. Here, instead of phosphate normally sulfate tetrahedra coordinate boron atoms in a tetrahedral fashion and form the terminal moieties coordinating metal atoms [236, 237]. Therefore, the effective coordination charge

$$c_{O-S} = 2 - \frac{6}{4} = 0.5 \tag{6.38}$$

is generally lower than in any other in the so far mentioned silicate-analogous compounds. $Gd_2[B_2(SO_4)_6]$ contains similar building units like the aforementioned strontium borophosphate, i.e., $[B(SO_4)_4]$. But two of them condense to build a vierer ring cyclosilicate-analogous anion. These anions coordinate the gadolinium atoms in a square-antiprismatic manner. Due to their similar size, Tb^{3+} or Eu^{3+} ions may be doped onto the gadolinium sites.

In *fluorooxoborates*, some terminal oxygen atoms are substituted by fluorine atoms. They reduce the coordination strength further. For instance, a terminal fluoride ion only achieves an effective coordination charge of

$$c_{\text{F-B}} = 1 - \frac{3}{4} = 0.25 \ (\text{c. n.} = 4), \quad c_{\text{F-B}} = 1 - \frac{3}{3} = 0 \ (\text{c. n.} = 3) \tag{6.39}$$

for BO_3F tetrahedra and BO_2F triangles. Moreover, the electronegativity of fluorine is higher than that of oxygen, which reduces covalent interactions to metal atoms. In $Ba[B_4O_6F_2]$, layers of condensed BO_3 and BO_3F moieties were reported, which coordinate the barium atoms. These are coordinated by nine oxygen and four flourine atoms yielding an extraordinarily high coordination number of thirteen with quite long distances. This situation reminds of the situation in $Ba[Si_7N_{10}]$ mentioned earlier. We will see in Part II how weak this coordination actually is when divalent europium atoms are doped on the barium sites in these two examples.

Part II: **Properties and Applications**

7 Transitions

Certainly, spectacular properties always fascinate people. Either "magic" behavior like the attraction of certain objects by others or a "miraculous" glow of cold objects upon exposure to light are properties for which also rare-earth materials are famous. Accordingly, in this second Part I will discuss selected optical and magnetic properties of such compounds. Due to the vast spectrum of meanwhile acquired property details and their applications, this selection has to be incomplete but aims to trigger interest and to enable the reader for further self-studies on these topics. Not only for the optical, but also for the magnetic properties, the knowledge about which transitions are feasible at room temperature is important. This is relevant to be able to assess a population of ground and excited states, which may influence the magnetic and optical properties.

7.1 Transition Probability

Between ground and excited states, electronic transitions are feasible. Such transitions may be excited optically or also thermally. Based on time-dependent perturbation theory, their intensity depends from the transition probability W between the states under consideration. This probability is proportional to the integral over space of the product of the wavefunctions of the excited (ψ_1) and the ground state (ψ_0). Moreover, the transition operator $\mathbf{O}$ has to be incorporated:

$$W \propto \int \psi_1 \mathbf{O} \psi_0 \mathrm{d}V \qquad (7.1)$$

From this equation, *selection rules* will be concluded, which tell us how intense optical absorptions, excitations and emissions will be. I will restrict the discussion herein to a rather concise background of the theory, its practical application and assessment of the results. The deeper background of the theory may be found in [238] or any other book dealing with an emphasis on electronic transitions.

7.2 Configuration, Terms and Levels

Basically, f–f transitions are parity forbidden. Therefore, electric dipole transitions are not possible here as the linear transition operator O requires different parities for ψ_0 and ψ_1. This will be explained in more detail in Chapter 7.3. But under certain local symmetries, partial mixing with states of different parity is allowed giving rise to *induced electric dipole transitions*. In the case of rare-earth elements states of different parity might be 5d states or charge-transfer states. For the following discussion on the Russel–Saunders coupling scheme, I choose as a prominent example trivalent europium, i. e., Eu^{3+}, that comprises the electronic *configuration* [Xe]4f^6 (Table 2.1). Within closed shells

https://doi.org/10.1515/9783110680829-007

like $1s^2$ or $5s^2p^6$, electrons can only be arranged in a single way. For the remaining six electrons within the seven 4f orbitals,

$$\binom{14}{6} = 3003 \tag{7.2}$$

arrangements, so-called *microstates*, are possible. These split energetically into 119 *terms* due to the interelectronic repulsion of the six electrons. These terms cover all microstates. The ground term considering *Hund's rule*[45] carries a total spin of

$$S = 6 \times \frac{1}{2} = 3 \tag{7.3}$$

giving a spin multiplicity of

$$M = 2S + 1 = 7 \tag{7.4}$$

called a septet term.[46] Further, the total orbital angular momentum quantum number amounts to

$$L = 3 + 2 + 1 + 0 + (-1) + (-2) = 3 \tag{7.5}$$

yielding an F *term*.[47] The relative orientation of spin and angular momentum is considered by the good quantum number J running from $L + S$ via $L + S - 1$ through $|L - S|$ defining the *levels*. Below half-occupation of the respective shell, the stability declines with increasing J and vice versa. In the case of a $4f^6$ configuration, the minimal J represents the most stable level in a given term. Thus, the ground term is a septet ^7F_J with rising level energies according to $J \in \{0, 1, 2, 3, 4, 5, 6\}$ covering in total

$$(2S + 1)(2L + 1) = 49 \tag{7.6}$$

microstates. The ground level is therefore $^7F_0$. Following at higher energies are the ^5D_J with $J \in \{0, 1, 2, 3, 4\}$ levels and another 117 terms; a listing of these may be found in [239]. Within a term, spin-orbit coupling leads to a splitting of the J levels. Finally, the ligand-field perturbs the J levels further up to $2J + 1$ *sublevels* depending from the local symmetry (Figure 7.1). To determine the number of lines, character tables containing irreducible representations are employed. Figure 7.1 displays the magnitude of splitting on the example of Eu^{3+}. In the spectrum accordingly the emission $^5D_0 \rightarrow {}^7F_0$ comprises

---

45 *Friedrich Hund*, German physicist (*1896 †1997).

46 The spin multiplicity $M = 2S + 1$ is named singlet ($M = 1$), doublet ($M = 2$), triplet ($M = 3$), quartlet ($M = 4$), quintet ($M = 5$), sextet ($M = 6$), septet ($M = 7$) and octet ($M = 8$).

47 The total orbital angular momentum quantum number is dubbed with letters S ($L = 0$), P ($L = 1$), D ($L = 2$), F ($L = 3$), G ($L = 4$), H ($L = 5$), I ($L = 6$), K ($L = 7$), L ($L = 8$) continuing with the alphabet.

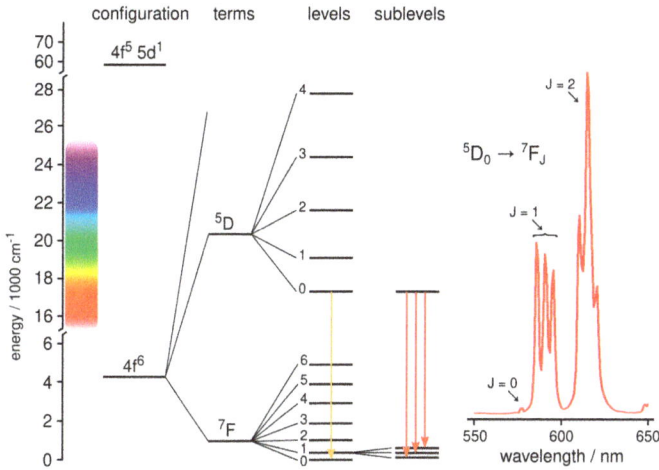

Figure 7.1: Basic energy diagram of Eu^{3+} illustrating the relative magnitudes of the splitting effects, the arrows indicate the observed transitions to the ground levels $^7F_0$ (orange) and $^7F_1$ (red); the spectrum shows the emissions of the transitions $^5D_0 \rightarrow {}^7F_J$ with $J \in \{0, 1, 2\}$.

one line, the transition $^5D_0 \rightarrow {}^7F_1$ three and one can guess that the transition $^5D_0 \rightarrow {}^7F_2$ indeed comprises five lines—although only three are obvious, another two are hidden in the shoulders of the band. Moreover, external magnetic fields may split these sublevels via a Zeeman[48] effect further. At least for optical properties, the last effect is usually neglected as its magnitude is very small compared with the other effects, which are given with approximate magnitudes in Figure 7.1. Contrarily, for the magnetic properties of trivalent europium these splittings and also the small energy differences between the ground levels cannot be neglected. With a rising temperature, the excited levels become increasingly populated altering the magnetic moment of Eu^{3+}. The ground state terms of all relevant rare-earth ions are given in Table 7.1.

Table 7.1: Terms of the ground states of the chemically most relevant lanthanide ions.

name	Ln^{2+}	Ln^{3+}	Ln^{4+}	name	Ln^{2+}	Ln^{3+}	Ln^{4+}
cerium		$^2F_{5/2}$	$^1S_0$	terbium		$^7F_6$	$^8S_{7/2}$
praseodymium		$^3H_4$	$^2F_{5/2}$	dysprosium		$^6H_{15/2}$	
neodymium		$^4I_{9/2}$		holmium		$^5I_8$	
promethium		$^5I_4$		erbium		$^4I_{15/2}$	
samarium		$^6H_{5/2}$		thulium		$^3H_6$	
europium	$^8S_{7/2}$	$^7F_0$		ytterbium	$^1S_0$	$^2F_{7/2}$	
gadolinium		$^8S_{7/2}$		lutetium		$^1S_0$	

─────

48 *Pieter Zeeman*, Dutch physicist, Nobel Prize in Physics 1902 (*1865 †1943).

If the electrons are distributed over two or more shells, the respective terms are derived per shell and subsequently coupled. For instance, in Eu^{2+} ions the excitation from its $4f^7$ ground state and ^8S ground term by bluish or higher energetic light usually leads to a $4f^6 5d^1$ configuration. Such transitions are called *interconfigurational transitions*. 4f and 5d states are sufficiently discrete in that their electrons will first couple within the shell and afterwards both terms will couple. Accordingly, the $4f^6$ shell yields a ground term ^7F, the $5d^1$ shell a ground term ^2D; the coupling of both might lead to spin multiplicities 6 or 8, L might total to 1 (P term) or 5 (H term)—with an octet ^8H term expected as lowest energetic excited one. This is indeed the case [240]. But due to transition selection rules, the most relevant excited term of the $4f^6 5d^1$ configuration is finally ^8P. The J levels were omitted in this brief discussion, but can certainly be included. In cases of interconfigurational transitions, such as 5d–4f, the J levels are normally omitted as these transitions yield broad and unresolved excitation and emission bands.

Throughout this book, the higher energy state is always mentioned first, the lower one second and the arrow in between shows which transition, either excitation and absorption or emission, is considered.

7.3 Electric and Magnetic Dipole Transitions

For a basic estimation whether a transition is allowed or forbidden, considering the symmetric behavior of the function under the integral in Equation (7.1) is sufficient. This transition will be allowed if the integral is positive and forbidden if it is zero. Any integral of an *even function,* which is symmetric with respect to a mirror plane localized in the origin like

$$x \mapsto x^2 \tag{7.7}$$

will be of the first type while any integral comprising an *odd function* being centrosymmetric with respect to the origin like

$$x \mapsto x \quad \text{or} \quad x \mapsto x^3 \tag{7.8}$$

is of the latter type. The symmetry behavior or so-called *parity* of respective functions will be indicated in this book in conformance with group theory as *ungerade* (German for odd, indicated with index "u") and *gerade* (German for even, indicated with index "g"), respectively. According to this definition, s orbitals are even ones, p odd, d even and f odd. Multiplying an odd with an even function yields an odd one, while the products of two odd or two even functions always are even.

Electrons interact with *electromagnetic* waves comprising perpendicular oscillating *electric* and *magnetic* fields propagating with the speed of light. Accordingly, optical transitions are classified as *electric dipole, magnetic dipole* and—considering higher-order interactions—*electric quadrupole* transitions depending on the respective transition operator. Considering their relative intensities only the first two are relevant and will be discussed here. Among these, electronic dipole transitions are normally several orders of magnitude (typically 10^3 to 10^5 times) more intense than magnetic dipole transitions.

The symmetry behavior of electronic states can be reliably determined via a careful analysis employing the character table of the local symmetry. Therein not only the states' symmetry is relevant but certainly also the symmetry of the transition operator. An electric dipole operator **D** transforms like a charge propagating along an axis, i. e., like a *translation* in a character table with odd symmetry [241, Chapter 4.5]. Hence, the characters of the product under the integral

$$W \propto \int \psi_1 \mathbf{D} \psi_0 \mathrm{d}V \tag{7.9}$$

with $\mathbf{D} = \mu\mathbf{r}$ are obtained as direct product of characters.[49] Therefore, the transition operator according to

$$\chi(\psi_1) \otimes \chi(\psi_0) \otimes \chi(\mathbf{D}) \tag{7.10}$$

can be calculated. The resulting representation has to contain the total symmetric representation of the point group. Only then a positive integral is obtained. Since the electric dipole operator is an odd operator, the parities of both wavefunctions have to be different. This is also known as *Laporte*[50] or *parity selection rule, i. e.,* $\Delta L = \pm 1$, and it is in accordance with the classic approach, the rule of conservation of all angular moments, as the emitted photon carries an angular momentum of $l = 1$. The Laporte rule requires the change of parity during transition. In case of accordingly forbidden transitions, these might become partially allowed if the surrounding allows for mixing with states of opposite parity. In a non-centrosymmetric surrounding, 4f states might mix with 5d states to some extent. This rule can be exemplified on a simple case considering the character table of point group O_h. In a centrosymmetric environment like the octahedral one, the set of d orbitals transforms like the symmetry races $e_g + t_{2g}$, the set of f orbitals transforms like $a_{2u} + t_{1u} + t_{2u}$.[51] Consequently, a mixing of f and d states is symmetrically forbidden. What changes upon lifting the inversion symmetry? In the resulting point group O, any set of d states transforms like $e + t_2$ and any set of f states like $a_2 + t_1 + t_2$. Thus, the states transforming like t_2 may mix. For those striving for further reading, I recommend [241].

Analogously, the significantly less intense *magnetic dipole* electronic transitions are treated. The magnetic dipole transition operator **M** corresponds to a rotating charge

---

49 The characters of the irreducible representation are found in the respective character table.

50 *Otto Laporte*, German and later American physicist (*1902 †1971).

51 $E = 7, 8C_3 = 1, 6C_4 = -1, 3C_2 = -1 \, (\| \, C_4), 6C_2' = -1$, the parity is certainly *ungerade*.

about an axis, i. e., it transforms like a *rotation*, also given in the respective character table (R_x, R_y, R_z), e. g., that of D_{3h} given in Appendix E. The product $\chi\,(\psi_{gr}) \otimes \chi\,(\psi_{ex}) \otimes \chi\,(M)$ has to contain the total symmetric representation to achieve a positive transition probability

$$W \propto \int \psi_1 M \psi_0 dV > 0. \tag{7.11}$$

Here, in contrast to electric dipole transitions no parity change is allowed between both states as M is an even operator and any change of parity would cause the integral to vanish. Thus, the selection rule for magnetic dipole transitions is $\Delta L = 0$. An exemplary calculation how to determine if a given pair of wavefunctions yields an allowed transition according to the discussion in this chapter is given in the Appendix E.

The frequency of the light under consideration influences the manner in which light interacts with matter as its oscillations may induce oscillations within the irradiated atoms. While low-frequent light enables the excitation of molecular rotations or vibrations of chemical bonds, the light corresponding to electronic transitions is orders of magnitude too fast for both. This is reflected in the *Franck–Condon principle*[52] based on the *Born–Oppenheimer approximation*;[53] this tells us that nuclear and electronic motions are independent. Thus, atomic movements can be neglected in the course of an optical electronic transition. According to the actual speed of transitions, electronic transitions occur on a femtosecond scale, vibronic ones on a picosecond scale with an approximate lifetime of excited states ranging from the nanosecond through to the millisecond scale for allowed and forbidden transitions, respectively.

7.4 Spin Selection Rule

Finally, the spin of both states has to be considered. The emitted or absorbed photon, on one hand, carries no spin. On the other hand, the spin must not change because the overall spin of the system is to be conserved. Therefore, the spin of both states must not be different. This is known as the *spin selection rule* $\Delta S = 0$. With increasing atomic mass, the Russel–Saunders coupling becomes less important compared with jj coupling. Thus, the quantum numbers L and S are less precisely defined. Accordingly, the spin and angular momentum selection rule changes smoothly to a selection rule regarding J, and the spin selection rule applies less strictly. This means that even at low temperatures transition intensity can be recorded for spin-forbidden transitions, although significantly

---

52 *James Franck*, German physicist, Nobel Prize in Physics 1925 (*1882 †1964)
Edward Condon, American physicist (*1902 †1974).
53 *Max Born*, German-British physicist and mathematician, Nobel Prize in Physics 1954 (*1882 †1970)
J. Robert Oppenheimer, American physicist, director of the Manhattan project (*1904 †1967).

Table 7.2: Summary of the selection rules discussed in this chapter.

electric dipole transitions	magnetic dipole transitions
$\Delta S = 0$	$\Delta S = 0$
$\Delta L = \pm 1$	$\Delta L = 0$
$\Delta J \leq 6$	$\Delta J = 0, \pm 1$
$\Delta J = 2, 4, 6$ if $J = 0$ or $J' = 0$	$(0 \longrightarrow 0'$ forbidden$)$

lower than for respective spin allowed transitions. A summary of all discussed selection rules is given in Table 7.2.

7.5 Jablonski Diagrams

During your studies on luminescence of rare-earth ions, you will frequently come across so-called *Jablonski diagrams*[54] like the one shown in Figure 7.2 (right). In such diagrams, all relevant electronic states including their vibrational sublevels are sketched relatively to energy. The energy is normally given on the ordinate against a rough radius or reaction coordinate on the abscissa. Within this scheme, transitions like optical absorptions as well as possible energy transfer mechanisms or energy loss pathways and feasible luminescence emissions either as fluorescence or phosphorescence are illustrated.

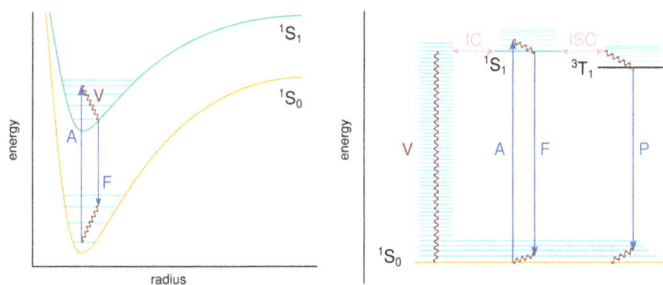

Figure 7.2: Potential curves of a ground (orange) and an excited state (green) with light blue vibrational sublevels and a Jablonski diagram displaying schematically relevant electronic states and vibrational sublevels; shown are also an absorption A, a subsequent vibrational relaxation V, a fluorescence F and a phosphorescence P emission.

Each electronic state should be viewed as a potential curve like that of an anharmonic oscillator with a series of vibronic sublevels starting from a vibrational ground

---

54 *Aleksander Jabłoński*, Polish physicist (*1898 †1980).

level; with increasing vibrational energy this potential curve develops increasingly anharmonically caused by interactions with adjacent atoms. While in harmonic oscillators, the energy gap between two arbitrary adjacent vibronic sublevels is the same; in real, anharmonic ones, this energy gap decreases with increasing energy as shown in Figure 7.2 (left). The width of the potential curve also illustrates the imagined radial interval within the object oscillates. You may imagine a child on a swing; this child will have the highest speed in the minimum of the potential curve where the whole potential energy is converted into kinetic energy. Moreover, the maximum probability density to find an oscillating object within any vibrational level is on the turning points where the object has speed zero and only possesses potential energy. Hence, in terms of the potential curve the probability density of an object will be highest near both ends of the vibrational level. This also means that electronic transitions will end at that sublevel with the highest probability density, and that will be close to the point where the ascending (absorption) or descending transition (emission) intersects the potential curve of the target electronic state.

In a Jablonski diagram, the states are also assigned their term symbols, which have been derived in Chapter 7.2. In our schematic diagram, we use the singlet states $^1S_0$ (electronic ground state), $^1S_1$ (excited singlet state) and an excited triplet state $^3T_1$, which is lower in energy than $^1S_1$ as the repulsion of electrons with opposite spin is smaller than of those with the same.

Electronic excitations like A in Figure 7.2 and relaxations like F or P are faster (approx. 10^{-15} sec) than any vibrational mode (approx. 10^{-13} sec) caused upon an optical transition. Accordingly, the average radius position of the electron would not change during the transition. Since on the abscissa the radius is plotted, only vertical arrows represent such optical absorptions and emissions. This is known as *Franck–Condon principle* based on the *Born–Oppenheimer approximation*. After such an optical transition, an excited vibrational state is yielded from which thermal relaxation until the vibrational ground state of the target electronic state occurs. This lasts approximately 10^{-13} sec. From an excited electronic state like $^1S_1$, a direct transition to the ground state $^1S_0$ can occur either by optical emission as fluorescence F, lasting approximately 10^{-9} sec, or radiationless as *internal conversion*, abbreviated IC. Internal conversion may occur, if an excited vibrational state of $^1S_0$ is within reach. For instance, if both potential curves touch close the vibrational ground state of $^1S_1$ as depicted in the Jablonski diagram. A further alternative is an *intersystem crossing* (ISC in Figure 7.2), which represents a radiationless energy transfer to a close state of different spin multiplicity like $^3T_1$. Energy transfers are generally depicted by a horizontal arrow since during this process in first approximation no energy is lost; as the energy is plotted on the ordinate, arrows have to be horizontal indicating the states from and to which the energy is transferred. A more thorough discussion on energy transfer mechanisms follows in Chapter 7.6. From the excited state $^3T_1$, the energy might be released as luminescence. This process is rather slow due to the spin selection rule, which lowers the transition probability significantly and which can only be bypassed via spin-orbit coupling. Luckily, this is in action for rare-earth atoms.

The thermal relaxations after optical excitation and relaxation lead to rearrangements of the system, and thus slight changes of the interatomic distances and energetic stabilizations. Consequently, during a whole cascade from optical absorption, via vibrational relaxation to final optical emission (either F or P) energy is lost and the emission occurs at longer wavelengths than the excitation. This energy shift is called the *Stokes shift* and is typically small for 4f–4f emitters and large for 5d–4f emitters.

7.6 Energy Transfer Mechanisms

Energy transfer processes cause several consequences. First, any energy transfer from an excited atom D, the donor, onto another atom A, the acceptor or activator, quenches feasible luminescence from D. Second, this may be beneficial if A takes advantage of this energy. Possibly A lacks of another efficient excitation process—and eventually luminesces. Third, A may collect more than one excitation and finally emits a single photon of higher energy than initially by D absorbed, i. e., *up-conversion*. Here, we will focus on those energy transfer mechanisms important for rare-earth compounds discussed in this book.

As depicted in Figure 7.3, initially the donor species D absorbs energy like a photon yielding an electronic excitation into an arbitrary vibrational sublevel of an excited electronic state. From here, it relaxes fast to the bottom sublevel of the excited state. Then the energy might either be emitted by fluorescence or it is transferred onto other nearby atoms A. This gives rise for quenching pathways, *sensitized luminescence* in the so-called *antenna phosphors* or up-conversion processes. Such energy transfers are only feasible if the energies of both transitions are very similar or if the tiny difference might be settled employing phonons. To distinguish between radiative and nonradiative transfers, one measures the respective lifetime of the excited donor state. If the transfer is radiative, this lifetime would be independent from the activators concentration; if the transfer is nonradiative, an increasing concentration of A would yield a shorter lifetime.

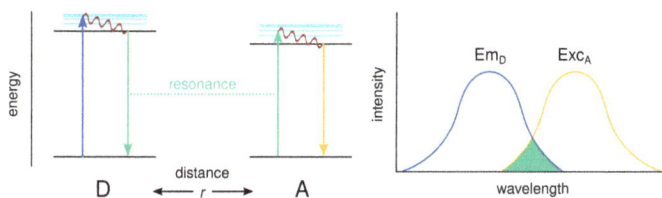

Figure 7.3: After excitation of donor D by a photon (blue) into an arbitrary vibrational sublevel (light blue) of the excited electronic state (upper black bar) fast vibrational relaxation (red) occurs; in case of resonance (green) with suited states on acceptor A the energy is transferred and finally emitted (yellow); the FRET requires an overlap of emission spectrum of D (Em$_D$) and excitation spectrum of A (Exc$_A$).

In general, the probability of an energy transfer scales with the spectral overlap of emission of D and excitation of A. Further, it is inversely proportional to the lifetime of the donor excited state, because the longer the excited state lives, the higher the chance of energy transfer from here. Practically, these energy transfers might occur by emission from D and direct absorption of this emitted light by A or the energy transfer occurs radiationless. There are two radiationless energy transfer mechanisms, the FRET and Dexter mechanisms.

7.6.1 FRET Mechanism

By the Förster[55] Resonance Energy Transfer (FRET), only energy is swapped between D and A [242]. This is possible as soon as there is an energetic overlap of D's emission spectrum and A's excitation spectrum because then both atoms can get into *resonance*. Based on a classic model, the excited pulsating dipole D causes oscillations on A, and thus transfers energy. According to quantum mechanics, this energy transfer occurs if there is a positive overlap of the wavefunctions of the excited state on D and the ground state on A. The type of transition may be electric dipole, magnetic or other multipole transition; the respective selection rules apply. In solids, additionally the relative orientation of both moments, donor and acceptor has to be considered. Classically treated, this is akin to proper positions of emitter D and antenna A. In liquids, these geometric considerations also apply but due to permanent reorientation of molecules and ions an averaged situation can be assumed. According to the type of transition, the transfer efficiency relates to the distance between D and A. For instance, in case of a dipole-dipole transition the efficiency decreases with r^{-6}. It certainly also scales with the overlap integral of D's emission spectrum and A's excitation spectrum shown as the green area in 7.3. Consequently, such energy transfers may occur via quite large distances up to approximately 100 Å.

7.6.2 Dexter Mechanism

Contrarily to the FRET mechanism, in the course of the Dexter[56] mechanism, a (blue) excited electron in Figure 7.4 on the energy-donor D swaps with a (green) ground state electron on the energy-acceptor A to exchange energy. Accordingly, this requires a direct overlap of orbital clouds, i. e., covalent interactions between both protagonists limiting the regime to distances of approximately 10 Å for such transitions. The efficiency of Dexter-type energy transfers typically decays exponentially with increasing distance r.

---

55 *Theodor Förster*, German physicochemist (*1910 †1974).
56 *David L. Dexter*, American physicist (*1924).

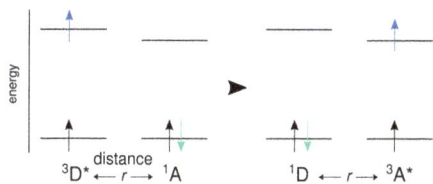

Figure 7.4: The excited triplet donor D* transfers its energy via swap of the excited (blue) electron versus a (green) ground state electron of the acceptor A to yield an excited A*.

Akin to the FRET mechanism, the emission and excitation spectra of D and A have to overlap. In contrast to FRET, a spin transfer can be achieved as shown in the example displayed in Figure 7.4, where the formal reaction

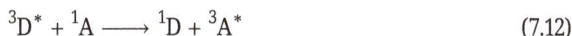

$$^3D^* + {}^1A \longrightarrow {}^1D + {}^3A^* \tag{7.12}$$

takes place. Thinking further, the now relaxed ground state ^1D may be excited again to $^1D^*$. A subsequent intersystem crossing to $^3D^*$ might yield another adjacent excited $^3A^*$. Given a sufficiently large lifetime of this triplet, state two neighboring $^3A^*$ can interact according to the formal equation

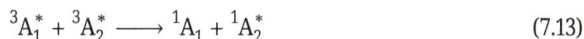

$$^3A_1^* + {}^3A_2^* \longrightarrow {}^1A_1 + {}^1A_2^* \tag{7.13}$$

to give singlet states. Thus, a so-called triplet triplet annihilation (TTA) results. The important consequence is that both excitations add up on a single A atom yielding a single emission with almost double frequency related to the excitation energy. Hence, an important up-conversion process is provided. Up-conversion luminescence is treated in more detail in Chapter 8.4.7.

7.7 Further Parameters

In this chapter, we will shortly discuss further parameters, which are relevant figures especially of luminescent materials like bandwidth, thermal quenching and some remarks on colorimetry.

7.7.1 Bandwidth of a Transition

The Jablonski diagram in Figure 7.2 might suggest that only distinct optical transitions are feasible, but actually also transitions to adjacent vibronic sublevels with somewhat lower transition probabilities occur. This leads to a bandwidth of the absorption and the emission spectrum as depicted in Figure 7.3 (right). This bandwidth depends from few main factors.

Ground and excited states are described by potential curves (Figure 7.2, left). The broader a potential curve, the smaller the energy gap between adjacent vibrational sublevels and the more transitions become probable. The excited states' potentials are normally shifted to larger radii, since increased electronic repulsion enlarges their radius. In the case of 4f–5d transitions, even another shell on a higher orbit becomes occupied. Moreover, states with a larger radius interact stronger covalently with neighboring atoms. This so-called *nephelauxetic effect*, described in Chapter 8.1, further increases the width of these states. Additionally, the diffusivity of states is reflected in the width of the respective potential curve. In the case of transitions from contracted 4f to diffuse 5d states, the latter not only show a strong radial shift but also a clearly broader potential leading to very broad excitation and emission bands. So both, increasing radial shift as well as increasing the width of excited states foster larger transition bandwidths.

With increasing concentration of the emitting ions in a luminescent material, their average distance decreases. Consequently, interactions among them become more and more relevant. Such interactions may further yield broader potentials and larger radial shifts of excited states. By the way, this also enhances nonradiative transition probabilities. These consequences lead to *concentration quenching* of selected emissions—via cross-relaxation as discussed for Tb^{3+} in Chapter 8.4.2—or of the luminescence at all.

Independently from these two effects, increasing temperature contributes to the bandwidth as higher and higher vibrational sublevels are populated; at 300 K, the thermal energy amounts to

$$k_B T \approx 0.026 \, \text{eV} \approx 208 \, \text{cm}^{-1} \qquad (7.14)$$

leading to a thermal equilibrium. This figure is relevant for the estimation whether levels above the ground level are thermally populated, and thus have to be considered accordingly.

7.7.2 Thermal Quenching and Concentration Quenching

With an increasing temperature also in the excited state, a thermal equilibrium is achieved yielding the population of higher vibrational sublevels. Generally, optical emission processes compete with radiationless internal conversion to states of the emitter himself as depicted in Figure 7.2. Alternatively, possibly existing charge-transfer states might be addressed as described in Chapter 8.2. The more and the higher vibrational levels are populated, the higher the probability of internal conversion or *thermal quenching* becomes. Consequently, the luminescence intensity I_T decreases with increasing temperature T according to

$$\frac{I_T}{I_0} = \frac{1}{1 + \text{const.} \cdot \exp\left(-\frac{E}{kT}\right)} \qquad (7.15)$$

with a materials constant and the activation energy E of the radiationless transition [243, 244]. Typically, the thermal quenching temperature T_Q is achieved reaching 50 %

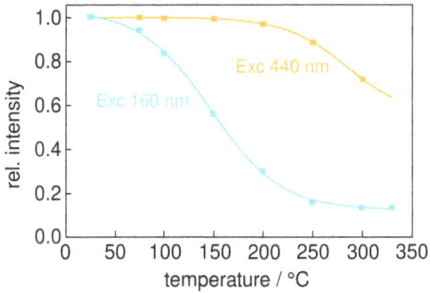

Figure 7.5: Thermal quenching behavior of $Sr_2[Si_5N_8]:Eu^{2+}$, excited at two different wavelengths, given is the integral emission intensity; data: [227].

of the maximum intensity. Figure 7.5 displays the thermal behavior of the luminescence of $Sr_2[Si_5N_8]:Eu^{2+}$ monitored around 620 nm according to Table 8.5 (Chapter 8.5.2). For the excitation at 160 nm, T_Q accordingly lies around 160 °C, for the excitation at 440 nm the thermal quenching temperature lies beyond 350 °C, apparently. Radiationless transitions occur in the regime of the intersection point of the potentials of ground and excited states. Therefore, T_Q depends from the relative *radial shift* and *differing widths* of both potentials. They can be analogously discussed as done in the chapter previously. Regarding the charge-transfer states, T_Q declines with decreasing transition energy. A similar effect is caused with increasing concentration of the emitting ion, called *concentration quenching*. Then a growing interaction between neighboring emitters can cause reabsorption of the emitted luminescence. Therefore, for every phosphor exists an optimum emitter concentration, which is the higher, the less allowed the emission transition is.

7.7.3 Color Coordinates

An important colorimetric figure are *color coordinates*. To standardize the color perception by the human eye, the *CIE*[57] developed a color diagram in 1931 comprising all visible colors, depicted in Figure 7.6. Based on the sensitivity curves of cone-cells of the human eye, three color coordinates can be calculated from any emission or absorption spectrum. Because

$$x + y + z \equiv 1 \tag{7.16}$$

a two-dimensional plot of two coordinates is sufficient. On the curved edge of the diagram, the pure colors are found while on the base-line the nonspectral colors like purple are situated. In the center, all colors add up to white. Complementary colors are arranged symmetrically with respect to the white point.

Practically, the CIE lists three sensitivity curves between 380 and 780 nm, which were meanwhile updated and can be downloaded from the internet, e. g., via http://

---

57 Commission International de l'Eclairage.

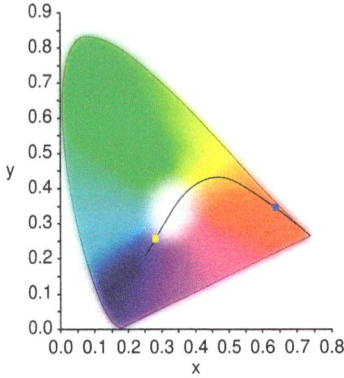

Figure 7.6: The CIE color diagram showing the color coordinates of a black body radiator at 1,000 (blue) and 10,000 K (yellow) on the black body radiator line.

cvrl.ioo.ucl.ac.uk. Then these data are multiplied with the experimental spectral data of a recorded spectrum and the color coordinates are obtained. For instance, for the aforementioned phosphor $Sr_2[Si_5N_8]:Eu^{2+}$ the coordinates $x = 0.63$ and $y = 0.37$ or for $Sr_6(BO_3)_3(BN_2):Eu^{2+}$ of $x = 0.47$ and $y = 0.53$ were reported [227, 245].

If a display is driven with three emitters (red, green and blue), the accessible color gamut corresponds to the triangle spanned by the three color coordinates. The same holds for phosphor-converted LEDs, where one color point is defined by the usually blue LED source and at least one, normally several phosphors converting blue into respective color points; this enables to tune the LED emission within the spanned color gamut by respective mixing of the contributing emissions.

7.7.4 Color Temperature

Closely related to the color coordinates is the colorimetric figure of the *color temperature* of light sources, which corresponds to the emission of a black body radiator at that certain temperature. The emission spectrum of a black body radiator was derived by Planck[58] where the spectral radiance at a wavelength λ is given via

$$B_\lambda = \frac{2hc^2}{\lambda^5} \frac{1}{\exp\left(\frac{hc}{\lambda k_B T}\right) - 1} \tag{7.17}$$

at the respective absolute temperature. A differentiation of the *Planck radiation law* then yields the emission maximum, which is known as the Wien[59] displacement law

$$\lambda_{max} \approx 2898 \ \mu m \ K \cdot \frac{1}{T} \tag{7.18}$$

---

58 *Max Planck*, German theoretical physicist, Nobel Prize 1918 (*1858 †1947).
59 *Wilhelm Wien*, German physicist, Nobel Prize 1911 (*1864 †1928).

Interestingly, a really hot black body radiator of 10,000 K emits already bluish light normally perceived as *cold* light, while a comparably cool black body of 1,000 K emits reddish light perceived as *warm* light (Figure 7.6). Daylight, which is essentially the reference for white light, corresponds to a color temperature of ca. 5,800 K. The color coordinates of a light source determine its color temperature.

7.8 Transitions within a 4f Configuration—Judd–Ofelt Theory

In Chapter 2.1, we discussed the special situation of orbitals like 1s, 2p, 3d and 4f, which are the first of their family, and thus lack radial nodes. We drew the conclusion that this leads to relatively contracted or localized orbitals. This makes it difficult to ionize electrons therein. In every case where electrons are hard to ionize, these apparently experience a quite strong effective nuclear charge. This furthermore yields a good shielding effect on more outwards localized orbitals. In the specific case of 4f orbitals, electrons therein screen next shell's orbitals 5s, 5p and 5d very well. Therefore, these are very diffuse (Figure 7.7), changing now the point-of-view onto an approaching nearby atom. This might, for instance, stem from a ligand or is just a single coordinating one. For this approaching atom, the 4f orbitals are hardly visible behind the completely filled $5s^2p^6$ shell—having in mind the image used in Figure 2.2. This filled octet shell shields the 4f orbitals very well against approaching atoms or ions as their electron density is repelled. Accordingly, the interaction of 4f electrons with adjacent atoms is weak, and the ligand-field splitting is small. This results in the remarkable situation that the energetic positions of 4f levels are almost independent from the chemical environment. Furthermore, also the energies of optical transitions within the 4f shell only show minor influence from ligands. This situation is beneficial for those who would interpret spectra of 4f–4f transitions. And it is beneficial for those looking for a suited dopant to achieve a specific luminescence emission since they do not have to care very much for the chemical surrounding as long as the band-gap is large enough and the site to be doped is appropriate. This will be a topic later, though, in Chapter 8.4.

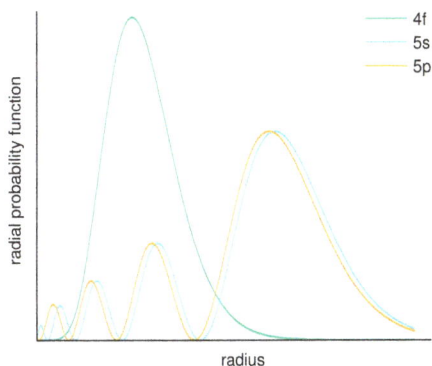

Figure 7.7: The radial probability densities of 4f orbitals versus the well against neighboring atoms shielding $5s^2p^6$ shell; the even more diffuse 5d states are not shown.

Due to this advantage, Dieke[60] collected the absorption and luminescence data of trivalent lanthanide ions published up to the early 1960s and set up the so-called *Dieke diagram* covering all experimentally recorded 4f levels up to 42,000 cm^{-1} [246]. He also marked all states from which emissions had been observed until then in respectively doped $LaCl_3$ with a filled semicircle like a hanging drop below each line representing a level. The width of these lines was chosen with respect to the observed splitting of the terms in the spectra of the respective $LnCl_3$. Since then, the energy regime of the Dieke diagram was extended to 70,000 cm^{-1} [247]. Finally, all 4f levels were calculated, ranging up to 200,000 cm^{-1}, many of them unobserved so far [248]. A complete list of atomic energy levels was collected in [240], which is available free of charge on the internet and which are the base of the term schemes shown in this book. These data can also be used for assigning 4f–4f transitions of R^{2+} or R^{4+} ions by choosing the respective electronic configuration, e. g., for Eu^{2+} the data of Gd^{3+} [249]. Due to the lower charge, the transition energies are reduced by 10 to 30 % according to data listed on https://physics.nist.gov/PhysRefData/ASD/levels_form.html.

In Figure 7.8, a selection of these collected data is depicted. Note that the terms of a considered $4f^n$ configuration are the same as those of the $4f^{14-n}$ configuration but with opposite order. This is obvious for the terms $^7F_0...^7F_6$ of Eu^{3+} and Tb^{3+}. Hence, the configurations $4f^1$ and $4f^{13}$ or $4f^6$ and $4f^8$ are named *spectroscopic twins*. Moreover, the energies of corresponding levels differ. This is due the increasing nuclear charge resulting in a stronger splitting of all states. The effect is further enhanced by a decreasing ionic radii going from left to right yielding a stronger interaction of the 4f electrons with the nuclear charge. For trivalent europium, also excited states assigned as *ligand-to-metal*

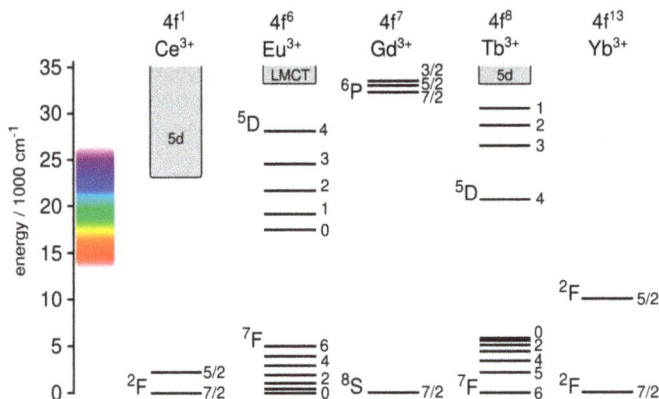

Figure 7.8: Energy levels of selected 4f configurations and the respective trivalent rare-earth ions based on data from [240]; in grey further excited states, 5d and charge-transfer states, relevant for the optical properties are depicted. The visible regime starting from the ground state energy is also shown.

---

60 *Gerhard Heinrich Dieke*, German and US-American physicist (*1901 †1965).

charge-transfer states (LMCT) are already indicated, which will be discussed in the next chapter.

You may wonder why I did not yet mention the next empty and basically available 5d states. In Figure 7.8, only for trivalent cerium and terbium the relative approximate position of 5d states are indicated. We will see in Chapter 8.3 that the *vacuum referred binding energies* of the 5d states are more or less constant for all rare-earth ions; the energy gap between the 5d and 4f ground states runs through a maximum for the $4f^7$ configuration, though. Figure 7.9 shows the energy differences of the lowest lying 5d and the 4f ground states of the free Ln^{3+} ions. Regarding this interconfigurational transition from 4f to 5d, the excited electron might retain or switch its spin. This situation is illustrated in the right part of Figure 7.9 on the example of Tb^{3+}. Hence, two different excited states with configuration $4f^7 5d^1$ arise. In the low-spin state (blue), a total spin of 6/2 results, in the high-spin state (orange) of 8/2. The ground state configuration $4f^8$ features a total spin of 6/2. Consequently, the excitation to the low-spin state (blue) is allowed with respect to the spin selection rule, and that to the high-spin state is forbidden (orange). Thus, the forbidden transition is very weak, if observed at all.

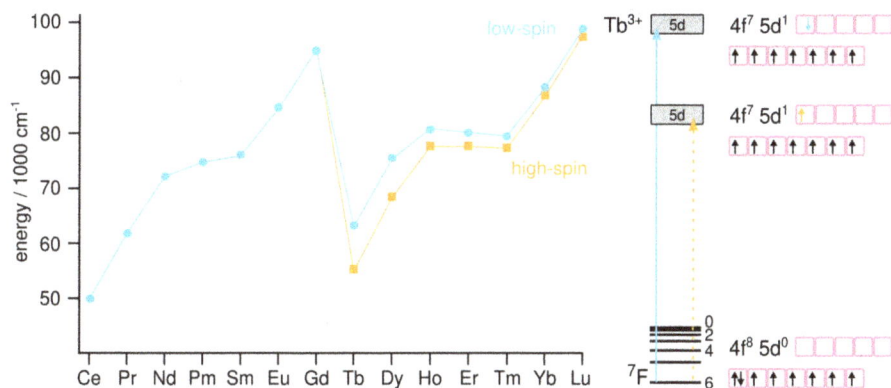

Figure 7.9: Energy differences of the lowest 5d states, discriminated according high and low spin, and the ground 4f states of the free Ln^{3+} ions calculated based on the data provided in [252]; illustration of the difference between high and low spin on the example of Tb^{3+}.

Generally, the high-spin state is more stable than the low-spin state. This can be rationalized by Hund's rule and recalling our discussion on shielding in Chapter 2.1. Electrons are better shielded by electrons of opposite spin and vice versa. Therefore, in the low-spin situation the 5d electron is shielded clearly better and might be released easier than in the high-spin situation. Thus, the low-spin state is situated at higher energy than the high-spin state. This effect declines with growing occupation along the series as obvious from Figure 7.9 since the shielding becomes increasingly similar for both situations. The discrimination of high- and low-spin states is practically only relevant for the second-half of the 4f series. Up to half-occupied 4f states the high-spin 5d–4f transitions are the spin-allowed ones and the only ones shown. In the second-half, the high-spin 5d–4f transitions are the spin-forbidden ones. So, the blue marked 5d states are the ones involved in spin-allowed transitions throughout Figure 7.9.

Absolutely, the energy differences between 5d and 4f states reflect the relative stabilities of the respective 4f configuration. These increase with growing nuclear charge as a general trend. Because of the specific stability of a semifilled 4f shell, the difference reaches a first maximum for Gd^{3+}. The eighth electron is shielded clearly better in Tb^{3+} (electronic configuration $[Xe]4f^8$) according to the discussion given just before, because it is the first with opposite spin. It is therefore clearly easier ionized than the seventh in Gd^{3+}. Thus, the 4f ground state of trivalent terbium is lifted significantly in energy. Hence, the energy difference between 5d and 4f is markedly reduced for Tb^{3+}. The development proceeds with a further increase until the end of the series where the second maximum is achieved. These energy differences can be measured very well by absorption spectroscopy. The basic values of the free ions were extracted from many compounds and led to estimated values for the free ions depicted in Figure 7.9. These values lie—apart from a few cases—far below the visible wavelength regime in the ultraviolet or even vacuum ultraviolet regimes.[61] In oxides, electrostatic and covalent interactions lower the 5d–4f differences up to $30,000\,cm^{-1}$ for trivalent lanthanide ions [250, 251]. Accordingly, the only exception where visible or u. v. luminescence may occur directly from 5d states in trivalent lanthanide ions are Pr^{3+} and Ce^{3+} discussed in Chapters 8.4.9 and 8.5.1. Excitation from the 4f ground states into the 5d excited states is basically possible for all trivalent ions and can be safely discriminated by their huge bandwidth compared with 4f–4f transitions as discussed later in Chapter 8.3. Figure 7.9 tells us, though, that even 5d←4f excitations in the near u. v. are restricted to Ce^{3+}, Tb^{3+} and Pr^{3+}. Here, I shall continue with a theory analyzing the intensities of 4f–4f transitions.

Within a few months in 1962, a theory was published independently by Judd[62] and Ofelt,[63] which addresses the transition probabilities, and thus the intensities of induced electric dipole transitions within the 4f shell [253, 254]. Their semi-empiric approach was inevitable due to the overwhelmingly vast number of states. Thus, ab initio calculations needed infinite memory and processor resources, which is still on the agenda today. The transition moments of magnetic dipole transitions can be calculated directly from the free-ion 4f wavefunctions, because the magnetic dipole operator is of even parity, and thus only states of same parity contribute (Equation (7.11)). Contrarily, the transition moments or probabilities of *induced electric dipole transitions* require a special parametrization as here appropriate mixing of the wavefunctions with those of opposite parity—odd-parity electronic states or vibrations—is required (Equation (7.9)). Such

---

61 VUV, electromagnetic radiation with a wavelength shorter than 200 nm or above $50,000\,cm^{-1}$.

62 *Brian R. Judd,* British physicist (*1931).

63 *George S. Ofelt,* American physicist (*1937 †2014).

an admixture of an odd-parity wavefunction ψ_{nl} *induces* the switch from a magnetic to an at least partially allowed electric dipole transition and generates an effective new wavefunction $|B\rangle$ with mixed parity. This may be written as

$$|B\rangle = \frac{|\ 4f_{JM} + (\langle 4f_{JM}\ |\ \mathbf{C}\ |\ \psi_{nl}\rangle)\ |\psi_{nl}\rangle}{E\ (4f_{JM}) - E\ (\psi_{nl})} \tag{7.19}$$

where a decreasing energy difference in the denominator enhances the mixing, so does a positive scalar product of the $4f_{JM}$ and ψ_{nl} wavefunctions with the crystal field operator **C** in the numerator. Due to this mixing—for instance with 5d or charge-transfer states—some parity change between the ground and excited state wavefunctions $|B\rangle$ and $\langle B'|$ is achieved. Then the intensity of this, otherwise according to the parity selection rule forbidden, pure 4f–4f transition is increased. A further discussion on the consequences and the ongoing scientific debate on mechanisms of such mixing will be found at the end of Chapter 8.3 and in Chapter 8.4.3 where *hypersensitive transitions* are treated.

The oscillator strength or transition probability P of an electric dipole transition with the dipole transition operator **D** is given by

$$P = \xi \cdot \frac{8\pi^2 m v}{3h\ (2J+1)} \cdot |\langle B\ |\ \mathbf{D}\ |\ B'\rangle|^2 \tag{7.20}$$

in which m represents the electron's mass, v the mean transition frequency, h Planck's constant and ξ considers the refractive index of the medium. Judd and Ofelt assumed for their approximation that the 4f levels under the ligand field are narrow enough to be well separated from others. So, the mixing of different J levels can be neglected. More-over, the energy differences employed in Equation (7.19) are thus equal for both 4f states; especially, the first approximation has to be kept in mind as J mixing certainly takes place to some extent. But nevertheless the results of Judd's and Ofelt's approach—now known as the Judd–Ofelt theory—are astonishingly precise. Considering the transition $B' \longrightarrow B$, its oscillator strength is then proportional to

$$P \propto \frac{v}{(2J+1)} \sum_{\lambda=2,\,4,\,6} \Omega_\lambda |\langle B\|U^\lambda\|B'\rangle|^2 \tag{7.21}$$

with calculated U^λ being a tensor of rank λ. Only for even λ, a nonzero contribution is obtained, which directly gives the selection rules for induced electric dipole transitions within this theory:

$$\Delta J \leq 6$$
$$\Delta J = 2,\,4,\,6 \quad \text{if } J = 0 \text{ or } J' = 0 \tag{7.22}$$

Because of some J-J' mixing, as mentioned before, these selection rules are somewhat weakened. For magnetic dipole transitions within a 4f configuration, these selection rules apply:

$$\Delta L = 0$$

$$\Delta J = 0, 1, \text{ but } 0 \rightarrow 0 \text{ is forbidden} \tag{7.23}$$

Here, you may recall the discussion on the Laporte rule in Chapter 7.3. Note also, that for the selection rules of the induced electric dipole transitions only the better quantum number J is mentioned. Therefore, all selection rules within the Russel–Saunders scheme are weakened as spin-orbit coupling cannot be neglected in this thorough approach [255]. The squared reduced matrix element terms $|\langle\psi_J\|U^\lambda\|\psi_{J'}\rangle|^2$—abbreviated as U^λ—for absorption and emission transitions are tabulated and describe the interelectronic interactions within rare-earth ions [256–258]. High U^λ values suggest thereby high transition probabilities, and the higher, the stronger apparently the average mixing with opposite-parity states. This behavior will be of further interest in the Chapters 8.3 and 8.4 on the chemical shift model and the 4f–4f emitters, respectively.

The intensity parameters Ω_λ ($\lambda = 2, 4, 6$) in Equation 7.21 are determined by a least-squares fit of the experimental data. They represent physically the square of the charge displacement due to the induced electric dipole transition. Mathematically, they represent the *radial part of the wave function*. For instance, in absorption spectra of trivalent europium ions Ω_2, Ω_4 and Ω_6 can be directly determined from the transitions $^5D_2 \leftarrow {}^7F_0$, $^5D_4 \leftarrow {}^7F_0$ and $^5L_6 \leftarrow {}^7F_0$. In these cases, all other matrix components U^λ are zero [257].

The Ω_λ parameters can also be obtained from emission spectra by integrating the areas under the emission bands of the respective transitions. Employing this approach a software is being developed, which currently covers Eu^{3+} emission spectra only [259, 260]. The obtained Judd–Ofelt parameters Ω_λ can be employed to calculate the transition probabilities of all transitions within the considered chemical situation. An elaborate description has to be considered to prepare the data properly for calculations and I suggest to refer to [239] and [261]. A detailed practical guide can be consulted in [262]. *BonnMag* is another program to calculate absorption spectra and temperature dependent magnetic susceptibilities of rare-earth ions. This emphasizes the relevance not only for the optical, but also the magnetic properties. You will find further details regarding *BonnMag* in [263] and the results of an application in [264]. Here, we shall now proceed with a concrete discussion of the optical properties of rare-earth elements.

8 Optical Properties

8.1 Nephelauxetic Effect

In the previous chapter, we discussed the Judd–Ofelt theory to estimate transition probabilities of electric dipole transitions within a 4f configuration. The 4f shell is very well shielded against approaching ligands, and thus interaction with these is very weak. Contrarily, the diffuse 5d states may interact directly and will interact strongly with electrons of ligands coordinating the rare-earth ions. Such covalent interactions cause the *nephelauxetic effect*. Any covalent interaction broadens the orbital's potential toward the covalently bound atom. Thus, the electron cloud of the metal ion moves to some extent toward its ligand as depicted in Figure 8.1. Consequently, the electron cloud *expands*. Jørgensen[64] apparently knew Greek—"cloud" means *nephos* and "to expand" translates to *auxanomai*. He accordingly suggested *nephelauxetic effect* for this phenomenon initially described for outer transition metal complexes. For the sake of better understanding, I shall explain this phenomenon on an as simple as possible example. The transfer on rare-earth chemistry follows at the end of this chapter. The stronger the nephelauxetic effect is, the stronger the radial shift of the potential curve becomes, strengthening the Stokes shift. Moreover, its shape becomes broader, and it is better stabilized. The respective curve is therefore lowered in energy. This covalent interaction increases with soft ligands according to the green scenario in Figure 8.1 and decreases with hard ligands (blue scenario). As consequence, a strong nephelauxetic effect yields in average larger electron-electron distances on the metal atom, and thus a reduced repulsion. For outer transition metals, this was quantified employing the Racah parameters. The *nephelaux-*

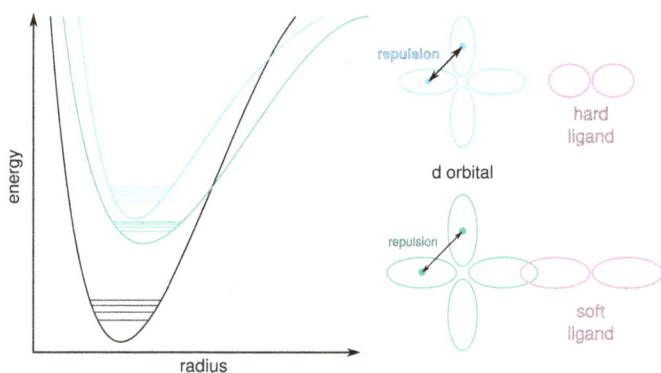

Figure 8.1: The nephelauxetic effect schematically displayed by the covalent interaction of a metal ion's d orbital with a hard (blue scenario) and a soft (green) ligand; on the left, the potential curves of the ground state (black) contrasted with those of the excited state with a hard and a soft ligand.

---

64 *Christian Klixbüll Jørgensen*, Danish inorganic chemist and spectroscopist (*1931 †2001).

https://doi.org/10.1515/9783110680829-008

etic ratio put the Racah parameter B of the considered system and of the noncoordinated, the naked atom in relation:

$$\xi = \frac{B_{\text{coordinated}}}{B_{\text{gaseous ion}}} \tag{8.1}$$

For transition metals, Jørgensen derived a quantitative *nephelauxetic series* [265]. To get an impression, in Table 8.1 the ξ values of ligands coordinated to the 3d and 4d transition metal ions Cr^{3+} and Rh^{3+} are listed. With decreasing ξ, the nephelauxetic effect of the respective ligand becomes stronger, the interacting states on the metal atom become diffuser. As expected, the latter holds vertically in Table 8.1. The ligands show stronger nephelauxetic effects as the coordinating atom becomes chemically softer. Here, also the atoms bound to these coordinating atoms are relevant, which can be nicely seen considering the row H_2O, $(NH_2)_2CO$, NH_3, $C_2O_4^{2-}$ and $(NH_2)_2(CH_2)_2$. This row also shows that the effect increases with increasing polarizability of the ligand, which is perfectly in line with the previous discussion [266].

Table 8.1: The nephelauxetic ratios ξ for homoleptic complexes of Cr^{3+} and Rh^{3+} with the given ligands (coordinating atom first in formula) based on data given in [265], additionally values for Pr^{3+} were calculated based on the data given in [267] (see text); $aq = OH_2$, $ur = $ urea $= (NH_2)_2CO$, $ox = $ oxalate $= O_4C_2^{2-}$, $en = $ ethylenediamine $= (NH_2)_2(CH_2)_2$.

	F⁻	aq	ur	NH₃	ox	en	SCN⁻	CN⁻	Cl⁻	Br⁻
Cr^{3+}	0.89	0.79	0.72	0.71	0.68	0.67	0.62	0.58	0.56	
Rh^{3+}		0.73		0.60		0.59			0.49	0.40
Pr^{3+}	0.95	0.93							0.94	0.94
Pr^{3+} ($^3P_0 \rightarrow {}^3H_4$)	0.95	0.93							0.93	0.93

You might wonder, why I am discussing this topic so thoroughly as the 4f states only show very weak covalent interactions with ligands? But as soon as 5d states enter the stage—in examples like Ce^{3+}, Eu^{2+}, Tb^{3+}, Yb^{2+} and others—the nephelauxetic effect dominates the energetic position of the 5d states with respect to the 4f ground states. Thus, the above drawn conclusions regarding the nephelauxetic effect are important as it will help us to estimate emission colors of phosphors containing rare-earth ions where 5d states play the crucial role. Furthermore, the detailed analysis of the optical properties of a single rare-earth ion delivers precise predictions of those of any other rare-earth ion. Also, for rare-earth ions the *nephelauxetic ratio* may be determined experimentally. The challenging task here is the calculation of the interelectronic repulsion parameters of the non-coordinated rare-earth ions, especially their separation from ligand-field effects and states of odd parity. Therefore, this cannot be treated precisely by Judd–Ofelt theory, as this applies an overall view on the ions. There has been quite a lot work done, but this is beyond the scope of this book as the results for nephelauxetic ratios for 4f states listed in Table 8.1 are almost negligible. Perhaps if you are interested in more details here, then I would recommend [268–271]. For Pr^{3+}, nephelauxetic ratios of a typical 4f–4f transition are given in Table 8.1 [267].

In general, the Judd–Ofelt parameters U^{λ} reflect the miscibility with states of opposite parity, and thus the effect of covalency on induced electric dipole transitions. Here both, 5d and charge-transfer states, give rise for covalent interactions with ligands. Approximate nephelauxetic ratios may also be determined by comparing the relative energies of excited states against a reference like the respective fluoride. For comparison purposes, I added the nephelauxetic ratios calculated by the relative transition energies of the prominent $^3P_0 \rightarrow {}^3H_4$ transition scaled to the precise calculation in the table. The figures very nicely quantify the weak covalent interaction of the inner transition metal ions R^{3+} with ligands compared with outer transition metals.

8.2 Charge-Transfer Transitions

8.2.1 General Aspects

During charge-transfer transitions, an electron moves upon optical excitation temporarily between two adjacent atoms. Such transitions can therefore essentially be understood as a redox reaction. The optical energy needed corresponds to the oxidation and reduction potentials of both partners. For instance, in the series of the isoelectronic colored ions, MnO_4^- and CrO_4^{2-}, the highly oxidized manganese and chromium atoms absorb visible light to catch an electron from a neighboring oxygen atom, according to the schematic reaction

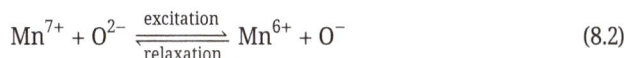

$$Mn^{7+} + O^{2-} \underset{\text{relaxation}}{\overset{\text{excitation}}{\rightleftharpoons}} Mn^{6+} + O^- \tag{8.2}$$

during which an electron is transferred from a ligand to a metal in the course of a *ligand-to-metal charge-transfer transition* (LMCT). In the case of MnO_4^-, blue, green as well as yellow light and in the case of CrO_4^{2-} ultraviolet as well as blue light is absorbed yielding the intensely violet and yellow ions, respectively. The absorption maximum of chromate lies in the u. v. but reaches until the bluish regime while permanganate's absorption peaks in the green regime at significantly lower energy. This coincides with the higher oxidation potential of permanganate compared with chromate in aqueous solutions.

Another vivid picture for the basic understanding may be to consider the charge gradient between both partners—which eventually urges a temporary redox reaction. The lower the stability of the high oxidation state, the lower the charge-transfer states are located energetically. Hence, the absorption energy is a measure for the stability of respectively oxidized species, and consequently, its *optical electronegativity*. Further prominent examples are transition metals in high oxidation states like tungstates, tantalates, titanates and vanadates. Also, in so simple anions like sulphates charge-transfer transitions are observed, normally deep in the u. v. regime reflecting their high stability.

Since an electron moves from one partner to a bonding state between both, the wavefunctions of both partners, they have to overlap. The potential curves of such *charge-transfer states* (CTS) are normally broad and considerably radially shifted, as they are positioned somewhere in between the partners. Both effects yield broad bands in absorption, excitation and emission spectra. Moreover, as these transitions are allowed, they are very efficient, and thus intense, if the other prerequisites like proper orbital overlap are met. We looked at these prerequisites in Chapter 7.6 on energy transfer mechanisms. For the sake of completeness, I would like to mention that charge-transfer transitions are also possible between two metal ions of different oxidation states like in $Pb^{[iv]}Pb_2^{[ii]}O_4$. These so-called *intervalence charge-transfer transitions* (IVCT) cause brightly colored or even black compounds.

The absorbed energy might be directly released as radiation again, certainly with a distinct Stokes shift considering the radial shift and broadness of typical CTS potentials. Additionally, the ion in its lower oxidation state is larger causing further steric relaxation. Alternatively—and this is the main reason why such excitations are important at least in this book—the energy might be transferred via a simple transfer mechanism as discussed in Chapter 7.6 onto another emitting ion such as a trivalent rare-earth ion. In this case, the moiety excited via a charge-transfer acts as an *antenna* to harvest optical energy efficiently due to its allowed nature. Consequently, it pumps the emitting states of emitters whose excitation is less allowed or even forbidden due to selection rules in action. Phosphors based on this principle are called *antenna phosphors* and will be discussed in Chapter 8.4.2.

8.2.2 View on Rare-Earth Ions

In some cases, though, even the emitting rare-earth ion may directly been excited via a *ligand-to-metal charge-transfer transition*. This occurs if the oxidation potential is high enough as another quite stable electronic configuration is achieved. Such an example is trivalent europium. Its electronic configuration is $4f^6$ and if it caught an electron from, say, a neighboring oxide ion according to the schematic reaction

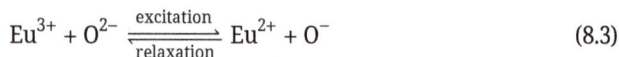

$$Eu^{3+} + O^{2-} \underset{\text{relaxation}}{\overset{\text{excitation}}{\rightleftharpoons}} Eu^{2+} + O^- \tag{8.3}$$

a semifilled f-shell is achieved. This is in excellent agreement with the relatively high stability of divalent europium ions. Similar situations are found for the rare-earth ions Yb^{3+} ($4f^{13}$), Sm^{3+} ($4f^5$) and Tm^{3+} ($4f^{12}$). For the remaining ions, normally significantly higher transition energies are required, a fact limiting practical relevance.

In spectra, charge-transfer transitions can be discriminated from 4f–4f transitions by two parameters. As mentioned, charge-transfer transitions show significantly broader bands compared with sharp intraconfigurational transitions. Second, charge-transfer bands are blue-shifted with decreasing temperature, and thus decreasing interatomic distances. The latter effect may also be achieved by doping the rare-earth ion under consideration onto smaller sites resulting in a blue shift of the charge-transfer band. Vice versa, if doped on larger sites, a red shift is observed. With declining distance, the effective charge difference—the driving force of charge-transfer transitions—is reduced and the energy of the charge-transfer states is elevated. Absolutely remarkable is, though, that almost regardless of the compound class, the difference of charge-transfer energies is constant between the lanthanide ions [252, 272]. This merits a closer look done in the next section after we shed some light on intervalence charge-transfer transitions in some rare-earth compounds.

If an element adopts different oxidation states within the same compound *intervalence charge-transfer transitions* are possible if the potentials of both overlap. Then an either thermally or optically activated electron transfer between both ions is feasible. The resulting potential curves of such IVCT states are very flat and broad. Therefore, also the absorption bands are expected to be broad. Theoretical calculations suggest that the IVCT state energy declines with decreasing electronegativity of the anions [273]. This behavior is the same as discussed for the LMCT transitions as potential curves become flattened with stronger covalent interactions. In the example of the red-brown mixed-valent oxide Eu_3O_4 and the respective black sulphide Eu_3S_4, Mössbauer spectroscopy confirmed the aforementioned trend. The oxide shows two signals representing distinct Eu^{2+} and Eu^{3+} ions. For the sulphide, only below $-60\,°C$ two signals were identified. At higher temperatures, the electrons are apparently already thermally excited to hop between adjacent europium atoms with an activation energy around 0.23 eV corresponding to roughly $1{,}800\,cm^{-1}$, well in the infrared regime [274, 275]. Accordingly, this europium sulphide is black.

Considering the chemical and physical properties of the rare-earth elements discussed in the first part of this book, it is not surprising that especially the pair Eu^{2+}/Eu^{3+} is relevant here. Another pair is Ce^{3+}/Ce^{4+} for which in the examples of intermediate oxides between Ce_2O_3 and CeO_2 and the phosphor $La(PO_4){:}Ce^{3+}$ IVCT transitions were successfully assigned [275, 276]. In the latter example, it became clear that the efficiency of the phosphor is reduced by the presence of significant amounts of Ce^{4+} if synthesized in air. Even worse, such states may quench luminescence via nonradiative relaxation paths. Unfortunately, in doped compounds such transitions are often elusive for direct measurements because of their extraordinarily broad and flat absorption bands. But sometimes they provide a possible explanation for unusually broad and red-shifted emissions [277].

In halides like $CaX_2{:}Eu^{2+}$ (X = halide) or the pure blue chlorides Eu_4Cl_9, Eu_5Cl_{11} and KEu_2Cl_6 or the violet $Na_5Eu_7Cl_{22}$, IVCT absorption bands were found in the visible regime. Their bands span more than $10{,}000\,cm^{-1}$ and yield only pale coloring in doped compounds. Certainly, they feature indeed bright coloring in the pure europium compounds [273, 278–281]. Also, the intense yellow color of the mixed valent europium borate $Eu_5(BO_{3-x}N_x)_4$ ($x \approx 0.5$) suggests the activity of IVCT transitions. But you have to be careful because other europium borates like $Eu_5(BO_3)_3F$—which exclusively contain divalent europium—may also be yellow. But a diligent look on the reflection spectra suggests the presence of an IVCT transition in the former example [282, 283].

8.2.3 Optical Electronegativity

The concept of electronegativity is closely related to ioniziation energies and, therefore, to oxidation and reduction potentials. So, it is not surprising that the charge-transfer energy relates also to the electronegativity of the ligand from which an electron is trans-ferred. The higher the electronegativity of a ligand, the lower lie its highest occupied and especially lowest unoccupied electronic states energetically. Assuming the same cation as partner, the charge-transfer energy from a more electronegative ligand will be higher than that of a less electronegative ligand.

The lanthanide halide oxides LnOX crystallize in similar structures (Chapter 6.2.4), where the R^{3+} are coordinated by both, oxygen and halide atoms. Therefore, the devel-opment of the charge-transfer energies can be assessed reliably in these compounds. Let us assume that a certain ion is doped onto sites coordinated by ligands of different electronegativity, e. g., Eu^{3+} with its valence electron configuration $4f^6$ doped on La^{3+} sites. In the halide oxides LaOX, the electronegativity of the ligands decreases along the halide series fluorine, chlorine, bromine and iodine. Careful studies show that exactly the postulated behavior is the outcome with the fluoride showing the highest charge-transfer energy. The data are depicted in Figure 8.2 where the charge-transfer energies of lanthanum halide oxides are plotted and illustrated by a blue line.

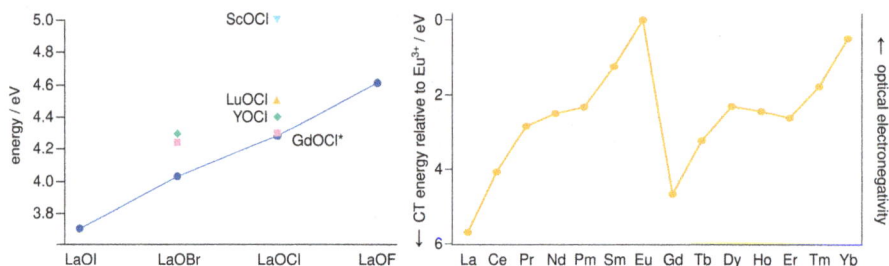

Figure 8.2: Charge-transfer energies of selected rare-earth halide oxides (left) and charge-transfer shifts relative to Eu^{3+} in reverse order to illustrate the trend of optical electronegativities of the R^{3+} (right); data: [272, 284].

Let us now assume that Eu^{3+} ions are doped onto sites of different size. This was done on the doped chloride oxides $ROCl:Eu^{3+}$ of lanthanum, gadolinium, yttrium, lutetium and scandium. Here, the R–Cl and R–O distances shrink smoothly from lanthanum to lutetium, and dramatically to the scandium compound.[65] The dramatic effect in ScOCl is due to a different structure. It adopts the FeOCl structure type in which Sc^{3+} is surrounded only octahedrally by four oxygen and two chlorine atoms. With decreasing distances between europium and the ligand atoms the overlap of orbitals is forcibly increased. You would conclude a stronger covalent interaction with simultaneously increased transfer of electron density from the ligands onto europium. This would reduce the effective charge difference as well as bring europium somewhat closer to a semifilled 4f shell. Hence, you would expect a growing charge-transfer energy from $LaOCl:Eu^{3+}$ to $ScOCl:Eu^{3+}$—and that is exactly what is observed and obvious from Figure 8.2. Moreover, the trends within a compound class are the same for all trivalent rare-earth ions. And additionally, the average differences between the charge-transfer energies of individual trivalent rare-earth ions are almost the same [252].

These insights opened the opportunity to actually measure the electronegativity of rare-earth ions by comparing their charge-transfer energies. Jørgensen[64] was the person to condense this into the equation

$$E^{CT} \approx 30{,}000 \text{ cm}^{-1} \cdot \{\chi_{opt}(X) - \chi_{opt}(M)\} \tag{8.4}$$

where the difference of so-called *optical electronegativities* of an electronegative partner X and an electropositive partner M yields an excellent guess of the charge-transfer energy, and vice versa [285]. The given 30,000 cm^{-1} correspond to 3.72 eV. Further, the obtained electronegativities resemble roughly the values of other common electronegativity scales. Figure 8.2 (right) displays the universally valid averaged differences of charge-transfer energies with respect to Eu^{3+} as well as the relative ion-specific optical electronegativities based on the discussion in [252, 272].

8.3 The Chemical Shift Model

Due to the weak interaction with host structures, the trivalent rare-earth ions behave almost independently from the local chemical environment. They essentially display the optical properties discussed later in this chapter. But the energies of their electronic states relative to the host structure's band structure is of crucial relevance. If they, for instance, came to lie close to the conduction or valence band, the rare-earth ion might be oxidized or reduced easily. Also, the relative position of the 5d states and charge-transfer states is important.

A typical scenario of rare-earth ions doped into an arbitrary host structure is depicted in Figure 8.3. Here, the energies of the *ground states of the configurations* of the lanthanide ions are plotted relatively to the conduction and valence bands. For instance,

---

65 Consulting Appendix A, the R–Cl and R–O distances decline from 303 to 256 pm and 260 to 213 pm, respectively.

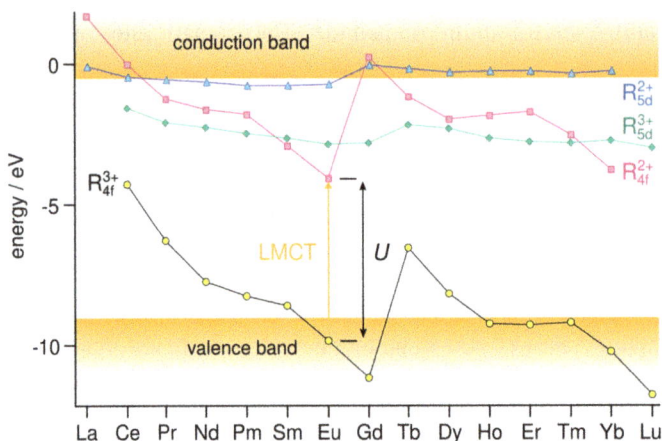

Figure 8.3: Relative vacuum referred binding energies of the 4f ground states and the 5d states as calculated by the chemical shift model of divalent and trivalent rare-earth ions on an arbitrary example and
scaled with a chemical shift $U \approx 6$ eV and a ligand-to-metal charge-transfer transition (LMCT) for Eu^{3+} of
approximately 5 eV.

the 4f states of the trivalent ions are given as R_{4f}^{3+}, and their 5d states are given as R_{5d}^{3+}.
The excited 4f levels are not shown, but certainly these are present. With the help of the
Dieke diagram, you can assess where the excited 4f levels are. The same is also displayed
for the divalent rare-earth ions. In a typical host structure, such as the halide oxides of
the previous chapter, the states of the ligand atoms form the valence band.

Upon a charge-transfer transition, formally an electron is transferred from a ligand
state onto, e. g., Eu^{3+}. This temporarily yields therefore a Eu^{2+} ion. Hence, the ligand-to-
metal charge-transfer energy of a trivalent rare-earth ion gives the approximate relative
position of the ground state of its R_{4f}^{2+} relative to the valence band. This is illustrated in
Figure 8.3 by an orange arrow on the example of europium.

Similarly, the relative energies of the R_{4f}^{3+} ground states with respect to the bottom of
the conduction band can be estimated from the charge-transfer transitions to the conduction band. Here, usually Ce^{3+}, Pr^{3+} and Tb^{3+} are favored. Their R_{4f}^{3+} ground states are
the ones closest to the conduction band. Accordingly, these charge-transfer transitions
are observed at lower energies than for all other R^{3+}.

As outlined in the previous chapter, the differences between charge-transfer energies within the same host structure are the same for all rare-earth ions. These considerations in mind, Pieter Dorenbos developed and introduced the basic, highly sophisticated
and theoretically founded *chemical shift model*. It employs the empiric finding that the
relative energetic setting within a given host structure always follows an at least similar
trend for all rare-earth ions. The energies given within this model are *vacuum referred
binding energies*. According to this model, the experimentally determined energies of a
single rare-earth ion's 4f ground state relative to the top of the valence band within a
certain chemical environment directly gives a reliable estimation for those of the other

rare-earth ions. For practical reasons, the *referred binding energy* of the valence electrons of an arbitrary R^{z+} with configuration $4f^n$ is estimated with respect to the anchor element europium via an on first view complicated equation:

$$E(R^{z+}) = E_0(R^{z+}) + E_{rep}(n) \cdot \xi(n) + U + \alpha(n) \cdot (r_{Eu^{z+}} - r_{R^{z+}}) \qquad (8.5)$$

Here, $E_0(R^{z+})$ is the basic energy of a valence electron. Then an interelectronic repulsion term $E_{rep}(n)$ multiplied with the nephelauxetic ratio (Equation (8.1)) follows; we recall here that the nephelauxetic ratio describes the decreasing interelectronic repulsion within a metal ion with increasing covalency toward a ligand. The *chemical shift parameter U* is defined by the energy difference between the 4f ground states of Eu^{2+} and Eu^{3+} in a considered host structure given by

$$E(Eu^{2+}) - E(Eu^{3+}) \equiv U. \qquad (8.6)$$

U reflects the different repulsion between the electrons of the ligands and the valence electrons of Eu^{3+} and Eu^{2+}, respectively. Typical values were determined to range between 6 and 8 eV for inorganic solid state compounds. Further, more recent figures and application examples are discussed in [252, 286, 287]. Finally, the parameter $\alpha(n)$ considering the lanthanoid contraction is multiplied with the difference of ionic radii, also relative to the anchor element europium.

All relevant data for such calculations are given in Appendix C. For the understanding, it is crucial to have the different interaction behavior of 5d and 4f states with host structure bands in mind. 5d states strongly interact and mix with the conduction or valence bands while the 4f states are very well shielded. Thus, in the example depicted in Figure 8.3, 4f–4f transitions within Gd^{3+} can be well recorded spectroscopically. The 4f ground state lies in the valence band, but does not recognize this situation due to the excellent shielding by the filled $5s^2p^6$ shell. In contrast, in this example a 5d←4f excitation of Yb^{2+} corresponds to an excitation from the 4f ground state into the conduction band because of the excellent mixing of $5d(Yb^{2+})$ with the conduction band states.

In Chapter 8.5, transitions between excited 5d and 4f ground states will be the center of interest. Also here, the chemical shift model provides helpful data. It allows the fairly precise prediction of the energies of 5d relative to the 4f ground states. These show an astonishingly stable energy difference relative to the well known behavior of the 5d–4f transitions in Ce^{3+}. So, regarding 5d–4f transitions Ce^{3+} acts as reference for the trivalent rare-earth ions and Eu^{2+} as reference for the divalent ones. These basic figures are given as $E_0^{fd}(R^{z+})$ in Appendix C. In addition, also the red shift of the energy upon interaction with ligands is the same. Interestingly, the absolute energies of 5d states only vary in comparably small energy windows below one electron volt. In Chapter 8.7, we will also dare a closer look on the 5d states as these split into a high-spin and a low-spin branch. In Figure 8.3, for the heavy lanthanide ions beyond a configuration $4f^7$ only the most relevant low-spin branch of the 5d states is depicted, because the transitions from and to there are spin-allowed.

Apparently, you need some data of a host structure to estimate the optical properties of rare-earth ions therein. Relevant are the host structure's band gap, the charge-transfer energy of Eu^{3+} to scale the 4f states relative to the valence band, the energy of the 5d←4f transition in this structure of Ce^{3+} as well as Eu^{2+} to scale the 5d states. Further relevant are estimates for the nephelauxetic ratio ξ and the chemical shift U. But also with less data, this model helps to get a good guess, whether certain transitions will be possible or not. Certainly it delivers interesting empirically obtained basic insights, which will also help us in the following chapters to understand the luminescence properties of rare-earth ions. Employing this knowledge, you can develop a feeling which type of host structure might be interesting for a certain application.

Figure 8.3 implies interesting consequences for transition probabilities of induced electric dipole transitions. Ions of high optical electronegativity like Eu^{3+} show a certain tendency to accept a further electron. Therefore, the charge-transfer states lie relatively low for trivalent europium. Ions with rather low ionization energies like Eu^{2+}, Ce^{3+} and Tb^{3+} feature relatively low lying 5d states. In cases where states of opposite parity like charge-transfer or 5d states are quite close to 4f levels significantly higher transition probabilities can be expected and vice versa. The chemical shift model assists you to estimate at which energies states of opposite parity can be expected. From these, you can estimate which 4f levels might mix with these according to Equation (7.19) to give induced electric dipole transitions. Thus, for ions where both, charge-transfer and 5d states, are far away from excited 4f states, very low transition probabilities are to be expected. Both effects are beneficial for respective properties. If you are interested in comparably efficient 4f–4f emitters, you will preferentially choose ions like Ce^{3+}, Eu^{3+} or Tb^{3+}; if you are interested in special emission properties like up-conversion or pumping of excited states and subsequent stimulated emission, you will choose one of the other ions like Er^{3+}, Ho^{3+}, Tm^{3+}, Gd^{3+} and Nd^{3+} providing low transition probabilities and long lifetimes of the excited states.

8.4 Emitters Weakly Interacting With Ligands

Based on the principles discussed in Chapter 7.1, the optical properties discussed herein can be well understood. In this chapter, all relevant trivalent rare-earth ions will be discussed where the luminescence is dominated by 4f–4f transitions. Since the 4f states show only weak interactions with any adjacent atom due to strong shielding by the filled $5s^2p^6$ shell, this chapter got its title. This weak interaction causes minor radial shifts of the excited 4f states, low Stokes shifts and comparably high thermal quenching temperatures. The small radial shifts also cause narrow emission bands, almost lines; thus, such ions are also named line emitters. As a result, the emitted colors are clean, and hence such emitters are well suited for lamps and monitors, which require three clean color spots of a red, a blue and a green source. Then they can emit many colors of the visible spectrum and this as brilliant as possible. Moreover, due to the weak interaction the emission colors only show marginal dependence from the host structure. For the sake of exhilaration, I will not only discuss the optical properties of the rare-earth ions but also do some excursions addressing applications for which the considered element is specifically interesting—but not necessarily exclusively, of course.

According to the selection rules discussed in Chapters 7.3 and 7.4 f–f transitions are parity forbidden, many also spin forbidden. The caused low transition probabilities yield pale colors of the compounds containing lanthanide ions—except the cases where other transitions are at work. These selection rules also cause limited luminescence efficiencies. However, in solids a low symmetric surrounding of such ions may enable partial mixing of 4f with 5d states resulting in significantly increased transition probabilities. Hence, in general, host structures providing low symmetric surroundings are often favored. Since the parity selection rule is based upon the presence of a local inversion center, by removing the latter better transition probabilities can be expected. Thus, host structures without inversion centers are attractive; this is fostered by the presence of non-centrosymmetric basic building units such as tetrahedra, and hence host structures containing anionic networks based on tetrahedral building units are beneficial such as silicate-analogous materials, which I will employ as examples on several occasions in this book.

The structure of the following chapters follows the principle of *spectroscopic twins*. Terms and levels of spectroscopic twins are the same but the levels occur in opposite order of stability. For instance, Eu^{3+} and Tb^{3+} ions are spectroscopic twins. Trivalent europium has a configuration of $[Xe]4f^6$, trivalent terbium $[Xe]4f^{14-6}$ equaling $[Xe]4f^8$. In the former case, six electrons, and in the latter, six holes occupy the 4f shell. As we will see, both are related to each other, helping to understand correlations of optical properties better. Therefore, the following sections start in the center with gadolinium and end with ytterbium directly leading to the following chapter on cerium.

Our journey starts with those rare-earth ions, which do not show any absorption or emission in the visible regime as these are suited as host structure cations, notably La^{3+}, Gd^{3+}, Y^{3+}, Lu^{3+} and Sc^{3+} given with decreasing ionic radius. They may be replaced partially during doping with emitting ions. They are suited as host structure ions because they can be chosen to fit the size and charge of doped rare-earth ions. This reduces malevolent defect formation and inadvertent significant host structure distortions. The stronger the ions differ in size, chemical hardness, and thus chemical and crystallographic behavior, the higher the risk of separation and a less homogeneous distribution of doped ions within the material. This would cause higher nonradiative transition rates and thermal as well as concentration quenching effects.

8.4.1 Gadolinium—Only Partially Innocent

Out of the aforementioned host structure cations, only Gd^{3+} features 4f–4f transitions all of which are found in the ultraviolet regime. The reason for the large gap of $32.2 \cdot 10^3$ cm^{-1} (311 nm ≈ 4.0 eV) between the $^8S_{7/2}$ ground and first $^6P_{7/2}$ excited state depicted

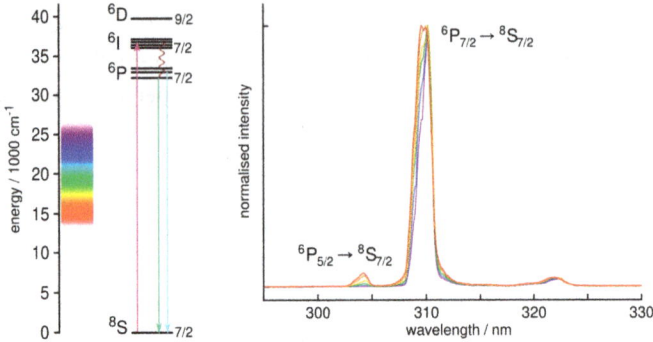

Figure 8.4: Term scheme showing the 4f states of gadolinium with the relevant transitions regarding the thermometry discussed in the text (left) and the resulting spectra recorded at temperatures between −180 and +120 °C with steps of 60 degrees (from blue to red, data: ref. [289]), the transition around 322 nm is due to vibronic fine structure (right).

in Figure 8.4 is the highly symmetric charge distribution of the exactly half-filled $4f^7$ shell. Also, a tiny change of the electron distribution like in the first excited state leads to a considerable drop in stability, and thus rise in energy. Nevertheless, as host structure cation Gd^{3+} may act as sensitizer ion absorbing radiation in the ultraviolet regime and transferring it onto doped ions or to transport energy via its excited 4f band. Although the transitions are spin and parity forbidden, the sheer mass of absorbers takes care that incoming radiation can be efficiently absorbed.

Let us look at an example. In GdF_3:Ce^{3+},Tb^{3+} even a double energy transfer is employed to convert mercury plasma radiation into visible light. According to

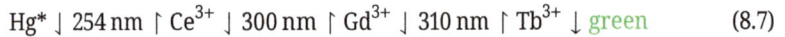

$$Hg^* \downarrow 254\,nm \upharpoonright Ce^{3+} \downarrow 300\,nm \upharpoonright Gd^{3+} \downarrow 310\,nm \upharpoonright Tb^{3+} \downarrow green \qquad (8.7)$$

the emission of a low pressure mercury plasma emission around 254 nm ($39.4 \cdot 10^3$ cm^{-1}) is absorbed by the efficient 5d←4f transition of doped Ce^{3+}. Trivalent cerium acts as an antenna here and emits radiation around 300 nm in the u. v. Trivalent gadolinium ions can absorb a sufficient number of these photons and subsequently supply doped ions like nearby doped Tb^{3+} ions with excitation energy. Eventually, the Tb^{3+} ions emit bright green luminescence [288].

Also, in the gadolinium silicate $Gd_2O(SiO_4)$:Ce^{3+}, a fast scintillation material (see Chapter 8.5.1), the Gd^{3+} ions at least contribute to the absorption of the high-energy radiation and their transfer onto the doped Ce^{3+} ions. You will find more examples and more details about such *antenna phosphors* in Chapter 8.4.2.

Thermometry Using Rare-Earth Ions

Regarding thermometry you probably think of mercury thermometers, pyrometers or thermocouple devices. In this section, we will look at a remote temperature recording approach based on the principle of the luminescence intensity ratio of two possible emissions. These emissions stem from excited 4f states of neighboring atoms, which are thermally coupled via common ligands. The material hosting these thermosensing atoms has to stay in close thermal contact with the medium under consideration—for instance, in biological systems or to monitor catalysis processes. The populations of both emitting states strive for a thermal equilibrium according to Boltzmann's law

$$R_{21} \propto \exp\left(-\frac{\Delta E_{21}}{k_B T}\right) \tag{8.8}$$

Here, the relative population depends from the energy difference ΔE_{21} of the two contributing states and the temperature. Since both states may emit, their intensity ratio resembles the relative population, and thus reveals the local temperature. This equilibrium may be achieved via a FRET mechanism (Chapter 7.6) or induced electromagnetic fields by collectively vibrating surrounding ligands. This equilibration competes with the emission or other nonradiative relaxation mechanisms. Moreover, the energy difference between both states determines the temperature range, in which such a *Boltzmann thermometer* might take action.

Here, we will take advantage of the fact that excited 4f states feature relatively long lifetimes since 4f–4f transitions are more or less spin- and parity-forbidden. Especially trivalent gadolinium is an interesting candidate as there are no intermediate states between the ground state and the excited states in the ultraviolet regime. Further, they are sufficiently apart to exclude any thermal interaction between ground and excited states. According to Figure 8.4, the very first adjacent excited levels of Gd^{3+}, ${}^6P_{7/2}$ and ${}^6P_{5/2}$, are approximately $600\ cm^{-1}$ apart, and thus may thermally equilibrate. Following the selection rules (Equations (7.22) and (7.23)), the transitions ${}^6P_{7/2} \rightleftharpoons {}^6P_{5/2}$ are magnetic-dipolar ones. In the herein chosen host structure, i. e., $Y_2[B_2(SO_4)_6]:Gd^{3+}$, the gadolinium ions are doped on the yttrium sites of a silicate-analogous borosulfate, where the cations are coordinated by sulfate tetrahedra, which themselves are covalently bound to borate tetrahedra. The tetrahedra of the anion may provide vibrational modes in the desired region of $600\ cm^{-1}$. As derived in Chapter 6.5, these anions exhibit only a weak coordination strength. Nevertheless, this is apparently sufficient to ensure fast thermal equilibration between the gadolinium ions. Their local surrounding shows no symmetry and this prevents further symmetry-based selection rules. The excitation occurs from the ground state ${}^8S_{7/2}$ into the excited levels ${}^6I_J$ from where relaxation occurs into the ${}^6P_J$ multiplet, followed by thermal equilibration. Figure 8.4 shows the evolution of intensities for the phosphor $Y_2[B_2(SO_4)_6]:Gd^{3+}$. Between -120 and $+120\ °C$, the emission ratio behaves as required for a Boltzmann thermometer [289].

Further emissive states suited for such investigations on thermalization equilibria and maybe applications are ${}^2H_{11/2}$ and ${}^4S_{3/2}$ of Er^{3+} as demonstrated in $YVO_4:Er^{3+}$. Here, the transitions between these two levels are of induced electric-dipolar nature, and thus much faster. [290] A further example are the states $^4F_{5/2}$ and $^4F_{3/2}$ of Nd^{3+} in $LaPO_4:Nd^{3+}$, which are of comparable speed of the states in Gd^{3+} [291].

8.4.2 Terbium—Bright Green or Also Blue?

The ground state configuration of trivalent terbium ions is $[Xe]4f^8$, just one electron above the very stable half-filled 4f shell. Therefore, its ionization energy is rather low as the terbium ions strive for a half-filled 4f shell either in Tb^{4+} or by excitation to a $4f^7 5d^1$ state. This is reflected nicely in Figure 8.3 where the 4f ground state $^7F_6$ lies at considerably higher energy than that of the neighboring Gd^{3+}. Since the 5d states of both ions are of similar energy, those of terbium are now located comparably close to the ground state. In the compound featured in this chapter, $Y_2[B_2(SO_4)_6]:Tb^{3+}$, these are recorded as broad bands around 212 and 254 nm (approximately 47 and $39 \cdot 10^3$ cm^{-1}) as shown in Figure 8.5 [292]. These transitions enable parity-allowed excitation via $4f^7 5d^1 \leftarrow 4f^8$ transitions. Because of the efficient absorption around 254 nm, also Tb^{3+} can be excited in applications like compact fluorescent lamps via the main emission line of a low pressure mercury plasma. Host structures doped with Tb^{3+} are colorless like terbium salts and show a bright green luminescence upon excitation in the ultraviolet regime. The luminescence is based on four dominant emission lines from the first excited state $^5D_4$ to the ground state multiplet ^7F_J ($J = 3 \dots 6$) represented by the green arrows in Figure 8.5. Apparently, Tb^{3+} does not actually emit a really clean green stemming from a single emission band, but bluish and yellowish side-bands yield something whitish, thus causing in total a bright green emission. This holds for doping concentrations above a few percent, which are necessary for a reasonable light output in the mentioned applications like compact fluorescent lamps for lighting.

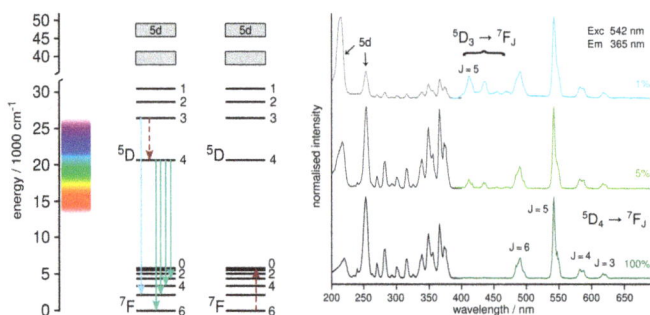

Figure 8.5: Term scheme showing relevant 4f and 5d states of terbium with optical transitions and nonradiative transitions (broken arrows) discussed in the text (left) and the resulting excitation (greyish, monitored at an emission of 542 nm) and emission spectra (colored, excited with 365 nm); given are the respective doping concentrations on the yttrium site in $Y_2[B_2(SO_4)_6]:Tb^{3+}$ in percent, the transitions are assigned (data: [292]).

Cross Relaxation

After excitation by ultraviolet radiation the terbium ions relax fast to the excited $^5D_3$ level. From there, direct emission is possible to the ground state multiplet. This competes with a nonradiative relaxation to the excited $^5D_4$ state, though; this relaxation is fostered by a so-called *cross relaxation* with a neighboring Tb^{3+} in its ground state. Such energy transfers according to

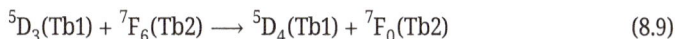

$$^5D_3(Tb1) + {}^7F_6(Tb2) \longrightarrow {}^5D_4(Tb1) + {}^7F_0(Tb2) \tag{8.9}$$

follow a FRET mechanism via a resonant interaction, and thus occur below an average interatomic distance of typically 4 Å (Chapter 7.6). At low concentrations, the average distances between adjacent Tb^{3+} ions are too large for an efficient cross-relaxation. Then also blue emissions from the $^5D_3$ level may be recorded as depicted in Figure 8.5 where the cross-relaxation transitions are marked by the red broken arrows.

Such relaxation is a type of concentration quenching of certain optical transitions and can be observed, if the energy gap between the excited states matches that between the ground state and a suited twin state. The critical concentration, above which only the green emissions of Tb^{3+} can be monitored depends from the average Tb-Tb distance and from the strength of interaction with the ligands—which is very weak in this compound here. A further example of cross-relaxation will be mentioned in the next chapter on Eu^{3+} ions.

Sensitized Luminescence—Antenna Phosphors

There are several ions featuring desired emission wavelengths for certain applications, but they unfortunately lack an appropriate efficiency due to the circumstance that their excitation and emission transitions are more or less forbidden. To overcome this problem, several decades ago sensitized luminescence was discovered and since then thoroughly investigated.

As you know from previous chapters, trivalent rare-earth ions normally show 4f–4f transitions, which are forbidden regarding the parity and may be even forbidden regarding the spin selection rule. In the case of Eu^{3+}, we will see that here at least an efficient excitation via a charge-transfer transition might help, and in the case of Tb^{3+} allowed 5d←4f excitations are feasible. Both efficient excitations only help for a certain application if the light source provides the necessary excitation wavelength. In many other cases, neither charge-transfer nor 5d←4f excitations are available; hence, here arises the need for alternative excitations via efficient absorbers, namely *sensitizers*, situated nearby the emitting ion and capable of transferring the absorbed energy onto the activator. Application of this approach is found in lanthanide complexes. There, ligands act as the antenna, which contain aromatic groups with a broad absorption band in the u. v. region. Upon excitation, this antenna transfers the energy onto the emitter as discussed earlier. Due to this antenna effect, the phosphors discussed here are also called

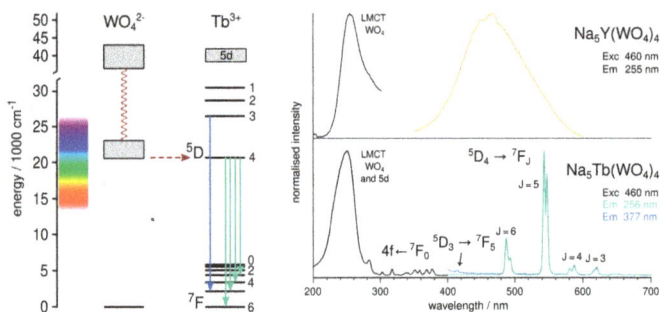

Figure 8.6: Term schemes showing the charge-transfer states of tungstate ions and selected relevant 4f states of Tb^{3+} including relevant radiationless relaxation (waved line), energy transfer onto terbium (red, broken) as well as observed transitions of Tb^{3+} (left); besides the resulting excitation (black) and emission spectra of WO_4^{2-} (orange) and $Na_5Tb(WO_4)_4$ excited at 256 nm (green) or 377 nm (blue) with assignments of transitions (right, data: [293]).

antenna phosphors. The great advantage of complexes is that the absorption of the antenna ligand can be tuned very finely via selective substitutions within the organic part. Unfortunately, these organic parts make them less stable against higher temperatures and also lead to typically lower thermal quenching temperatures. For many examples and a thorough discussion on such antenna complexes I recommend [294]. Here, we will continue with a purely inorganic solid antenna phosphor.

The principles are basically the same as depicted in Figure 8.6. Here, in $Na_5Tb(WO_4)_4$ the omnipresent tungstate anions can be efficiently excited via a charge-transfer transition. From there, three relaxation mechanisms compete. Either, it shows the shown broad-band emission of tungstate or it relaxes somehow radiationless back to its ground state. As a third possibility, Tb^{3+} offers suitedly lying excited states for a resonant energy transfer (FRET, see also Chapter 7.6). As both ions, WO_4^{2-} and Tb^{3+}, are usually in direct neighborhood, this transfer can become very efficient yielding the bright green luminescence known for trivalent terbium ions. This is shown nicely in Figure 8.6 where in the (green) emission spectrum upon excitation via the charge-transfer transition around 256 nm only $^5D_4 \rightarrow {}^7F_J$ transitions can be monitored. In contrast, if the same compound is directly excited into the level $^5D_3$ with u. v. light around 377 nm, also weak but significant emissions in the blue occur. The absolute intensity of the green emission curve is by far stronger than that of the blue curve; this can be deduced from the ratio of the intensities at 256 and 377 nm in the excitation spectrum. In $Na_5Tb(WO_4)_4$, certainly the LMCT states of tungstate overlap with the 5d states of terbium to some extent. Further investigations on the yttrium compound show that the lifetime of the LMCT state decreases upon slight doping with Tb^{3+}. Moreover, investigations on such phosphors with the same host structure, but larger cations like lanthanum can establish the expected intensity decay with increasing average distance between the donor WO_4^{2-} and the doped activator Tb^{3+}. One of the very first intentionally and successfully doped tungstates was

$Y_2(WO_6)$:Eu^{3+} by *Blasse*[66] in 1966 [295]. In the following paragraph, selected further examples are discussed.

Tuning the absorption wavelength is somewhat tricky in solids, so the diversity is surely more slender; in contrast, thermal quenching may occur at higher temperatures in solids. Of course, there are prominent examples for phosphors like the pair Sb^{3+}/Mn^{2+} where the efficient 5p←5s transition of antimony acts as antenna for the parity and spin-forbidden 3d→3d transitions within divalent manganese. The slightly different chemical surroundings in the doped *apatites*, also called *halophosphates*, $Ca_5(PO_4)_3$(OH):Sb,Mn and $Ca_5(PO_4)_3$(OH,Cl):Sb,Mn [296] allow to employ the different nephelauxetic effect on the absorption and emission of Sb^{3+}. In the latter, the weaker nephelauxetic effect yields a blue shifted emission. Also, in the latter the ligand field splitting is smaller and yields a blue shifted emission of the 3d^5 ion Mn^{2+}.[67]

Divalent manganese ions can also be sensitized via codoping with divalent europium ions. The thorough discussion of the optical properties of Eu^{2+} will be the topic of Chapter 8.5. In short, Eu^{2+} is normally excited via allowed 5d←4f transitions and shows also efficient emissions of the same type. It is also capable of transferring the excitation at least partially onto Mn^{2+}, which has been proven on many examples such as the polyphosphate α-Sr[PO$_3$]$_2$:Eu,Mn. There the combination of both, the europium and manganese based emissions, yield white fluorescence comprising a blue emission due to the 5d→4f transition of Eu^{2+} and an orange-red 3d→3d one of Mn^{2+} [297]. Both ions herein are typically doped with concentrations of 1% for the—compared with Sr^{2+}—slightly larger Eu^{2+} and 4% for clearly smaller Mn^{2+}. As mentioned before, the interatomic distance is crucial for an efficient energy transfer. Statistically, it is not very likely to find manganese and europium ions in a close neighborhood. Luckily, it was proven on codoped potassium chloride, KCl:Eu,Mn, by electron spin resonance that in highly symmetric arrangements of ions apparently a pairing of the introduced doped ions occurs. This minimizes the overall loss of lattice energy [298]. The same can be assumed for α-Sr[PO$_3$]$_2$, which is based on a diamond-like, and hence quite symmetric, arrangement of strontium. So, this approach is more promising than to offer distinct sites fitting the sizes of both ions. In summary, in an ideal case you need allowed transitions for the sensitizer who transfers the energy completely onto the activator with both being adjacent to each other. Allowed transitions are charge-transfer transitions like in the previously given examples $Na_5Tb(WO_4)_4$, $Y_2(WO_6)$:Eu^{3+} or further ones like $Na_2Eu(WO_4)(PO_4)$, $Na_2Tb(WO_4)(PO_4)$ and $EuKNaTaO_5$ [299, 300].

There are also examples of sensitized luminescence by Gd^{3+} described in Chapter 8.4.1. Another type of even more sophisticated antenna phosphors will be treated

66 *George Blasse*, Dutch solid state chemist (*1934 †2020).

67 This maybe an unexpected trend but is usual for 3d^5 systems (consult the respective Tanabe–Sugano diagram); decreasing interactions with ligands apparently increase the interelectronic repulsion in this electronically highly symmetric system.

in Chapter 8.4.7 where the absorbed energy of the sensitizer is added up by the activator during an up-conversion process.

8.4.3 Europium—Sometimes Hypersensitive

Trivalent europium ions are spectroscopically the twins of Tb^{3+}. While the latter come with a surplus electron with regard to the particularly stable half-filled 4f shell, the former lack one with its ground state configuration $[Xe]4f^6$. The energetic consequences are opposite since Eu^{3+} strives for a further electron, maybe supplied by an efficient charge-transfer transition. And indeed, trivalent europium features the lowest charge-transfer energies, and thus the lowest optical electronegativity of all trivalent rare-earth ions (Figure 8.7).

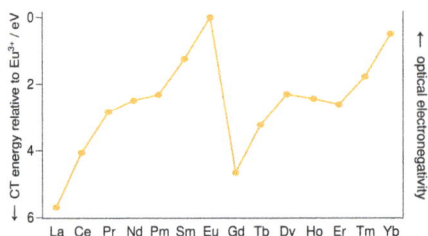

Figure 8.7: Charge-transfer energy shifts relative to Eu^{3+} in reverse order to illustrate the trend of optical electronegativities; data: [272].

Due to this symmetry, with regard to the configuration $4f^7$ the term schemes of Tb^{3+} and Eu^{3+} are very similar, only that the J levels occur in opposite order as displayed in Figure 8.8. The ^7F_J ground state levels of Eu^{3+} doped into LaF_3 lie within $5,000\ cm^{-1}$.

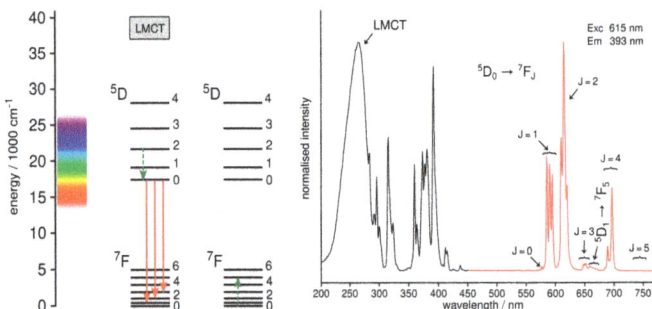

Figure 8.8: Term scheme showing relevant 4f states of europium with the transitions regarding the cross-relaxation (green), the characteristic emissions (red) and the resulting excitation (greyish, monitored at 615 nm) and emission spectra (red, excited with 393 nm) of $Eu_2[B_2(SO_4)_6]$ with assigned transitions; data: [292].

The first excited state $^5D_0$ is located around $17.3 \cdot 10^3$ cm^{-1}. While the ground state levels of Tb^{3+} span over 5,800 cm^{-1}, the first excited state $^5D_4$ is already located around $20.6 \cdot 10^3$ cm^{-1}. This reflects nicely the slightly larger ion size with decreased interelectronic repulsion combined with a lower nuclear charge of Eu^{3+}. Thus, for Eu^{3+} a smaller splitting of all terms and levels results. Like trivalent terbium, also Eu^{3+} is subject to cross-relaxation. After excitation via a charge-transfer transition or direct excitation of higher 4f states, e. g., the cross-relaxations

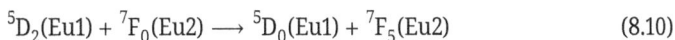

$$^5D_2(Eu1) + {}^7F_0(Eu2) \longrightarrow {}^5D_0(Eu1) + {}^7F_5(Eu2) \tag{8.10}$$

or

$$^5D_1(Eu1) + {}^7F_0(Eu2) \longrightarrow {}^5D_0(Eu1) + {}^7F_3(Eu2) \tag{8.11}$$

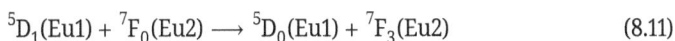

are feasible. Consequently, all relevant emissions start from the first excited state $^5D_0$ down to the ^7F_J levels of the ground state, typically causing an orange-red color impression of the emission. Among these, the basically forbidden $0 \rightarrow 0$ transition is sometimes hard to observe. Nevertheless, this transition is very useful as it is never split ($J = 0$). Therefore, the number of observed bands corresponds to the number of different chemical environments for trivalent europium in the sample. The magnetic dipole transition $^5D_0 \rightarrow {}^7F_1$ can be taken as a reference against the so-called hypersensitive $^5D_0 \rightarrow {}^7F_2$ transition; see the respective special chapter on this topic below. The transitions visible in the absorption spectrum depend on the temperature at which it is recorded. At room temperature, already higher J levels, normally $^7F_1$ and $^7F_2$, become thermally populated. Then also absorptions from these like $^5D_0 \leftarrow {}^7F_1$ or the hypersensitive one $^5D_1 \leftarrow {}^7F_1$ can be seen. These *hot bands* are invisible at low temperatures—these are not shown in Figure 8.8. Finally, a short *nota bene* from my experience as a teacher: the transitions ^5D_J-^7F_J are not 5d-4f transitions.

As mentioned above, Eu^{3+} features the lowest charge-transfer transition energies of all rare-earth ions due to the proximity of its configuration to the stable half-filled 4f shell. As outlined in Chapter 8.2, the energies lie around the famous 254 nm (approximately $39.4 \cdot 10^3$ cm^{-1}) emission of mercury plasma, which makes Eu^{3+} an important dopant for phosphors of respective lamps as discussed later in this chapter. The simple correlation of charge-transfer energies derived from Figure 8.3 allows also to estimate whether trivalent europium can be reliably doped into a certain host structure despite its limited chemical stability or if it presumably becomes reduced to the divalent state. This occurs if the estimated charge-transfer energy lies below 2.5 eV—which consequently depends from the electronegativity of the ligands [272]. Another aspect to be taken into account is that due to the low lying charge-transfer states also the thermal quenching temperatures of phosphors based on Eu^{3+} may be relatively low. The charge-transfer band of our example Eu$_2$[B$_2$(SO$_4$)$_6$] is significantly blue-shifted upon cooling. It is also blue shifted if Eu^{3+} ions are doped onto smaller Y^{3+} sites in Y$_2$[B$_2$(SO$_4$)$_6$]:Eu^{3+}. This corresponds with the experience that with decreasing average distance to the ligands the charge-transfer energy increases (Chapter 8.2). Since trivalent europium is one of the spectroscopically best investigated ions, a lot of literature is available. For a concise and helpful overview, I recommend [239].

Hypersensitive Transitions

Essentially, pure 4f–4f transitions are forbidden due to the parity selection rule. Accordingly, no electric dipole and only magnetic dipole transitions would be expected for the rare-earth ions discussed in this chapter; but also lined out earlier, low symmetric coordination lacking inversion symmetry gives rise for the so-called *induced electric dipole transitions* because under these symmetry conditions mixing with other states of different parity is feasible [301, 302]. Among these, *hypersensitive transitions* react especially sensitive—as the name implies—to the lack of local inversion symmetry, which fosters the intensity of such transitions by up to two orders of magnitude. Unfortunately, the reasons that make these transitions so special are not as yet fully understood and are still subject of scientific discussion. From the latter, we may conclude that there are more factors influencing the intensity of such transitions maybe adding up or even promoting mutually. Below, I will mention relevant aspects of the ongoing discussion. Anyhow, there is consensus about the empirically determined selection rules for transitions being possibly hypersensitive:

$$\Delta S = 0 \qquad \Delta L \leq 2 \qquad \Delta J \leq 2 \tag{8.12}$$

Accidentally, these selection rules resemble those of electric quadrupole transitions, but since such transitions are extremely weak ones, this presumably is indeed accidental. Nevertheless, you might sometimes come across the labeling *pseudo quadrupole transition*.

Most people hear about the concept of hypersensitive transitions for the first time when they face luminescence of Eu^{3+} where in every emission spectrum the prominent hypersensitive transition $^5D_0 \rightarrow {}^7F_2$ occurs. Certainly, also the respective excitation transition $^5D_0 \leftarrow {}^7F_2$ is hypersensitive and should be discussed as it is frequently employed as in situ probe for local inversion symmetry. This is done by comparing the intensities of this transition and the $^5D_0 \rightarrow {}^7F_1$ transition, which is a purely magnetic dipole transition and can thus act as a good reference. Moreover, several of such transitions have been identified for the trivalent rare-earth ions as listed in Table 8.2. Indeed, the aforementioned $^5D_0 \rightarrow {}^7F_2$ transitions of Eu^{3+} are the most investigated and for certain reasons most employed hypersensitive transitions. Within the Judd–Ofelt theory (Chapter 7.8) for this specific transition only U^2 is nonzero according to the tabulated figures of U^λ, so Ω_2 determines the intensity of the transition [258]. A wider view shows that this holds not only here, but generally for hypersensitive transitions, since the ligand impact declines from Ω_2 via Ω_4 to Ω_6 [303]. While the squared matrix elements U^λ ($\lambda = 2, 4, 6$) describe the

Table 8.2: Hypersensitive transitions of rare-earth ions with their respective approximate transition energies in wavenumbers.

R^{3+}	transitions	energy $/10^3$ cm^{-1}
Pr	$^3P_2 \leftarrow {}^3H_4$	22.5
	$^1D_2 \leftarrow {}^3H_4$	17.0
	$^3F_2 \leftarrow {}^3H_4$	5.2
Nd	$(^4G_{7/2}, {}^2K_{13/2}, {}^2G_{9/2}) \leftarrow {}^4I_{9/2}$	19.2
	$(^4G_{5/2}, {}^2G_{7/2}) \leftarrow {}^4I_{9/2}$	17.3
	$^4F_{5/2} \leftarrow {}^4I_{9/2}$	12.4
Sm	$(^6P_{7/2}, {}^4D_{1/2}, {}^4F_{9/2}) \leftarrow {}^6H_{5/2}$	26.6
	$(^6F_{1/2}, {}^6F_{3/2}) \leftarrow {}^6H_{5/2}$	6.4
Eu	$^5D_2 \leftarrow {}^7F_0$	21.5
	$^5D_1 \leftarrow {}^7F_1$	18.7
	$^5D_0 \leftarrow {}^7F_2$	16.3
Dy	$^6F_{11/2} \leftarrow {}^6H_{15/2}$	7.7
	$(^4G_{11/2}, {}^4I_{15/2}) \leftarrow {}^6H_{15/2}$	23.4
Ho	$^3H_6 \leftarrow {}^5I_8$	27.8
	$^5G_6 \leftarrow {}^5I_8$	22.1
Er	$^4G_{11/2} \leftarrow {}^4I_{15/2}$	26.4
	$^2H_{11/2} \leftarrow {}^4I_{15/2}$	19.2
Tm	$^1G_4 \leftarrow {}^3H_6$	21.3
	$^3H_4 \leftarrow {}^3H_6$	12.7
	$^3F_4 \leftarrow {}^3H_6$	5.9

interelectronic interactions within the levels of the rare-earth ion under consideration, the intensity parameters Ω_λ represent physically the square of the charge displacement due to the induced electric dipole transition and mathematically the radial part of the wave function (Chapter 7.8). Accordingly, the latter are influenced by the surrounding in terms of symmetry and polarization.

On the other hand, you cannot neglect the respective squared reduced matrix element U^λ as this apparently determines the sensitivity toward any symmetry change, which allows to override the parity selection rule and is a measure for miscibility with states of opposite parity. Accordingly, it is not further surprising that high values here are beneficial for hypersensitive transitions, already noted by Judd [253].

Jørgensen and Judd found a correlation between hypersensitive behavior and an inhomogeneous dielectric environment of the respective rare-earth ion caused by an asymmetric arrangement of dipoles [301]. They concluded that this surrounding may lead to strongly enhanced pseudo quadrupole transitions. Judd later added certain symmetry restrictions. Also, a nexus to the basicity of the ligands was identified and discussed [303]. This seems quite reasonable as this approach, known also as *covalency model*, identifies charge-transfer states as a further contributing factor. Their intensity grows with a decreasing energy difference between the forbidden 4f–4f and the allowed charge-transfer transitions.

The *dynamic coupling model* or *ligand polarization model* was developed by Mason[68] and extends the *static coupling model*, i. e., the Judd–Ofelt theory. There the ligands are assumed to be passive (or *static*) with regard to dynamics of charge distribution within the rare-earth ion upon excitation. Here, in this theory these dynamics induce electric dipoles on the individual ligands—they are polarized. In turn, the dynamics foster the probabilities of transitions forbidden due to the parity selection rule [304, 305]. According to this theory, such polarization effects contribute only to Ω_2 if an inversion center is absent and are stronger the higher the polarizability of the ligands is.

Unfortunately, to all of these rules and theories exceptions or arguments were found which not necessarily falsify these hypotheses but make clear that at least many of these aspects may contribute to the hypersensitivity of transitions. But none of them is solely capable of explaining the phenomenon. Eventually, not even the same Ω_2 value of any two Eu^{3+} emission spectra yield the same intensities of the hypersensitive transitions. Therefore, it was postulated that also the refractive index n may play a role since a growing n seems to correlate with higher intensities [306].

Compact Fluorescent Lamps

During the mid-19th century, and during the same time when the incandescent light bulb was developed, many people experimented with electricity driven lighting approaches. After evacuating glass tubes and applying high electric tension on both ends, the tube lit up—and the *Geissler*[69] *tube* was born. The Geissler tube was a gas-discharge lamp filled with neon, air or mercury. Today we still use this concept in a slightly modified form in *compact fluorescent lamps*, also called *energy saving lamps*. This name implies that it consumes less electricity compared with incandescent light bulbs and this is indeed the case. Out of one watt, a light bulb generates approximately 2 % visible light, corresponding to a luminous efficacy of 15 lumen per watt. The same energy employed in a fluorescent tube yields approximately 100 lumen per watt or an efficiency of up to roughly 20 %. White light emitting diodes manage even more, up to 40 %. These will be discussed in Chapter 8.6. Nowadays fluorescent tubes are still driven by a low-pressure mercury plasma. Each tube contains a few miligrams of the toxic metal, which upon switching on forms a plasma. Mercury was chosen due to its high energy conversion efficiency of 75 % into u. v. light with three main emissions around 185, 254 and 365 nm [307]. To get white light out of the tube, a manifold of phosphors was developed with excellent absorption around the main line of 254 nm. Such lamps are normally designed on a three-color-band concept, where phosphors emit mainly three colors red, green and blue adding up to white—according to Figure 8.25 in Chapter 8.6. A typical emission spectrum is shown in Figure 8.9 besides a schematic picture of such an energy saving lamp. A compact

---

68 *Stephen Finney Mason*, British chemist and scientific historian (*1923 †2007).
69 *Heinrich Geißler*, German glassblower and physicist (*1814 †1879).

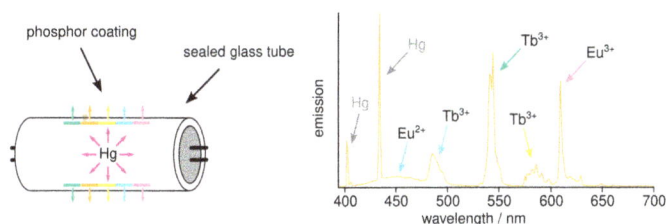

Figure 8.9: Principle of a compact fluorescent tube and a typical emission spectrum, selected emissions are assigned to rare-earth ions and mercury plasma.

fluorescent tube consists of a glass tube containing the mercury. The tube is coated on the inner surface with a phosphor mixture. These phosphors convert the u. v. light as completely as possible into visible light. A typical phosphor coating consists of red emitting Y_2O_3:Eu^{3+} (YOX), green emitting $La(PO_4)$:Ce^{3+},Tb^{3+} (LAP) and $Ce[MgAl_{11}O_{19}]$:Tb^{3+} (CAT) as well as blue emitting $Ba[MgAl_{10}O_{17}]$:Eu^{2+} (BAM). Frequently, besides these rare-earth based phosphors also halophosphates like $Ca_5(PO_4)_3(F,Cl)$:Sb^{3+},Mn^{2+} are employed which contribute broader bands in the blue and orange regime [93]. Coming back to the emission spectrum depicted in Figure 8.9, the relevant emissions of the contributing rare-earth ions can be identified. Apparently, there is a preference for Tb^{3+}, contributing in particular green light, Eu^{3+}, contributing red light, and Eu^{2+}, contributing blue light to the in total white emission. The trivalent ions thereby show the emissions already discussed. The very versatile divalent europium is employed here with a famous phosphor described in the course regarding the 5d–4f emitters in Chapter 8.5.2.

To convert the plasma's u. v. light as efficiently as possible, terbium is sensitized by trivalent cerium ions, which feature broad-banded 5d←4f excitations. This absorbed energy can then be transferred onto the doped trivalent terbium ions according to a FRET mechanism. For a thorough understanding on the principles of such antenna phosphors, a glance into Chapter 8.4.2 is recommended. Trivalent europium is the emitter of choice for the red emission because the efficient ligand-to-metal charge-transfer transition in oxidic environments manifests itself in a broad excitation band around the main mercury plasma emission at 254 nm. Finally, Eu^{2+} is efficiently excited via allowed and broadbanded 5d←4f excitations. Thus, at least the excitation path is solved efficiently for the employed emitters fostering a quite reasonable overall energy efficiency.

To get rid of the toxic mercury, alternate plasma generating elements such as xenon are feasible. Unfortunately, the excitation occurs at clearly higher energies in the vacuum u. v. regime, which then requires a higher energy efficiency of the employed phosphors—and here quantum cutting comes into play, which is treated later in Chapter 8.4.9.

Although the fluorescent lamps emit white light, this white light does not fully resemble that of the black body radiator. Therefore, it is recognized as deficient and colored objects might look different under illumination compared with incandescent light bulbs or high quality white light LEDs. An application where distinctly colored emissions are a compulsory requirement are *plasma display panels.*

Plasma Display Panels

For quite a time, *plasma display panels* (PDPs) were commercially used for flat, large televisions until the first decade of our century. They were quickly replaced by brighter and overall cheaper concepts such as liquid-crystal displays and OLED screens needing less energy for operation. Nevertheless, a few remarks should be made here on PDPs, especially because this technology is closely related to fluorescent tubes and may deliver interesting ideas for possible future applications. In PDPs, normally a noble gas plasma like xenon was employed, which upon ignition emits in the VUV regime with two prominent emissions at 147 and 172 nm. Each pixel comprises a plasma chamber coated with a phosphor emitting the respective color. Certainly, also these phosphors had to absorb efficiently in the VUV and to emit as fast as possible to avoid ghostly images. Therefore, the host structures need large band gaps, and the phosphors require very efficient and allowed emissions [307, 308]. Accordingly, most of the inorganic phosphors are colorless, thus a coating looks whitish. This is a disadvantage for a display, which requires a dark background of the nonilluminated pixels. For this purpose, colored phosphors were developed carrying a dark body color such as the red emitting red powder of $Zr(SiO_4):Pr^{3+},Pr^{4+},P^{5+}$ discussed in Chapter 8.4.9 or the blue emitting blue phosphor $BAM:Eu^{2+},Co^{2+}$, i. e., $Ba[MgAl_{10}O_{17}]:Eu^{2+},Co^{2+}$, where blue cobalt ions achieve the same effect [309]. Moreover, the strong VUV irradiation led to the decomposition of phosphors. Consequently, the image burn-in was a big problem. Here, long-term display of a certain image yielded persisting shadow images.

8.4.4 Samarium and Burning Holes

The electronic configuration of trivalent samarium ions is $[Xe]4f^5$. It features the same states like the smaller Dy^{3+} but with opposite order and a smaller splitting. According to its configuration, Sm^{3+} shows only a weak tendency to get reduced to the divalent state—and thus features high lying charge-transfer states. Moreover, it also shows only a weak tendency to get oxidized—thus the 5d states are situated at high energies. This can also be concluded from the chemical shift model; see Figure 8.3. Both tendencies lead to lower probabilities of mixing with states of opposite parity according to the Judd–Ofelt theory (Chapter 7.8) and subsequently to low transition probabilities. Moreover, the emission spectra like that depicted in Figure 8.10 show a manifold of lines in the visible regime and normally do not comprise clean colors preventing it from many applications—unlike the prominent red line in Eu^{3+} or the green emission of Tb^{3+}.

Due to the absorptions in the violet-blue regime, solutions containing Sm^{3+} look yellowish, certainly pale as the transition probabilities are low. The special situation of

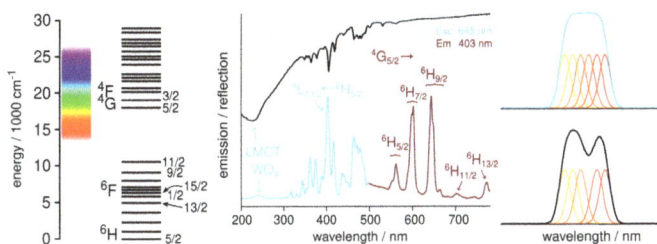

Figure 8.10: Term scheme of Sm^{3+}, excitation (blue, monitored at 645 nm), emission (red, excited with 403 nm) and reflection spectra (black) of $Na_5Sm(WO_4)_4$ (left); right: scheme of an inhomogeneously broadened emission band (blue) with individual bands (colored) without (top) and with a burnt hole (bottom); data: [293].

Sm^{3+}, somehow undecided between being reduced and being oxidized easily, makes it attractive for permanent data storage based on *spectral hole burning* as discussed in the following paragraph.

Spectral Hole Burning

Spectral hole burning may sound somewhat weird, but it describes the approach to save information employing the excitation of absorbing or luminescent ions, such as trivalent samarium. In any material, any excitation band of transition metal or rare-earth ions shows a certain width caused by thousands of slightly different local surroundings or interactions with adjacent doped ions. To minimize vibrational broadening, the samples are normally cooled. The observed band with its *inhomogeneous* linewidth comprises the slightly different extremely sharp individual absorptions of all ions of *homogeneous* linewidth as depicted schematically in Figure 8.10. The ratio of inhomogeneous to homogeneous widths within an absorption band typically spans several orders of magnitude, values of six are mentioned in literature. If an individual ion is excited by a laser beam, it leaves behind a tiny, local but recordable dip in the absorption band at the wavelength and with a depth proportional to the intensity of the laser beam, which may be regarded as a *bit*. Thus, at a specific spot more than one bit may be saved by employing a tunable laser beam and different beam intensities. The local width of the bit is determined by the size of the beam. In total, optical data densities up to terabits per square centimeter are imaginable [310].

Several materials and ions were investigated, e. g., $Y_2(SiO_4)O:Er^{3+}$, where the ratios of inhomogeneous to homogeneous widths amounts to more than 10,000. The problem here is the lifetime of the stored bit as after excitation trivalent erbium relaxes quite fast to the ground state. If you now imagine an ion, which might be ionized temporarily, so that the excited electron is trapped in more or less shallow traps in the conduction band, this information will be stored permanently as long as you would not release the trapped electron by suited radiation. Such an ion should be relatively stable in two oxidation states—not switch too easily, though. Further, it should certainly tend to very sharp individual absorptions like a 4f–4f absorber. The samarium ions Sm^{2+} and Sm^{3+} fulfill these preconditions—like Pr^{3+}/Pr^{4+} or Eu^{2+}/Eu^{3+}.

Accordingly, a host structure like BaFCl is doped with divalent samarium ions. BaFCl is suited because it tends to the formation of many defects on the halide sites. It adopts the structure of *matlockite*, PbFCl, and comprises layers of fluorine and chlorine. This will be also subject of the next chapter on long lasting luminescence. Initially, BaFCl is doped with the stable Sm^{3+} ones, which are subsequently reduced by irradiation with ionizing radiation like X-rays. The generated excited electrons in the conduction band are transferred onto the Sm^{3+} ions, which are then locally reduced to the divalent state. Moreover, the barium ions in BaFCl may be replaced by strontium ions to generate a broader spectrum of different local surroundings. This gives rise for significant inhomogeneous line broadening of the typically 0.2 % doped Sm^{2+} ions. It can be further enhanced by doping nanosized crystals of systematically tuned compositions $Ba_{1-x}Sr_xFCl:Sm^{2+}$, which are physically mixed afterwards. The authors of a recent study employed the basically forbidden but nevertheless present emission $^5D_0 \rightarrow {}^7F_0$ of Sm^{2+}. Sm^{2+} is isoelectronic with Eu^{3+}. Compared with Eu^{3+}, where this $^5D_0 \rightarrow {}^7F_0$ transition occurs below 600 nm, the emission is recorded at significantly lower energies for Sm^{2+}, peaking around 690 nm; this is due to the lower nuclear charge leading to a relative expansion of the isoelectronic ion and accordingly smaller splitting of all states. Due to better mixing with states of opposite parity, this transition is quite intense in divalent samarium. And it is not split—we recall the short discussion in the chapter on Eu^{3+}. So, every slightly different local surrounding generates its own sharp emission. As depicted in Figure 8.10, all these emissions sum up to a broader hat-shaped band out of which upon laser radiation dips of different depth are taken out—or holes are burnt [310].

8.4.5 Dysprosium—As Long-Lasting as Possible

The electronic configuration of trivalent dysprosium ions is $[Xe]4f^9$ and accordingly corresponds via $4f^{14-n}$ to that of its spectroscopic twin Sm^{3+} ($4f^5$) with just reversed order of levels but the same order of terms. The series of weak absorptions in the blue and u. v. regime (Figure 8.11) yields a pale yellowish color of the ions. Compared with Sm^{3+}, the same transitions occur in dysprosium at higher energies (see Figure 8.10) reflecting the higher nuclear charge here. The luminescence of three transitions in the visible regime by Dy^{3+} adds up to a whitish or yellowish color impression.

Long Persistent Luminescence
Introduction and Safety Application

For safety applications, like the power-independent marking of emergency routes in buildings, you need phosphors, which are ideally excited during daytime and show a long visible emission. Then during night-time or when it is dark, the signs can be read.

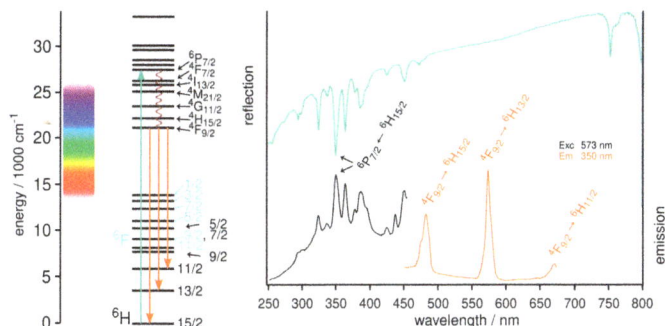

Figure 8.11: Term scheme of trivalent dysprosium with an excitation (green), relaxation (red) and the emissions in the emission spectrum (orange); right: reflection (green), excitation (black, monitored at 573 nm) and emission spectra (orange, excited with 350 nm) of $DyH[PO_3]_4$; data: [230].

Such phosphors are called *long persistent phosphors* and they are applicable in several fields. Besides the application as safety signs, they are also employed for the detection and quantification of high-energy radiation such as X-rays or u. v. light.

To illustrate the principle, I will focus on a long persistent phosphor employed for marking of emergency routes, and here eventually trivalent dysprosium ions will come into play. For this application, a bright green emission is desired since the human eye is most sensitive in the green regime. Moreover, it is believed that green yields an additional soothing effect. A famous green phosphor showing *afterglow*—another phrase for persistent luminescence—is $Sr[Al_2O_4]:Eu^{2+}$ where divalent europium ions are doped on Sr^{2+} sites embedded in a silicon dioxide type network structure comprising AlO_4 tetrahedra [311]. Further details of the luminescence will be discussed in Chapter 8.5.2; the structure was already described in Chapter 6.5.3.

To explain the afterglow effect, several mechanisms have been suggested, some of which appear to be improbable as they propose the participation of Eu^+ species, for instance. The in-conclusion most reasonable explanations so far consider the nature of *electron traps* and their role in the mechanism of long persistent luminescence [312–316]. Accordingly, in pure $Sr[Al_2O_4]$ and $Sr[Al_2O_4]:Eu^{2+}$ presumably exclusively oxygen vacancies act as traps. The depth of such traps can be evaluated by thermoluminescence measurements. Hereby the sample is irradiated during cooling to temperatures, which prevent the thermal release of the excited electrons. Subsequently, the sample is slowly heated, the emission spectra are recorded simultaneously, certainly without further excitation. This method can also be used to evaluate the age of ancient fireplaces by analyzing the thermoluminescence of stones featuring color centers filled by radiation over time. For problems and practical aspects of evaluating thermoluminescence measurements, refer to [315].

The basic persistent luminescence mechanism as displayed in Figure 8.12 starts with optical excitation of the activator, here Eu^{2+}, from its ground state of $[Xe]4f^7$ configuration ($^8S_{7/2}$) beyond the excited states into the conduction band of the host (blue arrow)

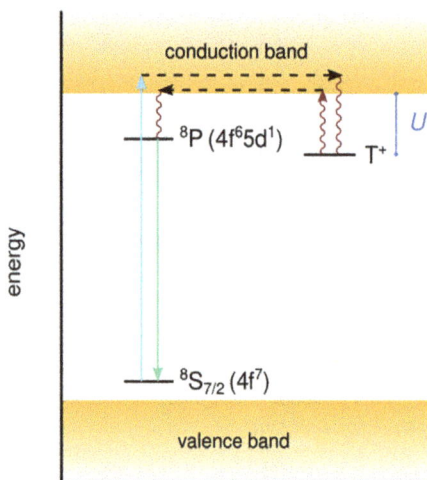

Figure 8.12: Term scheme showing relevant states of long-persistent phosphors like $Sr[Al_2O_4]:Eu^{2+}$.

giving a quite freely moving electron and a hole—Eu^{3+}. Although there is some discussion whether the transfer occurs via the conduction band or directly, the evidence is more on the former side. The excited electron can then be trapped in a nearby oxygen vacancy T^+ of depth U suited to host a negative charge because it is surrounded by divalent cations stabilizing negative charges according to

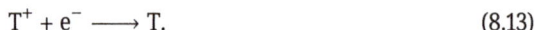

$$T^+ + e^- \longrightarrow T. \tag{8.13}$$

Such defects are usually generated during the synthesis of any phosphor.[70] Depending on the depth of the vacancy potential this electron will then be released thermally or optically via

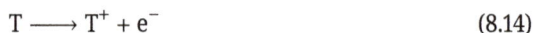

$$T \longrightarrow T^+ + e^- \tag{8.14}$$

For applications around room temperature, trap depths of approximately 0.6 eV are ideal as then the smooth thermal release of the trapped electrons occurs. If released, the electron finds an aforementioned Eu^{3+} hole where it yields an excited Eu^{2+} ion, which eventually emits the green photon around a wavelength of 520 nm upon relaxation to the ground state (green arrow). As the thermal excitation from traps only occurs spontaneously and strongly depends from the temperature, the emission might be delayed for many hours. The intensity and duration of this persistent luminescence depends from the number and the trap depth of the vacancies. The former can be increased significantly by doping with ions on the metal site with a differing and stable charge, such as Dy^{3+}.

---

[70] To avoid afterglow effects, phosphors for other applications are subsequently annealed under a suited atmosphere to heal these defects.

Dysprosium suits well for this role as the trivalent state is very stable. A reduction to the divalent or oxidation to the tetravalent state proves to be challenging. Thus, the charge introduced—with all consequences for a desired defect formation—will not be affected by surplus or lack of electrons. In $Sr[Al_2O_4]:Eu^{2+},Dy^{3+}$, two doped Dy^{3+} replace three Sr^{2+} ions, and apparently cause even more oxygen defects around the vacant cation sites. Presumably, most importantly for the long persistency of the luminescence, the Dy^{3+} ions stabilize the oxygen vacancies due to their higher charge and increased U. For the time being, it remains unclear whether here exclusively oxygen vacancies as proposed by *Aitasalo* or also dysprosium traps forming intermediately Dy^{2+} as suggested by *Dorenbos* contribute; but this latter detail does not actually abate the basic understanding of these phosphors. Moreover, even if excited somehow, Dy^{3+} also emits greenish and yellowish light, typically around 572 nm, enhancing the luminescence further. Certainly, also in the polyanion doping with fluorine or by partial replacement of aluminium by—for instance—boron, which can also be happy with only three adjacent oxygen atoms, is feasible to increase the number of defects on oxygen sites.

Such phosphors releasing the trapped energy smoothly over time by thermal activation may also be employed for photocatalysts to enable catalysis also in a dark setting. If the emitted wavelength peaks in the biologically transparent window between 650 and 1350 nm, also an application as detector for bioimaging and biodetection is feasible as here a high-intensity irradiation to release the trapped electrons can be avoided [317].

Application as Detector Material

Another prominent application of long-lasting luminescence is the quantitative recording of radiation such as X-rays in detectors of diffractometers or medicine. For this purpose, the trap depth is slightly deeper to prevent thermal relaxation. A material of choice is $BaFBr:Eu^{2+}$ where the trap depth U in Figure 8.12 is reported to be around 2 eV [318]. The doped Eu^{2+} ions are excited with energies above 6 eV (approx. 200 nm) to yield Eu^{3+} and a nearby trapped electron. The latter is released during read-out employing a red laser with 2 eV energy and then recombines with Eu^{3+} to give an excited Eu^{2+}. Via allowed $5d^1 4f^6 \rightarrow 4f^7$ transitions, a blue-violet emission around 3.2 eV (approximately 400 nm) will be recorded—and the stronger the emission intensity here, the stronger the irradiation by an X-ray diffracted beam was at this considered spot. Such—very sensitive—detectors are named *imaging plate detector* (IPD) or imaging plate detector system (IPDS).

8.4.6 Holmium—A Chameleon

Holmium compounds frequently show an interesting color change effect when watched under daylight where they look pale yellowish and pink under the artificial light of fluorescent lamps, for instance. Therefore, it might be named a chameleon. This interesting behavior will be discussed in the second part of this chapter. As the 5d states lie at very high energies in the vacuum ultraviolet regime, holmium does not show any 5d←4f absorptions very well deduced from the chemical shift model in Figure 8.3. The charge-transfer transitions are found at even higher energies than the 5d states. Both confirm the high stability of the electronic configuration of $[Xe]4f^{10}$ of trivalent holmium ions. The transitions between the terms depicted in Figure 8.13 feature accordingly only low transition probabilities for induced electric dipole 4f–4f transitions. Among these, the transitions $^5G_6 \leftarrow ^5I_8$ and $^3H_6 \leftarrow ^5I_8$ are *hypersensitive* (Chapter 8.4.3), which were recorded as intense absorption bands in the depicted spectra around 360 and 450 nm. Since Ho^{3+} shows low transition probabilities, the lifetime of excited states is long opening the door for applications based on optically pumped states, such as solid state lasers (Chapter 8.4.8) or up-conversion phosphors (Chapter 8.4.7).

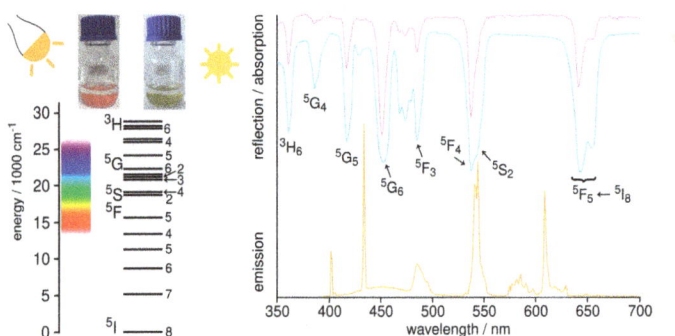

Figure 8.13: Term scheme of Ho^{3+}, the appearance of $Ho_2(SO_4)_3 \cdot 8\,H_2O$ illuminated by a compact fluorescent lamp and by daylight (left); reflection and absorption spectra of powdered (blue) and a solution of $Ho_2(SO_4)_3 \cdot 8\,H_2O$ (red) with assigned transitions compared with the emission spectrum of a compact fluorescent lamp.

Alexandrite Effect

The *Alexandrite effect* occurs if a solid shows different appearance under illumination with different light sources. Historically, this effect was named after tsar Alexander II, who reigned from 1855 to 1881. In the 1840s, the mineral *alexandrite* was discovered, namely a *chrysoberyl* of the composition $BeAl_2O_4:Cr^{3+}$; it looks greenish under daylight and reddish under campfire. The reason for this behavior are two strong absorption bands of the doped chromium ions peaking in the blue and yellow regime of visible light. While the emission of the sun—approximated as a black body radiator of 5500 °C—peaks in the green, the visible emission of campfire—a black body radiator of roughly 1200 °C—lies in the red. Consequently, the human eye, more sensitive to green than to red light, recognizes the dominating green reflection or transmission under daylight. Under the light of a campfire, the human eye gets a reddish color impression. To end this story, one should know that the colors of the imperial army of that time were green and red, so the discovered mineral was named after the tsar.

A similar color changing effect can be monitored for many compounds containing trivalent holmium ions if watched under daylight and the light of a fluorescent lamp. This can be well understood by looking at the absorption spectrum of Ho^{3+} ions and the emission spectrum of a fluorescent lamp shown in Figure 8.13. Apparently, the green emission of the fluorescent lamp around 550 nm is absorbed by the intense $(^5F_4,{}^5S_2)\leftarrow{}^5I_8$ transition. Accordingly, mainly the remaining lines in the blue and red regime yield the purple color impression of Ho^{3+} ions. Under sunlight, which consists of a continuous spectrum peaking in the green regime, the absorptions cause a yellow color appearance. The emissions of the fluorescent lamp are, by the way, dominated by emissions of rare-earth ions and are assigned in Figure 8.9 in Chapter 8.4.3. Such Alexandrite effects occur in every case where several absorptions in the visible regime selectively efface emissions of a light source as in the case of compounds containing Nd^{3+} where the color impression varies from reddish via purple to bluish.

8.4.7 Erbium—Putting Photons Together

Trivalent erbium ions usually form pink salts, and with their electronic configuration $[Xe]4f^{11}$ they are like the holmium and dysprosium ones that are very stable, which is reflected in their intermediate optical electronegativity (Figure 8.2). Thus they neither strive for further electrons nor they are interested in loosing one; accordingly they show high charge-transfer energies, also high lying 5d states and lack efficient excitation and relaxation pathways. This certainly limits the application of Er^{3+} in simple phosphors as is pointed out in a thorough discussion on the consequences of Equation (7.19) in Chapter 8.3. Remarkably though, the gaps between their excited 4f states are quite similar. In combination with the former facts, this makes them highly interesting for *up-conversion* as well as *signal amplification phosphors*. Both applications require long lifetimes of the excited states given the low transition probabilities for the subsequent absorption of more than one photon on the same Er^{3+} on one side. On the other, the long lifetime of excited states enables excited states to get pumped for stimulated emission in laser applications.

This goal is achieved by the basic properties like large energetic distance of states of opposite parity and charge-transfer states as outlined above. Moreover, doping Er^{3+} into host structures, which show very weak interactions, such as fluorides or oxides enhances these low transition probabilities further.

Over long distances—imagine for instance a transatlantic telecommunication cable—the intensity of the optical signals in glass fiber cables drops significantly. This requires regular modules where the intensity of the signal is amplified. And here phosphors or glass fiber sections doped with trivalent erbium ions come into play. In such a module, an infrared laser diode providing a 980 nm excitation wavelength pumps

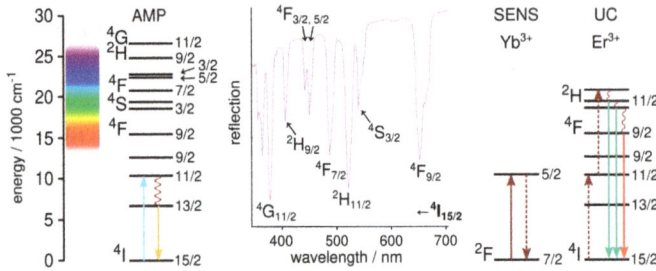

Figure 8.14: Term scheme of trivalent erbium for signal amplification (AMP) and reflection (left); reflection spectrum of powdered $Er_2[B_2(SO_4)_6]$ with assignments of transitions (middle) and relevant terms of an Yb^{3+} sensitizer (SENS) for up-conversion in Er^{3+} (UC, right).

the transition $^4I_{11/2} \leftarrow {}^4I_{15/2}$. Subsequent radiationless relaxation yields the excited state $^4I_{13/2}$ as depicted in Figure 8.14. The incoming weak signal of approximately 1,500 nm or 6,700 cm^{-1} causes a strong stimulated emission of many photons via the transition $^4I_{13/2} \rightarrow {}^4I_{15/2}$, and thus yields an amplification by a factor of up to three orders of magnitude. Due to relatively low transition probabilities, I here only display the reflection spectrum in Figure 8.14 where the hypersensitive transitions $^4G_{11/2} \leftarrow {}^4I_{15/2}$ and $^2H_{11/2} \leftarrow {}^4I_{15/2}$ are the strongest besides the lowest energetic one in the visible regime.

Upconversion

The subsequent absorption of more than one photon and final emission of a single photon of higher energy than the absorbed ones is called an *up-conversion* or *anti-Stokes process*—which just states that the emitted photon comprises a substantially higher energy than the exciting one. Nonsignificant processes in this context are the absorption of thermal energy, which might lead to a slightly higher-energy emission. Up-conversion may take place as subsequent absorption of photons of the ground and then of an excited state on the same ion, i. e., *excited state absorption* occurs. Alternatively, the second excitation occurs by assistance of a sensitizing ion, which transfers its excitation energy via one of the mechanisms considered in Chapter 7.6 onto the emitting ion, i. e., *energy transfer up-conversion* or other cooperative processes of low probability as outlined in [319]. Such up-conversion processes were discovered independently by Auzel[71] and Feofilov[72] in 1966. Since then, up-conversion has been described for several outer transition metal ions like Cr^{3+}, Mo^{3+} or Re^{4+}, but especially for lanthanide ions; compared with other nonlinear processes up-conversion is about eight orders of magnitude more efficient than second harmonic generation or a so-called two-photon absorption [319].

---

71 *Francois Auzel*, French engineer (*1938).

72 *Petr Petrovich Feofilov*, Soviet physicist (*1915 †1980).

The activator or emitter itself should be present in low concentration to avoid clustering and interactions between adjacent emitter ions as these lead to radiationless relaxation, such as cross-relaxation. The intensity of the up-converted emission scales via

$$I_{\text{up-conversion}} \propto I_{\text{exc}}^p \tag{8.15}$$

with the power of p subsequently absorbed photons of the exciting radiation intensity I_{exc}, if the energy transfer probabilities are low; if they are high, p converges to one, especially when the concentration of the sensitizer is sufficiently large. By the way, in this context such phosphors may also be named *antenna phosphors* as discussed in Chapter 8.4.2.

A proven efficient material for up-conversion from the near infrared around 980 nm to the visible regime—and efficient means a yield of 10 % photon conversion efficiency— is β-Na[YF$_4$]:Er^{3+},Yb^{3+}.[73] The process basically also works with trivalent erbium ions alone. But especially the higher excitations yielding also green emissions require a further sensitizer ion, namely Yb^{3+}. This is doped with a significant amount of around 20 %—which I would call rather a mixed crystal with yttrium than doping. Er^{3+} is present on up to 3 % of all trivalent cation sites; if erbium sensitizes itself the concentration is certainly higher within the range of 20 to 30 %. Because of the almost equidistant excited states, Er^{3+} is such a well-suited ion for up-conversion. It may be excited by photon energies around 980 nm and around 1,500 nm and. This makes them interesting for the enhancement of solar cell efficiencies as then infrared radiation below the band-gap of the respective semiconductor may be up-converted and also being harvested [308, 320].

In the example shown in the right part of Figure 8.14, the sensitizer ions Yb^{3+} absorb radiation of approximately 980 nm and transfer it onto neighboring Er^{3+} ions. Due to the long lifetime of the excited erbium state $^4I_{11/2}$, it may absorb another photon transferred from ytterbium via the transition $^4F_{7/2} \leftarrow ^4I_{11/2}$. From there, several radiationless relaxations compete with two green emissions from the states $^2H_{11/2}$ and $^4S_{3/2}$ as well as a red one from $^4F_{9/2}$. The intensity ratio of these visible transitions depends from the doping concentrations of emitter and sensitizer and certainly also from the host structure into which both were doped. Of course, further excitations are basically possible but these are even more improbable and were not only for the sake of better clarity omitted here. Further suited ions for such up-conversion processes based on ytterbium and erbium are Ho^{3+} and Tm^{3+}. These enable blue emissions; in combination with red and green emitters this opens the door for *transparent displays* based on up-conversion phosphors. In trivalent thulium, the same sensitizing mechanism as in the previously discussed example occurs from the 980 nm transition of Yb^{3+}. Via the excitations $^3H_5 \leftarrow ^3H_6$ followed by an intermediate radiationless relaxation, the second excitation $^3H_4 \leftarrow ^3F_4$ follows. From here, a further direct excitation $^1G_4 \leftarrow ^3H_4$ yields the final

---

73 The theoretical maximum value is 50 % for a two-photon up-conversion process.

emitting state causing the blue emission of $21.3 \cdot 10^3$ cm^{-1} (ca. 470 nm) to the ground state via the hypersensitive transition $^1G_4 \rightarrow {}^3H_6$ (see Figure 8.17).

Another up-conversion system, capable of converting visible into u. v. light, is Pr^{3+}, where the up-conversion process runs via a first excitation into the ^3P_J multiplet (see Figure 8.16 in Chapter 8.4.9) as intermediate state and further by absorbing another photon to reach the high lying 5d states. From here, an allowed transition to the ground state occurs as demonstrated on $Y_2(SiO_4)O:Pr^{3+}$ [321]. For further up-conversion systems on lanthanide ions, I would refer to [319] where up-conversion cascades are described also for Nd^{3+}, Gd^{3+}, Dy^{3+} and Tm^{2+}. Moreover, such up-conversion processes give rise for application in lasers or in commercial product identification systems via specific emission footprint patterns.

8.4.8 Neodymium—A Perfect Match for Stimulated Emission

Trivalent neodymium is the spectroscopic twin of trivalent erbium. Thus, Nd^{3+} belongs to the same group of rare-earth ions like Er^{3+}, which are in the pleasant situation to be fairly happy with their electronic configuration; here [Xe]4f^3. They are neither looking for a further electron—accordingly quite distant charge-transfer states—nor looking for getting rid of an electron yielding quite distant 5d states around at least 5 to 6 eV. The terms are the same as already seen for trivalent erbium ([Xe]4f^{11}), but with opposite order of J levels leading to the $^4I_{9/2}$ ground state. Therefore also, the transition probabilities for induced electric dipole transitions are relatively small, and luminescence is surely not on the top of neodymium's interests. Figure 8.15 depicts the relevant states for one of the most important applications of Nd^{3+}, i. e., in *solid state lasers*. It also displays all states in the visible regime important for the reflection spectrum shown and its violet color impression. Moreover, the different intensities of spin-allowed and spin-forbidden transitions may be pointed out here. To emphasize this, I left out some

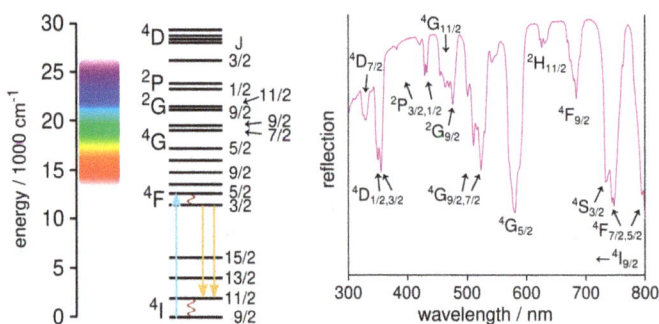

Figure 8.15: Term scheme of Nd^{3+} for the application in lasers (blue: optical pumping via an external source, red: thermal relaxation processes, yellow: laser emission), reflection spectrum of powdered Nd$_2$[B$_2$(SO$_4$)$_6$] with assignments of transitions (right).

transitions from the elusive manifold of transitions possible in the visible regime and selected those with the relatively highest intensities according to the largest Judd–Ofelt parameters U^λ (Chapter 7.8). Also here, the *hypersensitive* transition ${}^4G_{5/2} \leftarrow {}^4I_{9/2}$ is very intense. It was found that the shape and intensity of this transition as well as those of the other hypersensitive transitions of Nd^{3+} and Pr^{3+} elucidate details about the interaction of Nd^{3+} with complex ligands, especially biomolecules [322].

Solid State Lasers

In a phosphor like the aforementioned $Eu_2[B_2(SO_4)_6]$ individual Eu^{3+} ions may be excited via 4f–4f or charge-transfer transitions (see Chapter 8.4.3). After an arbitrary time, they relax back to the ground state by *spontaneous* emission of a photon. Since 4f–4f transitions are parity forbidden, the lifetime of the excited states is fairly long. If the excitation transition is an at least partially allowed one, and the finally achieved excited state only shows forbidden transitions to the ground state, there might occur a situation in which several ions are simultaneously excited. If, under these circumstances, an emitted photon interacts with another excited ion, *stimulated* emission of a further photon of same energy, same polarization and same phase occurs, i. e., *light amplification by stimulated emission of radiation*.

The first laser, invented in 1960, was a so-called three-level laser, where the final excited state ${}^2E$ of Cr^{3+} was reached via the spin-allowed ground state excitation ${}^4T_{2g} \leftarrow {}^4A_{2g}$ and a subsequent relaxation to the ${}^2E$ term. The emission back to the ground state ${}^4A_{2g}$ from here is spin-forbidden. Hence, by *optical pumping* a population inversion on the ${}^2E_g$ term is achieved. The emission to the ground state ${}^2E_g \rightarrow {}^4A_{2g}$ can then be stimulated by the emission of adjacent chromium ions. The disadvantage of this setup is that it is necessary for efficient laser operation to excite more than half of all chromium atoms at the same time; otherwise, no population inversion with respect to the ground state is achieved. Further, the laser emission occurs only pulsed, as the emitted pulse effaces the population inversion.

A better setup is a so-called four-level laser in which the target state of the stimulated emission—the fourth state in the game—is slightly above the actual ground state. This fourth state should be sufficiently high not to be populated thermally and makes sure that already at low excitation rates a population inversion is achieved. Hence, such a laser may be operated in a continuous mode.

And here neodymium enters the stage. Considering Figure 8.15, neodymium can be excited from its ground state, ${}^4I_{9/2}$, via a relatively efficient hypersensitive transition to the ${}^4F_{5/2}$ term and many higher ones. Further, a quite stable excited state ${}^4F_{3/2}$ with a lifetime around 200 µs, and finally the desired fourth state ${}^4I_{11/2}$ approximately 2100 cm^{-1} above the ground state exists—sufficiently prevented from thermal occupation (see Equation (7.14)). Therefore, we see here a four-level system fulfilling the aforementioned preconditions for such a four-level laser—optical pumping and easy population inversion on ${}^4F_{3/2}$ with respect to the ${}^4I_{11/2}$ term. The optical pumping is also possible via an interconfigurational and, therefore, allowed transition in the 5d states localized at around five to six electron volts.

As the host structure, many oxidic materials are feasible providing suited sites for doping with typically 1 % trivalent neodymium ions like yttrium aluminium garnet (YAG, $Y_3Al_2(AlO_4)_3$), $CaWO_4$ or glasses. Accordingly, the Nd^{3+} solid state laser may be optically pumped via charge-transfer transitions of sensitizers or direct excitation of neodymium's 5d—yielding a poor energy efficiency. But nowadays either lamps emitting white light, which can be absorbed by the manifold of lines in the visible regime, or laser diodes, which directly address the hypersensitive transition $^4F_{5/2} \leftarrow {}^4I_{9/2}$ are employed. The emitted wavelength usually peaks around 1,064 nm (9,400 cm^{-1}). Green laser pointers are frequently based on YAG:Nd^{3+} lasers including a frequency doubling crystal yielding the green emission at 532 nm—certainly this gives also rise for further frequency-doubled lasers with an emission of ca. 265 nm for various applications.

YAG:Er^{3+} lasers are suited for medical applications as the emission wavelength of 2.80 μm (3,570 cm^{-1}) based on the $^4I_{11/2} \rightarrow {}^4I_{13/2}$ transition (Figure 8.14) is strongly absorbed by matter containing water. Thus, thermal damage of adjacent tissue when removing, e. g., cancerous tissue can be reduced. Within the targeted tissue, the water evaporates vigorously, and thus the tissue is locally destroyed. Such lasers are employed in medicine also for eye surgeries or dental and skin treatments.

Table 8.3 lists selected reported lanthanide lasers with their respective transitions. Apparently, some ions allow for several laser transitions; their suitability depends from the host structure and filters employed fostering specific transitions. The wavelengths

Table 8.3: Selected solid state laser transitions of lanthanide ions with their respective approximate transition wavelength (Data: [323]); the target states, which are not the ground state are marked by *.

R^{3+}	transition	wavelength / μm	R^{3+}	transition	wavelength / μm
Pr	$^3P_0 \rightarrow {}^3H_4$	0.49	Ho	$^5S_2 \rightarrow {}^5I_5*$	1.40
	$^3P_1 \rightarrow {}^3H_5*$	0.53		$^5F_5 \rightarrow {}^5I_6*$	1.49
	$^3P_0 \rightarrow {}^3H_6*$	0.62		$^5I_7 \rightarrow {}^5I_8$	2.05
	$^3P_0 \rightarrow {}^3F_2*$	0.64		$^5F_5 \rightarrow {}^5I_5*$	2.37
	$^1D_2 \rightarrow {}^3F_4*$	1.04		$^5I_6 \rightarrow {}^5I_7*$	2.90
	$^1G_4 \rightarrow {}^3H_4$	1.04	Er	$^4S_{3/2} \rightarrow {}^4I_{15/2}$	0.55
Nd	$^4F_{3/2} \rightarrow {}^4I_{9/2}$	0.93		$^2H_{9/2} \rightarrow {}^4I_{13/2}*$	0.56
	$^4F_{3/2} \rightarrow {}^4I_{11/2}*$	1.06		$^4F_{9/2} \rightarrow {}^4I_{15/2}$	0.67
	$^4F_{3/2} \rightarrow {}^4I_{13/2}*$	1.35		$^2H_{9/2} \rightarrow {}^4I_{11/2}*$	0.70
	$^4F_{3/2} \rightarrow {}^4I_{15/2}*$	1.85		$^4S_{3/2} \rightarrow {}^4I_{13/2}*$	0.85
Eu	$^5D_0 \rightarrow {}^7F_2*$	0.61		$^4S_{3/2} \rightarrow {}^4I_{11/2}*$	1.30
Tb	$^5D_4 \rightarrow {}^7F_5*$	0.54		$^4I_{13/2} \rightarrow {}^4I_{15/2}$	1.60
Dy	$^6H_{13/2} \rightarrow {}^6H_{15/2}$	3.02		$^4S_{3/2} \rightarrow {}^4I_{9/2}*$	1.70
Ho	$^5S_2 \rightarrow {}^5I_8$	0.55		$^4I_{11/2} \rightarrow {}^4I_{13/2}*$	2.80
	$^5S_2 \rightarrow {}^5I_7*$	0.75	Tm	$^3H_4 \rightarrow {}^3H_6$	1.97
	$^5F_5 \rightarrow {}^5I_7*$	0.98		$^3H_4 \rightarrow {}^3H_5*$	2.30
	$^5S_2 \rightarrow {}^5I_6*$	1.01	Yb	$^2F_{5/2} \rightarrow {}^2F_{7/2}$	1.03

span a quite wide range between 0.49 and 3.02 μm, and certainly even more are accessible by using frequency doubling crystals.

8.4.9 Praseodymium—Occasionally a Knife for Photons

The electronic configuration of trivalent praseodymium ions of $[Xe]4f^2$ (ground state: $^3H_4$) is significantly less stable as that of the previously discussed ions Nd^{3+}, Er^{3+} and Ho^{3+} (Figure 8.3). Already, the averaged composition of the oxide, $Pr_6O_{11} = Pr_2O_3 \cdot 4\,PrO_2$ with a brown body color, suggests a significant stability of the tetravalent state; thus, 5d states are localized at comparably low energies, here around 200 nm. So, Pr^{3+} may efficiently be excited via the parity allowed absorption into the 5d states by VUV radiation.[74] Radiationless relaxation occurs to the excited and relatively long-lived $^1S_0$ state if it is sufficiently separated from the 5d states. This is the case in weak coordination surroundings with a weak nephelauxetic effect; such a situation is basically attractive for the application as a so-called *quantum cutter* or *down-conversion phosphor*—discussed in the second part of this chapter.

Normally, the $^1S_0$ state will lie within the 5d states, and thus the fast radiationless relaxation propagates down to the emitting state $^3P_0$. From here, the usual luminescence of Pr^{3+} yields visible red emissions mainly into the $^3F_2$ and $^3H_6$ levels and further in the orange, yellow and green regimes. Such clean emissions might be suited for xenon plasma driven lamps or displays. There the excitating plasma wavelengths of 148 and 172 nm fit nicely to the efficient $5d \leftarrow {}^3H_4$ transition. Let us now consider a single pixel of such a display. For the sake of a better contrast, it would be advantageous if the phosphor particles of the respective color point would not be white but colored—for instance, in the same color, which is emitted. For Pr^{3+} doped host structures this might be feasible if traces of tetravalent praseodymium ions were also present. These usually colorize compounds dark violet or red likewise the mentioned oxide Pr_6O_{11} because of ligand-to-metal charge-transfer (LMCT) transitions according to

$$Pr^{4+} + O^{2-} \overset{LMCT}{\rightleftharpoons} Pr^{3+} + O^-. \tag{8.16}$$

Whether or not this was an accidental discovery, but finally the researchers of a company developed the red phosphor $Zr(SiO_4):Pr^{3+},Pr^{4+},P^{5+}$. Essentially, *zircon* was doped with trivalent praseodymium. Due to the quite high doping concentration of up to 15 %, after absorption in the VUV efficient relaxation to the $^3P_0$ level occurs, from where a dominating red emission can be recorded via the spin-allowed transition $^3P_0 \rightarrow {}^3H_6$. During the synthesis starting from SiO_2, ZrO_2 and Pr_6O_{11}, $Zr(SiO_4)$ is simultaneously doped with Pr^{3+} and Pr^{4+} yielding a dark red powder. During the following treatment with the

---

74 VUV = vacuum UV light: 100–200 nm.

mild reducing agent, carbon monoxide, most of the tetravalent praseodymium ions were reduced, but some survived on the zirconium sites. These feature a significantly smaller ionic radius compared with that of Pr^{3+} and Pr^{4+} ions; it was occasionally observed that the site size within the host structure influences the stability of the better fitting ion. Relevant ionic radii of Zr^{4+} and the rare-earth ions with different coordination numbers are listed in Appendix A. Here, the small sites suited for Zr^{4+} stabilize Pr^{4+} regarding size and charge. This stabilization occurs at the cost of the desired emitter, here trivalent praseodymium. Moreover, every introduced Pr^{3+} in exchange for a Zr^{4+} generates a defect in the anion lattice; therefore, the authors undertook a further doping with PO_4^{3-} for SiO_4^{4-} by adding sodium phosphate during the synthesis to minimize the number of defects [325].

Quantum Cutting

As soon as the $^1S_0$ term is sufficiently decoupled from the 5d states, *quantum cutting*, a *down-conversion* process, is basically possible. To achieve this situation, a very weak coordination around Pr^{3+} is required. This is most easily the case in fluorides, such as $Sr(AlF_5):Pr^{3+}$ (0.3 %) where the rare-earth ion substitutes for strontium [324]. In the crystal structure, the strontium sites are coordinated in a ninefold irregular fashion by helical chains of AlF_6 octahedra [326]. Accordingly, the effective coordination charge of the fluoride ions amounts to only 0.5. Hence, the 5d states of Pr^{3+}, which interact strongly with ligands, experience a very weak nephelauxetic effect and also weak electrostatic interactions. After excitation into the 5d states, the $^1S_0$ term is thermally populated from where at least two subsequent emissions according to Figure 8.16 occur, a so-called cascade emission. This effect was discovered 50 years ago by Sommerdijk[75] and Piper[76] for $YF_3:Pr^{3+}$ [327, 328].

In such cases, one absorbed photon causes emission of approximately two photons. These phosphors are interesting for energy efficient lighting based on xenon plasma excitation, which emits radiation in the VUV at 148 and 172 nm. Compact fluorescent lamps are usually based on a mercury plasma; since mercury is highly toxic, a substitute such as xenon is desired. Unfortunately, the excitation occurs at clearly higher energy, which then requires a higher energy efficency of the phosphors—and here quantum cutting urges onto the wish list. Further examples for quantum cutting are $Cs_2KYF_6:Pr^{3+}$ or, based on other rare-earth ions, $Gd_2O(SiO_4):Eu^{3+}$ and $Ba_9Lu_2Si_6O_{24}:Tb^{3+}$ [329–331]. For even more examples and for further reading, I recommend [332]. In the chapter on trivalent ytterbium, we will see another quantum cutting effect including energy transfer and of actual interest.

---

[75] *J. L. Sommerdijk*, Philips Research Eindhoven, The Netherlands.
[76] *W. W. Piper*, General Electric Schenectady, USA.

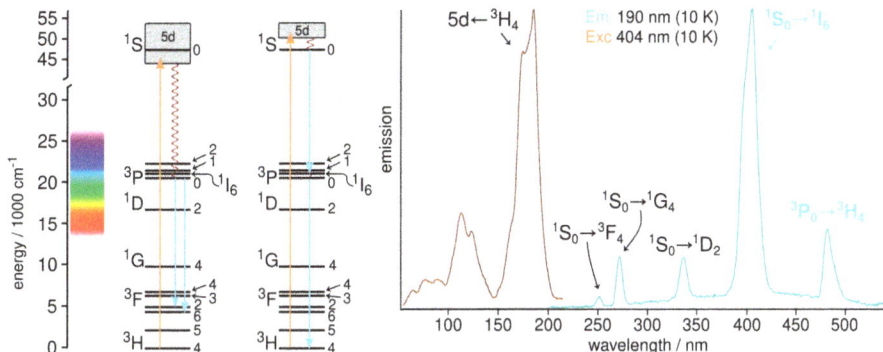

Figure 8.16: Term schemes showing relevant 4f states of praseodymium for its usual and cascade luminescence (left) and the luminescence spectra of $Sr(AlF_5):Pr^{3+}$ with assignments of transitions (right); data: [324].

8.4.10 Thulium—Normally Blue

With its electronic configuration of $[Xe]4f^{12}$ (ground state: $^3H_6$) the trivalent ion of thulium is the spectroscopic twin of Pr^{3+}. While these had a certain tendency to get oxidized to the tetravalent states, Tm^{3+} has a certain tendency to get *reduced* to the divalent state. It is like Pr^{3+} and less stable compared with its neighbors Er^{3+} and Ho^{3+}. This is also obvious from the chemical shift model (Figure 8.3) and the Frost diagram (Figure 3.8). Tm^{3+} apparently strives toward a fully occupied 4f shell while praseodymium did the same toward an empty one—both eventually are thus aspiring toward a closed-shell scenario. Also here, as already discussed for the previous twins, the terms span a significantly broader energy regime. For instance, the gap between the terms $^3H_4$ and $^3H_6$ amounts to roughly 4,500 cm^{-1} in Pr^{3+} and to more than 12,700 cm^{-1} in Tm^{3+}. Still, the states of opposite parity, the 5d and charge-transfer states, are localized at high energies. Therefore, the transition intensities are relatively low. The most prominent emission of Tm^{3+} is the dominating blue one due to the $^1D_2 \rightarrow {}^3F_4$ transition around 450 nm as shown in Figure 8.17.

Trivalent thulium is mainly used for rare-earth based lasers [333]. Our chosen example here is the YAG:Tm^{3+} laser, mainly employed for medical applications. There, a *two-for-one cross-relaxation mechanism* occurs by which initially the transition $^3H_4 \leftarrow {}^3H_6$ is pumped by a diode laser with an energy of 12,800 cm^{-1} (780 nm, orange arrow in Figure 8.17, right). If $Y_3Al_2(AlO_4)_3$ (YAG) is doped moderately with Tm^{3+} like 2 %, a cross-relaxation occurs according to

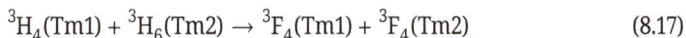

$$^3H_4(Tm1) + {}^3H_6(Tm2) \rightarrow {}^3F_4(Tm1) + {}^3F_4(Tm2) \tag{8.17}$$

illustrated by the broken blue arrows. In the aftermath, two Tm^{3+} ions are excited into the $^3F_4$ term, which explains the phrase *two-for-one*. Finally, the stimulated emission

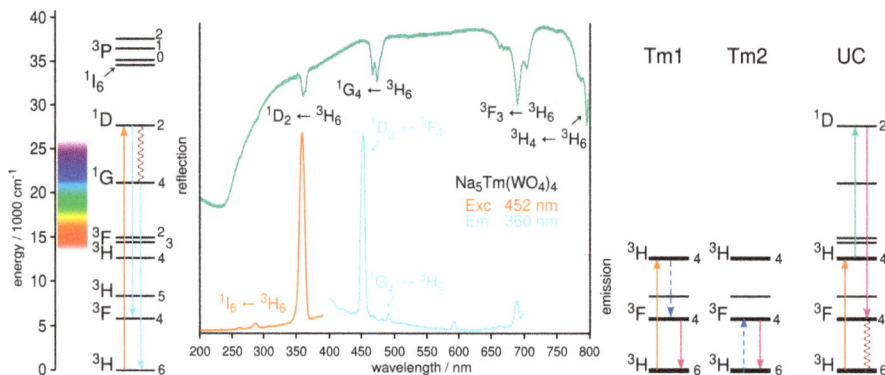

Figure 8.17: Term schemes showing relevant 4f states of Tm^{3+} for its usual luminescence including typical luminescence spectra of $Na_5Tm(WO_4)_4$ [293] (left) and the applications in lasers and up-conversion phosphors (right).

occurs from here to the higher J sublevels of the ground term $^3H_6$ with a wavelength of 1.8 µm. This term $^3H_6$ shows a significant sublevel splitting, so effectively a four-level laser system is realized here. Since the vitreous body of the human eye fully absorbs the emitted wavelength, this laser system can be employed quite safely in medical applications.

Because of similar reasons alike Er^{3+} also Tm^{3+} shows efficient up-conversion illustrated in the term scheme in Figure 8.17 (right). In an *up-conversion laser* system like YLF:Tm^{3+}(1 %), where YLF stands for yttrium lithium fluoride Y(LiF_4), two dye lasers subsequently pump according to

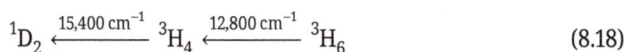

$$^1D_2 \xleftarrow{\text{15,400 cm}^{-1}} {}^3H_4 \xleftarrow{\text{12,800 cm}^{-1}} {}^3H_6 \tag{8.18}$$

a term from which blue emission around 450 nm in the term $^3F_4$ occurs, thus adding up two photons of different energy by up-conversion. Also here, the eightfold coordination of the yttrium sites is very weak. Y(LiF_4) crystallizes in the *scheelite* type, which derives from *fluorite* by a distortion yielding a tetrahedral surrounding of lithium. Thus, the effective coordination charge of fluorine amounts to 0.75 here.

8.4.11 Ytterbium—Solar Cells are Pleased

The ground state of trivalent ytterbium is $^2F_{7/2}$ in accordance with an almost fully occupied $[Xe]4f^{13}$ shell. Therefore, ytterbium alike europium shows a tendency to get reduced to the divalent state. Trivalent ytterbium is also the spectroscopic twin of Ce^{3+}, which will be discussed in the next chapters. Again, the splitting between the terms and states is larger for Yb^{3+}, and thus here the 5d states do not play that crucial role than

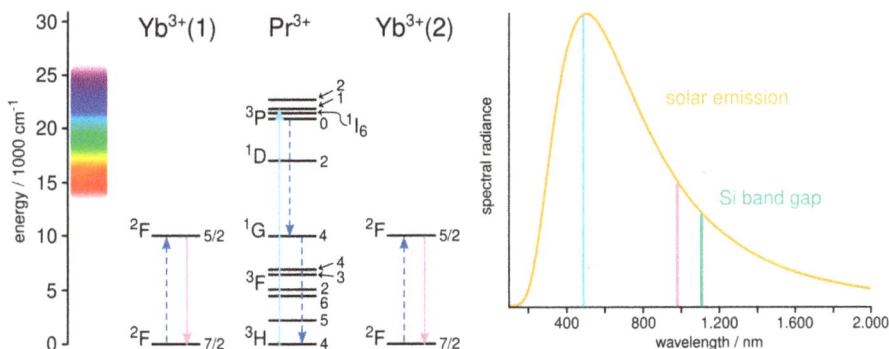

Figure 8.18: Term scheme of trivalent ytterbium showing all 4f terms relative to those of Pr^{3+} and the solar emission spectrum calculated with Equation (7.17) where the band gap of silicon is indicated in green, the relevant blue excitation and red emission transitions from the left section are also depicted.

in Ce^{3+} as we will see there. Hence, we have to consider only two 4f terms here, the ground term and the excited term $^2F_{5/2}$, separated by approximately 10,000 cm^{-1} in the infrared regime—yielding colorless Yb^{3+} ions (Figure 8.18). Besides a role as a sensitizer ion for up-conversion processes as mentioned in Chapter 8.4.7 trivalent ytterbium ions may also act as an emitter in down-conversion processes.

Considering the term scheme, only 4f–4f emissions in the infrared regime are expected from trivalent ytterbium ions. This emission can be well absorbed by semiconductors typically employed in solar cells. For instance, silicon exhibits a band gap around 9,000 cm^{-1} or 1,100 nm. Looking at the solar emission spectrum (Figure 8.18), it is obvious that the sun emits strongly in the blue, but even a blue photon only excites a single electron in such a semiconductor. Therefore, the idea arose to cut blue photons into two infrared ones, and thus enhance the light to electricity conversion efficiency of solar cells [334]. In a *quantum cutter* phosphor like SrF_2:Pr^{3+}(0.1 %),Yb^{3+}(5 %), the praseodymium ions absorb blue light and then a two-step cross-relaxation illustrated by the broken dark blue arrows in Figure 8.18 occurs, where two neighboring Yb^{3+} ions are excited. These then emit infrared light, which can be absorbed by the solar cell. This phosphor shows an efficiency of 140 %, so apparently in average 1.4 photons are generated from one absorbed blue photon. Both praseodymium and ytterbium are doped on the strontium sites in SrF_2, which crystallizes in the *fluorite* type. The charge mismatch between R^{3+} and Sr^{2+} generates cation defects in the material. It is well known that in such defect structures the doped ions form local clusters to minimize the overall loss of lattice energy by defect formation [335, 336]. Such clusters, though, make sure that the sensitizing praseodymium ions are in a close neighborhood of the ytterbium emitters onto which the energy has to be transferred. What a nice coincidence!

8.5 Emitters Strongly Interacting With Ligands (5d–4f Transitions)

While the 4f states are well shielded by the full $5s^2p^6$ shell, the 5d states are fully exposed to the impact of the ligands. Applying the chemical shift model (Chapter 8.3), from the diagram shown in Figure 8.3 typical positions of 5d states with respect to the 4f ground states can be deduced. For most of the trivalent rare-earth ions, the 5d states are found at very high energies, except for Ce^{3+}. In the case of sufficiently stable divalent ions, Eu^{2+} is the most interesting one with 5d states in reach. In this section, we will focus first on Ce^{3+}, discuss relevant factors and parameters influencing the optical properties, which essentially also apply to divalent europium discussed thereafter.

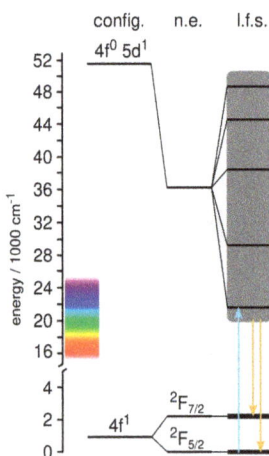

Figure 8.19: Schematic energy level scheme of YAG:Ce^{3+} illustrating the consequences of nephelauxetic effect (n. e.) and ligand-field-splitting (l. f. s.).

As depicted in Figure 8.19 in trivalent cerium ions, the 5d states in the naked ion are situated around $51{,}500 \text{ cm}^{-1}$ above the ground state level. Already, the covalent interaction with the ligands' electron clouds, the *nephelauxetic effect*, results in a stabilization of roughly $15{,}000 \text{ cm}^{-1}$. The electrostatic interactions yield an enormous ligand-field splitting of all five 5d orbitals spanning approximately $30{,}000 \text{ cm}^{-1}$ [251]. So, the centroid shift of the 5d states' barycenter is caused by the nephelauxetic effect already discussed in Chapter 8.1 and combined with the ligand-field splitting the lowest lying 5d states enter the visible regime. Therefore, YAG:Ce^{3+}, i. e., $Y_3Al_2(AlO_4)_3$:Ce^{3+} doped with trivalent cerium on the yttrium sites, can be excited already by blue light and subsequently emits yellow light. Most importantly, 5d→4f transitions are allowed according to the parity selection rule regardless of the local surrounding. Thus, the excited states may be relatively short-lived; moreover, such transitions may be very intense with quantum efficiencies up to 100 %.

Because of the strong nephelauxetic effect on the 5d states, and almost none on the 4f ones, the relative shift of the potential curves of 4f and 5d states is considerable. This situation was already depicted in Figure 8.1. Accordingly, many transitions from many different vibrational levels may occur, which leads to broad bands for both, the excitation and emission transitions with fairly large Stokes shifts.

The chemical shift model also suggests to distinguish between transitions into the low-spin and high-spin states of the 5d configurations as transitions there may be spin-forbidden [252]. At least for the here discussed ions Ce^{3+} and Eu^{2+}, this is not relevant. If considering divalent thulium or other ions of the second-half of the lanthanide series, this is another story addressed in Chapter 8.7.

8.5.1 Trivalent Cerium—the Most Efficient Ln^{3+}

Because of its electronic ground state configuration $[Xe]4f^1$, trivalent cerium is essentially the spectroscopic twin of Yb^{3+}. However, Ce^{3+} ions are clearly larger, and hence the spacing between the two 4f terms declines from 10,000 observed for Yb^{3+} to just above 2,100 cm^{-1} here. Moreover, the spacing between ground state and 5d states is with roughly 100,000 cm^{-1} out of reach for practical relevance in case of trivalent ytterbium. According to the chemical shift model, the 4f states of Ce^{3+} are situated at relatively high energies compared to the other trivalent rare-earth elements as shown in Figure 8.3. This very nicely reflects the fair stability of the tetravalent oxidation state. Looking at trivalent ions, the 5d–4f spacing is the smallest here and roughly around 51,500 cm^{-1}.

Nephelauxetic effect and ligand-field splitting then yield spacings between 4f and 5d states in the visible and u. v. regimes. YAG:Ce^{3+} (YAG:Ce) was already discussed before. Figure 8.20 shows two further compounds, an yttrium compound doped with cerium, $Y_2[B_2(SO_4)_6]$:Ce^{3+} (yttrium borosulfate YBS:Ce), and the pure cerium compound $Ce_2[B_2(SO_4)_6]$ (CeBS), both of which adopt the structure of $Gd_2[B_2(SO_4)_6]$ described in Chapter 6.5.6. Y^{3+} ions are smaller than Ce^{3+} ions. Therefore, the coordinating oxide ions approach the doped cerium ions closer in YBS:Ce than in CeBS. Hence, the electrostatic

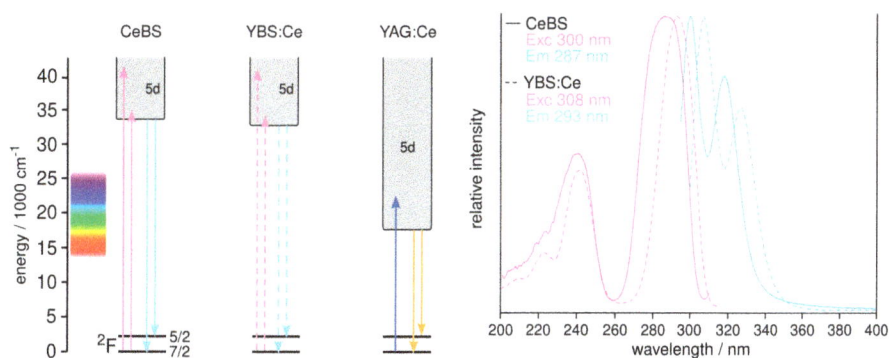

Figure 8.20: Energy level schemes of $Ce_2[B_2(SO_4)_6]$, $Y_2[B_2(SO_4)_6]$:Ce^{3+} and $Y_3Al_2(AlO_4)_3$:Ce^{3+} containing Ce^{3+} ions (left) and excitation (red) and emission spectra (blue) of the first two (right).

interactions are stronger in YBS:Ce compared with CeBS. The same holds for the co-valent interactions, thus the nephelauxetic effect is stronger in YBS:Ce. This situation is depicted in the schematic energy level diagrams shown. Consequently, the $5d \leftrightarrows 4f$ transitions should occur at lower energies and longer wavelengths for YBS:Ce. This is indeed the case considering the spectra in Figure 8.20. Moreover, we see two emission transitions toward the two 4f terms $^2F_{7/2}$ and $^2F_{5/2}$.

Comparing further YBS:Ce and YAG:Ce, the size of the yttrium site is comparable. But considering the next coordination spheres in YAG and YBS, in the former Si^{4+} and Al^{3+} are bound to the coordinating O^{2-} anions while in the latter S^{6+} are bound to oxide. Thus, the effective coordination charge is clearly lower in the latter case and the ligand-field splitting as well as the nephelauxetic effect are significantly smaller in YBS. Hence, the transitions in YBS:Ce occur solely in the UV, those in YAG:Ce already in the visible regime.

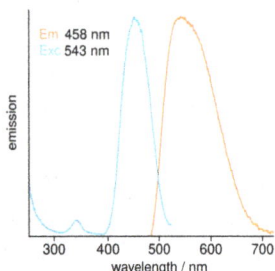

Figure 8.21: The excitation (blue) and emission (orange) spectrum of YAG:Ce^{3+}.

The most prominent cerium phosphor is indeed YAG:Ce. As depicted in Figure 8.21, it can be excited efficiently in the blue regime and converts approximately 90 % of all absorbed photons into a broad and unresolved single yellow emission band, i. e., a quantum efficiency of roughly 90 %. Its main applications are in white light emitting diodes, discussed in Chapter 8.6, or compact fluorescent lamps (Chapter 8.4.3). A typical synthesis runs via a coprecipitation where solutions of the respective nitrates form according to

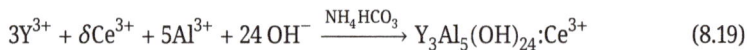

$$3Y^{3+} + \delta Ce^{3+} + 5Al^{3+} + 24\,OH^{-} \xrightarrow{NH_4HCO_3} Y_3Al_5(OH)_{24}{:}Ce^{3+} \tag{8.19}$$

a precipitate in a basic solution, which is then reacted at high temperatures via

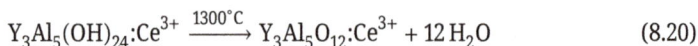

$$Y_3Al_5(OH)_{24}{:}Ce^{3+} \xrightarrow{1300°C} Y_3Al_5O_{12}{:}Ce^{3+} + 12\,H_2O \tag{8.20}$$

to give phase pure $Y_3Al_2(AlO_4)_3{:}Ce^{3+}$. Since YAG:Ce^{3+} is such an important phosphor, a manifold of different synthesis approaches are conducted depending on the respective application. The powder itself is bright yellow.

Extending our consideration, this means that the transition energies of trivalent cerium may be tuned over a quite large regime by an appropriate chemical variation of the coordination environment. Ce^{3+} emissions are observed from the deep ultraviolet through the orange-red regime, excitations from the deep ultraviolet until the blue regime. Accordingly, also the Stokes shifts are extraordinarily large. But certainly with increasing Stokes shift also the thermal quenching temperature decreases. Prominent applications dealing with large Stokes shifts are white light emitting diodes, where YAG:Ce is used, or as *scintillation phosphor*.

Scintillation and Storage Phosphors

Scintillation phosphors convert ionizing radiation immediately into ultraviolet or visible radiation. They are employed in detectors and wherever a quick and easily visible detection of such radiation is required. *Storage phosphors* also detect ionizing radiation but store and collect the absorbed energy until a respective physical impulse like a laser or heat leads to emission of the converted radiation. The ionizing radiation, such as X-, α-, β- and γ-rays excites the activators in these phosphors either directly via 5d←4f transitions or via ionization of the host structure. In the latter case, the generated very fast electrons are slowed down by ionizing further electrons, and finally the formation of a multitude of excitons. An *exciton* is an excited electron loosely bound to the respective hole forming a quasiparticle carrying an integral spin. In the case of scintillation, these generate excited Ce^{3+} ions, which then cause luminescence. Hence, the number of excited cerium ions is then proportional to the incoming intensity, the same holds for the emitted luminescence [337–339]. An example for a storage phosphor will be described in the second part of this chapter.

The basic requirements for such phosphors are a short relaxation time, an emission fitting for the respective application and a high radiation stability. To detect γ-rays efficiently, a high density and a high atomic number are also necessary [339]. The short relaxation time is required for a good time resolved detection. If the phosphor shall be employed for the visible detection of ionizing radiation during, for instance, alignment works, it should emit in the green regime as the human eye is most sensitive here. Regarding the excitation the phosphor has to provide suited excitation transitions in the regime of the radiation to be detected. And finally, a good chemical stability against the radiation is necessary to allow for a long-term application and for reproducible detection results. In a few examples listed in Table 8.4, these principles will be discussed.

Trivalent cerium ions are among the best candidates for scintillation as the 5d⇋4f transitions are allowed during excitation and emission for parity and spin. This ensures high absorption and a quantitative response scaling with the incoming radiation energy. Moreover, the transitions are fast and the luminescence properties are fairly straightforward to record as no further 4f–4f emissions are possible, and thus have not to be considered. The situation is more complicated employing other rare-earth ions like Pr^{3+} and Eu^{2+} where also 5d–4f transitions are relevant, as here further 4f–4f transitions might

Table 8.4: Selected parameters of scintillation phosphors; relaxation time, emission wavelength, radiation stability (Gy = Gray in J per kg, the energy dose absorbed by matter) and respective applications (for meaning of XCT and PET see text); data: [311].

compound	relax. time / µs	emission / nm	rad. stability / Gy	application
NaI:Tl$^+$	0.23	415	10	XCT, PET
CsI:Tl$^+$	1.05	550	10	XCT
CsI	0.01/1	305/>400	10^2	
Gd$_2$O(SiO$_4$):Ce^{3+}	0.06	430	>10^6	PET
Lu$_2$O(SiO$_4$):Ce^{3+}	0.04	420	10^6	PET
LaBr$_3$:Ce^{3+}	0.03	370	not known	PET

play a role. Further, in Ce^{3+} only one emission type has to be recorded in contrast to pure salts like CsI where two emissions with differing lifetimes compete.

The chemical stability of the host structure is enhanced with increasing bond energy, which is achieved by stable covalent bonds or high lattice energies. Accordingly, simple halides are generally less stable compared with oxides; silicates are often more stable than simple salts. This can be seen watching the values of the radiation stability in Table 8.4 where both silicates Gd$_2$O(SiO$_4$):Ce^{3+} and Lu$_2$O(SiO$_4$):Ce^{3+} show a clearly higher stability than the halides. Both compounds have the same structure as discussed in Chapter 6.5. Lu^{3+} are smaller than Gd^{3+} ions, thus the coordinating oxygen atoms are situated closer to the doped Ce^{3+} ions. Hence, ligand-field splitting and nephelauxetic effect increase from the gadolinium to the lutetium compound resulting in a shorter excitation and emission wavelength of the latter. Because of the slightly smaller size of the rare-earth site, the lutetium compound is slightly more sensitive against oxidation of the larger Ce^{3+} to Ce^{4+}, which would fit better on this site; therefore, the radiation stability of this compound is slightly lower than that of the gadolinium compound. A further advantage of the gadolinium compound is the presence of 4f states up to very high energies providing further excitation pathways. Moreover, the response and relaxation times of the cerium doped compounds are significantly shorter than those of classical scintillators like Tl$^+$ where allowed p–s transitions are at work. There, both states are broad yielding broad excitation and emission bands but both show stronger interaction, and thus longer lifetimes.

As mentioned before, several applications employ scintillation phosphors. In medicine and materials science, XCT (X-ray computed tomography) allows for fairly fast obtained three-dimensional images of matter with—at least for medical application—comparably low X-ray intensities. Another three-dimensional tomography is PET (positron emission tomography), where radioactive nuclei like $^{11}_6$C or $^{18}_9$F decay under formation of positrons which in turn annihilate with nearby electrons to give two γ-photons propagating in exactly opposite direction.

A further material, basically also capable of acting as scintillation phosphor, is BaFBr:Eu^{2+} [340]; it is isostructural with PbFCl. BaFBr:Eu^{2+} is much better known for

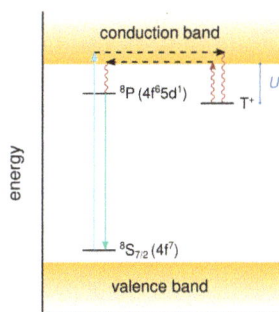

Figure 8.22: Schematic energy level scheme of a storage phosphor like BaFBr:Eu^{2+} with retarded luminescence.

its famous properties as a *storage phosphor*. The basic energy level scheme is depicted in Figure 8.22 and shows the storage mechanism. Like every inorganic compound also BaFBr contains many defects; because of the lower lattice energy contribution, mostly bromide defects are present here [318]. Upon irradiation with ionizing radiation, the Eu^{2+} ions doped into BaFBr are oxidized to the more stable trivalent state. These electrons migrate via the conduction band to nearby *electron traps*, i. e., especially bromide vacancies. Here, the excited electrons are stored. These traps are too deep for the electrons to be released by thermal energy. So, after excitation the irradiating energies are stored likewise a hidden image. To read out this information, a red laser excites the trapped electrons again back into the conduction band from where they are collected by adjacent Eu^{3+} ions. These are basically too small for the barium sites and strive for reduction back to Eu^{2+}, which fits much better. Hence, these electrons reduce the Eu^{3+} and yield an excited Eu^{2+} ion, which then immediately emits its excitation energy. Because of the presence of europium, indeed a defined single emission is caused. In case of BaFBr:Eu^2, this energy is emitted as blue light. Since blue light is higher in energy than the red laser light used for read-out, it can be excluded that other sources than the hidden image contribute to the emitted radiation. Eventually, all traps are emptied by white light to erase any stored information prior to a new detection cycle.

The phosphor under consideration here, BaFBr:Eu^{2+}, is employed in so-called *imaging plates* for the detection of X-rays. There, the irradiated matter causes a hidden image. This image can be the diffraction pattern of a crystalline sample or the transmission image of a human body. Accordingly, it is necessary here to store the image until it is read-out. The same mechanism occurs in *persistent phosphors*, treated in Chapter 8.4.5, and in natural rocks where excited electrons are trapped in defects. There a read-out may give information about the age of ancient fire places because the strong heat then released any trapped electron thermally in contrast to the adjacent rocks not heated sufficiently. The excitation here is caused by natural ionizing radiation. It should certainly be noted that usually no suited doping with rare-earth ions is present so the luminescence behavior is complicated [311, 341].

8.5.2 Divalent Europium—a True Chameleon

While in Ce^{3+} only a single 4f electron screened the 5d states in its ground state and none in its excited state, in divalent europium with its ground state configuration $[Xe]4f^7$ ($^8S_{7/2}$), there are seven and in its first excited state $[Xe]4f^65d^1$ still six shielding 4f electrons. Hence, the 5d states in the latter are better shielded, and thus more diffuse than those in trivalent cerium. Further, the ionic radius of divalent europium is larger than that of trivalent cerium. The ionic radii of different coordination numbers are listed in Appendix A. Consequently, all these factors lead to a stronger interaction with coordinating atoms and, therefore, the Stokes shift is in comparable surroundings normally even larger, the thermal quenching temperature lower in Eu^{2+} compared with Ce^{3+}.

Comparing Eu^{2+} with the isoelectronic Gd^{3+} ions, the effective nuclear charge is lower in europium. Therefore, the splitting between the 4f states is lower here. In Eu^{2+}, the transition $^6P_{7/2} \leftrightarrows ^8S_{7/2}$ occurs around 27,800 cm^{-1}, in Gd^{3+} around 32,200 cm^{-1}. However, this transition is rarely visible in emission spectra as the 5d–4f transitions are allowed, and thus orders of magnitude faster and more intense.

Due to the half-filled 4f shell, divalent europium ions are the most stable divalent rare-earth ions, and so is the 4f ground term considering the chemical shift model discussed in Chapter 8.3. Like for trivalent cerium, also in divalent europium the energies of the 5d states can be tuned via the *nephelauxetic effect* causing a *centroid shift* of the 5d barycenter and the *ligand-field splitting,* which we discussed in detail in the beginning of this chapter. Because of the lower charge, the gap between 4f ground state and the 5d states is clearly smaller in Eu^{2+}, roughly 39,800 cm^{-1} for the spin-allowed transition to the ^8P term; for Ce^{3+} we remember the value of 51,500 cm^{-1}. How to derive the term symbols of the excited 5d states of divalent europium was explained in Chapter 7.2; the J levels can be omitted as the 5d states show a very large width. The energetically higher lying ^8P term dominates for the excitation over the lowest lying ^8H term around 33,900 cm^{-1} because of the transition selection rule

$$\Delta L = \pm 1 \tag{8.21}$$

Nevertheless, the emission occurs—with slightly longer relaxation time—from the lowest lying term. The larger Stokes shifts allow for tuning the emission color of Eu^{2+} from the ultraviolet through the whole visible spectrum until deep red emissions. Figure 8.23 depicts schematically the energy levels of selected phosphors and their emission spectra; Table 8.5 lists selected data of these phosphors. As for Ce^{3+} and also for Eu^{2+}, very high quantum efficiencies close to 100 % are reported. In this section, we will take a respective journey based on the phosphors displayed here.

Our journey starts with a phosphor $Ba[B_4O_6F_2]:Eu^{2+}$ where the divalent europium ions are doped on a thirteenfold coordinated barium site, which provides only a very

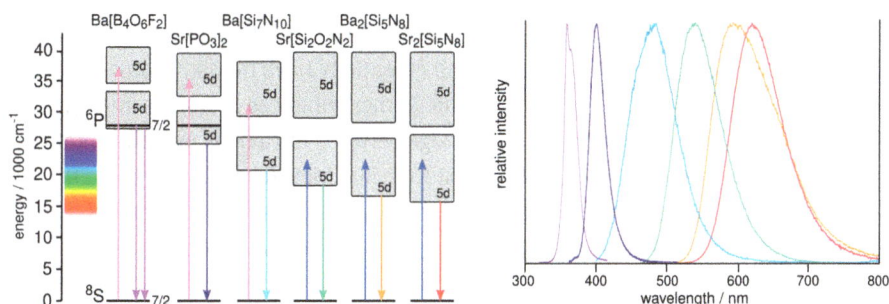

Figure 8.23: Energy level schemes of compounds containing Eu^{2+} ions (left) and their emission spectra, the colors illustrate the approximate emission color (right, data: [227, 235, 297, 343]).

Table 8.5: Excitation and emission wavelengths and energies as well as Stokes shifts of the phosphors discussed in this section; as Stokes shift the energy difference between the most efficient excitation of the emission was determined (data: [227, 235, 297, 311, 342–346]).

compound	excitation / nm	emission / nm	emission energy / 10^3 cm^{-1}	Stokes shift / 10^3 cm^{-1}
$Ba[B_4O_6F_2]:Eu^{2+}$	290	360	27.8	6.7
$Sr[PO_3]_2:Eu^{2+}$	323	404	24.8	6.2
$BAM:Eu^{2+}$	310	450	22.2	10.0
$Sr_6[B(PO_4)_4][PO_4]:Eu^{2+}$	348	480	20.8	7.9
$Ba[Si_7N_{10}]:Eu^{2+}$	330	484	20.7	9.6
$Sr[Al_2O_4]:Eu^{2+}$	410	520	19.2	5.2
$Sr[Si_2O_2N_2]:Eu^{2+}$	400	540	18.5	6.5
$Ba_2[Si_5N_8]:Eu^{2+}$	430	590	16.9	6.3
$Sr_2[Si_5N_8]:Eu^{2+}$	430	620	16.1	7.1
$CaO:Eu^{2+}$	490	733	13.6	6.8
$CaH_2:Eu^{2+}$	475	764	13.1	8.0

weak coordination as derived in Chapter 6.5 on silicate-analogous compounds. This example was chosen because here the coordination environment is so weak that even at room temperature both emissions, the 5d→4f as well as the 4f→4f transition, are monitored. The latter is expected to be a sharp, the former a comparably broad emission band. And indeed in Figure 8.23 (right) the violet emission band peaking around 360 nm shows the overlap of a broad and a sharp band. This weak coordination is mainly caused by the long coordination distances as Ba^{2+} is clearly larger than Eu^{2+} on one hand and the low effective coordination charges of the ligand oxygen and fluorine atoms. In such situations, it is likely that the europium atoms do not center the coordination polyhedron but are attached more or less to one side leaving the backside almost noncoordinated. Emissions in this regime are usually of interest for water disinfection or as invisible product safety markers since the whole emission band is located in the ultraviolet regime and the emission is excited with u. v. radiation around 290 nm [235].

In the second example, α-Sr[PO$_3$]$_2$:Eu^{2+}, the europium ions find themselves on strontium sites perfectly fitting for Eu^{2+} since both essentially have the same ionic radii. The coordination here is thus clearly stronger and further enhanced by a surrounding of oxide only. Hence, the emission occurs on the edge between ultraviolet and blue light with a maximum at 404 nm (Figure 8.23). The excitation is achieved best at 323 nm. In the second coordination sphere phosphorus atoms with oxidation state +V catch significantly electron density from the coordinating oxygen atoms thus weakening their coordination strength. To move further into the blue regime, this can be enhanced by formally replacing pentavalent phosphorus by trivalent aluminium atoms. As we will see later, this alone would shift the emission already into the green regime like in Sr[Al$_2$O$_4$]:Eu^{2+}. Therefore, to stay in the blue, the strontium atoms are formally replaced by the larger barium ones yielding the famous phosphor BAM:Eu^{2+}. BAM stands for Ba[MgAl$_{10}$O$_{17}$]. The host structure was described in Chapter 6.5.3. This blue phosphor is a prominent blue emitter used in compact fluorescent lamps and emits around 450 nm when excited in the ultraviolet regime. Another interesting phosphor is Ba[Si$_7$N$_{10}$]:Eu^{2+}, which emits already turquoise (Figure 8.23), on the brink toward green. It is the first nitridosilicate during this journey. Compared with BAM, also here the Eu^{2+} ions are doped on barium sites but instead of an aluminate surrounding the coordination is in summary slightly strengthened by an increasing charge in the second coordination sphere going from Al^{3+} to Si^{4+} and in the first coordination sphere going from O^{2-} to N^{3-}. Thus, the nephelauxetic effect as well as the ligand-field splitting represented by the effective coordination charge increase slightly, the emission wavelength is accordingly red-shifted to 484 nm. A comparable emission is achieved in Sr$_6$[B(PO$_4$)$_4$][PO$_4$]:Eu^{2+}—you might guess yourself how to explain this similarity. But due to the directly coordinating more electronegative oxygen atoms here the Stokes shift is significantly smaller.

Moving on to the green to red regime, a stronger coordination is required. Sr[Al$_2$O$_4$]:Eu^{2+} emits greenish light around 520 nm as implied before. This phosphor emits in the regime where the human eye recognizes light best, so it is employed in safety applications. A thorough discussion on such *long lasting* or *persistent phosphors* is given in Chapter 8.4.5. In green emitting Sr[Si$_2$O$_2$N$_2$]:Eu^{2+}, again the europium ions are doped on strontium sites, but due to the presence of nitrogen atoms in the coordination sphere the nephelauxetic effect and the effective coordination are stronger here. Switching to a pure nitrogen surrounding strongly fosters the coordination strength like in red emitting Sr$_2$[Si$_5$N$_8$]:Eu^{2+}. And again, to shift the emission to the orange regime, the respective barium compound is chosen. Ba$_2$[Si$_5$N$_8$]:Eu^{2+} shows an emission peaking at 600 nm from yellow through red and looking bright orange. Such phosphors like the latter two are employed in white light emitting diodes since they show excellent absorption in the blue regime and contribute the missing colors for white light [343]. Further red-shifted phosphors are only reasonable if the emission bands are narrow enough to avoid significant emission in the infrared, the regime unaccessible to the human eye perception. Here, an indeed very interesting and important finding were the

new nitride phosphors $M[LiAl_3N_4]:Eu^{2+}$ (M = Ca, Sr). Therein, not only the nephelauxetic effect of the coordinating nitride ions is stronger than in $Sr_2[Si_5N_8]$, due to a higher
coordination charge, as expected from our discussion so far. Remarkable is the almost
cubic coordination of the alkaline-earth sites on which the divalent europium ions are
doped. This apparently results in a fairly narrow potential curve and a narrow emission band peaking in the deep red region around 650 and 670 nm for the strontium and
calcium compounds. It is believed that this cubic environment hinders the vibrational
relaxation of the doped europium ions in ground and excited state, and thus further
fosters the narrow potential curve assumed here [228, 347].

Regarding lighting applications we are at the end of our journey here as a further
red shift of the emission yields an increasing contribution of infrared light [348]. Then
the energy efficiency of lamps drops as invisible light is produced. So, the red emissions
of $CaO:Eu^{2+}$ and $CaH_2:Eu^{2+}$ are hardly visible with the naked eye as they lie predominantly in the infrared regime with 733 and 764 nm, respectively. The trick, to shift the
emissions even further into the red regime, is to force the relatively large divalent europium ions on the clearly smaller sites of Ca^{2+}. Hence, the ligands approach the Eu^{2+}
ions closer leading to a stronger coordination. The higher ligand-field splitting of the
higher charged oxygen atoms in calcium oxide with *rock salt* structure is apparently
outperformed by the stronger nephelauxetic effect of hydride in CaH_2 with *cottunite*
structure of $PbCl_2$, nicely reflected in the larger Stokes shift here. But, as mentioned before, the stability of the divalent oxidation state is further reduced by doping on small
sites as then the formation of Eu^{3+} is favored even stronger. Hence, such phosphors show
a lower stability against oxidation. In case of the hydride, the reducing surrounding prevents an easy oxidation here.

8.6 On White LEDs—a Short Story

Global warming is a topic of our days and it fostered the necessity of saving natural
resources. So, preventing our earth from overheating is also about saving energy wherever possible. In 2013, the United Nations reported that 20 % of the worldwide electricity
consumption and 6 % of the global CO_2 emissions were caused by lighting. Lighting predominantly was achieved by incandescent lamps developed in the 19th century. And
with an estimated further increase of 60 % energy consumption for lighting by the year
2030—if not addressed properly—the United Nations demanded to phase-out conventional incandescent lamps to substitute them by either compact fluorescent lamps or
white LEDs [349]. In this section, I will give a rough overview on the concept of *white light
emitting diodes* and some historic aspects regarding blue light emitting diodes (LEDs).

The concept of incandescent lamps is simple. An electrical current flows through
a tungsten wire. The resistivity of tungsten then generates heat, and the tungsten wire
becomes so hot that it glows and produces light based on the principle of a black body radiator. Most of the radiation is invisible infrared radiation, though, circa 98 %. Tungsten

is chosen for this application because of its high melting point of approximately 3,400 °C allowing for the generation of white light. By the way, this light is usually recognized *natural* as it resembles the light of the big black body radiator named *the sun*. By far, more efficient are compact fluorescent lamps, already addressed in Chapter 8.4.3. Therein, a mercury plasma by electricity is generated, which then excites phosphors. And we saw there, most of them contain rare-earth elements. White light emitting diodes can save up to 80 % of energy used for conventional incandescent lamps. Thus, these will play a decisive role in the more efficient consumption of electricity for lighting applications.

The story on white light emitting diodes began already in the 1960s when blue diodes based on silicon carbide with a perfectly blue emission peaking around 465 nm corresponding to a band gap of 2.7 eV were developed. Unfortunately, this band gap is an indirect one yielding only poor emission efficiencies. A similar material, featuring the same valence electron concentration, is gallium nitride. Both SiC and GaN crystallize in the *wurtzite* structure type (see Appendix D). The band gap of 3.4 eV of gallium nitride is a direct one and lies just below the brink between u. v. and blue light. Theoretically, GaN would have been a perfect match for the concept of a white light emitting diode based on a blue semiconducting emitter and a phosphor converting this blue emission partially into yellowish light. According to the color diagram shown in Figure 8.25, the combination of blue and yellow then yields white light. But for several practical reasons there were considerable problems leading to low efficiencies here. It took approximately 30 years to solve these problems and to develop an efficient blue LED based on gallium nitride, which was rewarded with the Nobel Prize in Physics in 2014. This latter fact demonstrates how challenging the problems, and how elegant their solution was. Certainly, also other materials like zinc selenide were on the agenda but finally gallium nitride made the race. The prize was awarded to *Akasaki*,[77] *Amano*[78] and *Nakamura*[79] literally "for the invention of efficient blue light-emitting diodes, which has enabled bright and energy-saving white light sources."

Before we get to the realization of an efficient blue LED, we should shed light on a further important invention, which was awarded with the Nobel Prize in Physics 2000, the *double heterostructure junction*, which made LEDs substantially more efficient. Here, the relevant section of the LED where the light generation occurs, where holes and electrons combine was modified in a way that the band gap was slightly reduced. In this junction, between n- and p-type semiconductor, holes and electrons are collected since the holes strive for higher, the electrons for lower energies as depicted in Figure 8.24. This approach was employed in the actual blue LEDs by introduction of an indium-doped section. Indium nitride's band gap is somewhat smaller than that of gallium nitride and, therefore, the heterostructure was tuned to a band gap of 2.9 eV corresponding to bright

---

77 Isamu Akasaki, Japanese physicist (*1929 †2021).
78 Hiroshi Amano, Japanese physicist (*1960).
79 Shuji Nakamura, Japanese and American physicist (*1954).

Double Heterostructure Junction   **Two-Flow MOCVD**

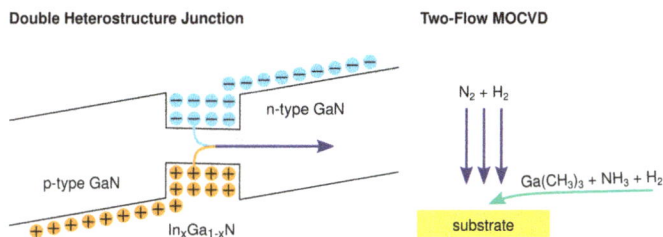

Figure 8.24: Principles of a double heterostructure junction where the blue electrons and orange holes are collected and the two-flow MOCVD process.

blue light of 430 nm—and an emission band lying fully in the visible regime. A pure gallium nitride LED would have an emission predominantly in the u. v. with 365 nm. So, the blue light output was optimized here, as well as the efficiency of the LED as such.

The two main problems to solve where the lattice mismatch of gallium nitride with the *sapphire* substrate and the p-doping of gallium nitride. The lattice mismatch amounts to 16 % between wurtzite-type GaN and sapphire, i. e., Al_2O_3. Consequently, during crystal growth many defects are generated impeding efficient semiconducting properties. To overcome this defect formation, a novel *two-flow metal-organic chemical vapor deposition* (MOCVD) process was developed. Here, the CVD process starting from trimethylgallium and ammonia according to

$$Ga(CH_3)_3 + NH_3 \xrightarrow{1050\,^\circ C} GaN + 3CH_4 \qquad (8.22)$$

is improved by introducing a secondary flow of chemical vapor perpendicular to the substrate plane according to Figure 8.24. This applies pressure to the tangential flow against the surface. Hence, an almost perfect gallium nitride or aluminium nitride buffer layer grows on the substrate [350, 351].

The second problem, the p-doping, was chemically even more challenging. Every contamination with oxygen yields an n-doping according to GdN:O where each oxygen atom replacing nitrogen contributes a surplus electron. Therefore, as-made gallium nitride is n-doped. The actual p-doping is usually made by doping the gallium sites with an element of group 2 such as magnesium. An important discovery was that due to the presence of hydrogen during the CVD process actually a codoped GaN:Mg,H with additional hydrogen atoms was formed, and that under a subsequent annealing in nitrogen/hydrogen atmosphere these hydrogen atoms remained in the material, which clarified one of the main remaining problems on the way to an efficient blue LED [352]. Since then, the material is finally annealed in a pure nitrogen instead of a nitrogen/hydrogen atmosphere above 700 °C, which dramatically enhanced the hole mobility as the doped hydrogen just evaporates. This marked a crucial breakthrough, the successful p-doping according to GaN:Mg was accomplished [353]. A typical doping concentration amounts to $2 \cdot 10^{-18}$ cm^{-3} or approximately 23 ppm magnesium. The same result was achieved by electron beam irradiation of the GaN:Mg,H raw material [354].

The last step was the realization of a double heterostructure junction by employing trimethylindium to obtain an indium-gallium nitride section, i. e., In$_x$Ga$_{1-x}$N. Also here, the development of the two-flow MOCVD solved the problem of an active layer rich in defects, which until then prevented any luminescence at room temperature [355]. Eventually, in 1994 the first tunable bright blue In$_x$Ga$_{1-x}$N LED was presented [356]. It is still tricky to tune the band gap of this emitting section finely. So, from LED to LED the band gap may vary yielding somewhat different blue tones.

So, 1994 was the actual starting point, from where efficient white light emitting diodes came on the agenda. To realize such a white LED, phosphors were needed to convert the blue LED light partially into the remaining colors of the visible spectrum. Covering the blue LED with a thin layer of Y$_3$Al$_2$(AlO$_4$)$_3$:Ce^{3+} (YAG:Ce^{3+}) yielded a quick approach to the first white LED following the color scheme in Figure 8.25. YAG:Ce^{3+} absorbs efficiently in the blue and shows a very broad-banded emission from yellow to orange (Figure 8.21). Figure 8.25 shows the color coordinates as defined in Chapter 7.7.3 of a blue LED (a) and YAG:Ce^{3+} (b). The broken connection line between both points represents all possible colours such a coated LED might display. It apparently intersects the bold line of the black body radiator in the regime of white light. An even broader regime is accessible if the blue LED is coated by a green emitting (c) and an orange emitting phosphor (d) since then the whole triangle spanned by the broken lines may be displayed. Examples for such phosphors like Ba$_2$[Si$_5$N$_8$]:Eu^{2+}, Sr[LiAl$_3$N$_4$]:Eu^{2+} or Sr[Si$_2$O$_2$N$_2$]:Eu^{2+} were mentioned in the last section [343].

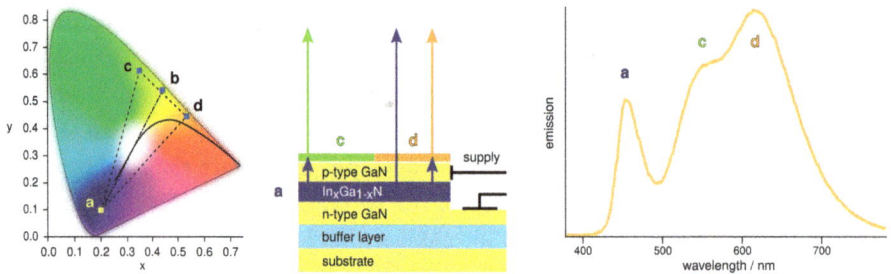

Figure 8.25: The CIE color diagram showing the color coordinates of blue LED (a) and typical phosphors Y$_3$Al$_2$(AlO$_4$)$_3$:Ce^{3+} (b), Sr[Si$_2$O$_2$N$_2$]:Eu^{2+} (c) and Ba$_2$[Si$_5$N$_8$]:Eu^{2+} (d) as well as the black body radiator line; principle of a white LED and typical emission spectrum of the visible regime (400–700 nm); data: [343].

With the former concept based on blue LED and YAG:Ce^{3+} only white LEDs generating *cold* white light with high color temperatures corresponding to the black body radiator can be realized while the latter approach enables the more reddish *warm* white light generation with lower color temperatures. Moreover, the shown emission spectrum of such a white LED more or less resembles the solar emission spectrum and should be perceived as *natural* [348]. Further developments might lead to u. v. LEDs based on aluminium gallium nitride, $Al_xGa_{1-x}N$, which then has to be coated with three efficient phosphors converting u. v. into visible light. Because more energy would have to be supplied for this approach, it is less energy saving than the former. However, this design would also solve the problem that each blue LED shows a slightly different blue hue as then solely the phosphor mixture determined the emission color of the white LED. But on the other hand, the goal of the United Nations to save even more energy would be harder to achieve.

8.7 Further Divalent Rare-Earth Ions

First reports about divalent rare-earth ions date back to the 1930s. Certainly, primarily divalent europium was identified as an efficient emitter, but also Yb^{2+} and Sm^{2+} were mentioned and believed to be the only stable ones beside Eu^{2+}. Later, also divalent thulium was thoroughly investigated [357, 358]. Interestingly, but not surprising, these are indeed the most stable ones. The synthesis of such phosphors can be challenging in most cases. But occasionally, even syntheses under ambient atmosphere were reported, such as the doping of $Sr[B_4O_7]$ with Eu^{2+}, Yb^{2+}, Sm^{2+} and even Tm^{2+} on the strontium site [359]. Here, presumably the comparably large site size for trivalent ions stabilized the larger R^{2+}. Usually, the reduction even with hydrogen fails and a metallothermic reduction yields the respective divalent halide.

   Because of the lower charge, the electronic situation for the divalent cations is quite different to that of the trivalent ones. Figure 8.26 shows an enlarged extract of the chemical shift model scrutinized in Chapter 8.3 based on the same data. In Figure 7.9, we thoroughly discussed the difference of low- and high-spin 5d states. The same holds for the divalent lanthanide ions as depicted here also in the chemical shift model. Important for possible applications is that in the second-half of the rare-earth series the lowest lying high-spin 5d states feature comparably long lifetimes as the relaxation to the 4f

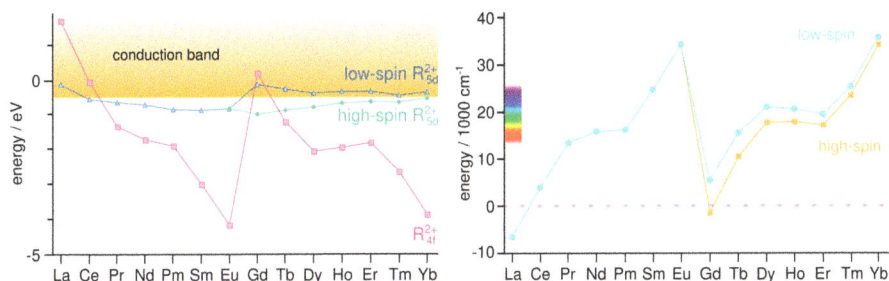

Figure 8.26: Chemical shift model of the divalent states only (left) and the energy differences of the lowest high- and low-spin 5d states and the ground 4f states of the free Ln^{2+} (right) calculated based on the data provided in [252].

ground state is spin forbidden. Therefore, like the late trivalent R^{3+}, up-conversion is a topic here. For the interpretation of spectra, it is also important, that the lower charge of R^{2+} yields a smaller splitting of the 4f levels, approximately 10 to 30 % lower than the figures given in the Dieke diagram as these were set up for R^{3+}.

The trends of relative stabilities are vividly reflected by the 5d–4f transition energies displayed in Figure 8.26. The most stable two, divalent europium and ytterbium, with the highest 5d–4f transition energies for the naked ions contrast the most unstable ones, cerium, gadolinium and lanthanum. Some figures are even negative; so, in these cases in divalent compounds an electron enters the conduction band forming a hidden trivalent ion according to $R^{3+} \cdot e^-$.

In aqueous solution, only Eu^{2+} survives a longer treatment. But in the solid state several divalent, and I mean really divalent, cations could be stabilized. Certainly, also divalent europium endures the handling of most of its solid compounds in air. Further, reasonably stable ions are Yb^{2+}, Tm^{2+} and Sm^{2+}. As a rule of thumb may act the color of their halides. The most stable $EuCl_2$ and $YbCl_2$ are colorless and apparently absorb only in the u. v. regime. This corresponds to a high stability of R^{2+} since the excitation of an electron into the conduction band yielding a R^{3+} requires a comparably high energy; the less stable a R^{2+} the less energy is needed for such transitions as observed for the following examples. $TmCl_2$ and $SmCl_2$ are described as dark red. Hence, they show absorption in the green. Further black chlorides—absorbing the whole visible regime—are $DyCl_2$ and $NdCl_2$ [112, 360].

The maybe most important trivalent rare-earth ion, Eu^{3+}, finds its sister in Sm^{2+} being isoelectronic ($[Xe]4f^6$). The lower charge generally leads to a lower splitting of all 4f states as described in Chapter 7.8; accordingly, the sharp 4f–4f emission $^5D_0 \rightarrow {}^7F_0$ occurs around 690 nm as discussed in the course of the possible application in optical memories (Chapter 8.4.4). Moreover, it can be well excited in the u. v. via 5d←4f excitations. Because of the energetically closer 5d states a mixing of the excited state, $^5D_0$ seems reasonable. Hence, the essentially forbidden $^5D_0 \rightarrow {}^7F_0$ transition becomes quite intense in Sm^{2+}, in striking contrast to its sister ion. Sm^{2+} has almost the same size than Eu^{2+} and it can therefore be applied for the investigation of host structures doped with Eu^{2+} (Appendix A). Because the transition $^5D_0 \rightarrow {}^7F_0$ always occurs as a single line, it can be employed to determine the number of chemically different sites occupied by europium. These facts already suggest that the 5d–4f emission, for which the ions are expected to be prominent, is not as frequent or relevant as for Ce^{3+} and Eu^{2+}.

Divalent thulium is isoelectronic with Yb^{3+} ($[Xe]4f^{13}$), and thus basically interesting for up-conversion processes. This was indeed proven on several examples, e. g., in $SrCl_2:Tm^{2+}$ where up-conversion could even be realized employing a xenon lamp, and in $CsCaI_3:Tm^{2+}$, where it reached an efficiency of 11 % [361, 362]. This extraordinarily high efficiency for an infrared to visible up-conversion became possible because of participation of the highly efficient 5d←4f excitation into high-spin 5d states. Quite recently, the doped halides $NaI:Tm^{2+}$, $CaBr_2:Tm^{2+}$ and $CaI_2:Tm^{2+}$ were identified as possible solar luminescent concentrator for the application on windows. Sputtered as thin films on normal window glass, these materials absorb a part of the incoming light and convert it into infrared radiation. Divalent thulium absorbs here via a $5d \leftarrow 4f$ excitation. The then emitted infrared radiation around 1140 nm remains trapped in the glass and is totally reflected therein until it is converted at the edges of the window into electricity by solar cells embedded there. Since Tm^{2+} absorbs uniformly over the entire visible spectrum, the otherwise inevitable coloring of the glass by selectively absorbing phosphors can be avoided [363].

Trivalent ytterbium is the spectroscopic twin of Ce^{3+}. In contrast to the latter, no 5d–4f emissions are relevant here. Adding an electron, yielding Yb^{2+} with a fully occupied $4f^{14}$ shell, only 5d–4f transitions are possible. The 5d–4f energy difference is slightly larger for Yb^{2+} compared with Eu^{2+} (Figure 8.26). Further, the ligand-field and nephelauxetic effects on the 5d states are very similar for all Ln^{2+} as they are for Ln^{3+}. Hence, the luminescence properties of both, Eu^{2+} and Yb^{2+}, are quite similar. The critical difference is the high-spin 5d states acting somehow as traps and generally leading to long lifetimes and slightly lower efficiencies, though. For instance, for $Sr[Si_2O_2N_2]{:}Yb^{2+}$ a good quantum efficiency with a red emission around 620 nm was reported [364]. For $Sr[Si_2O_2N_2]{:}Eu^{2+}$, a green emission around 540 nm was reported (p. 165 and [343]). The initial remarks would suggest a slightly blue-shifted emission of the ytterbium compound due to the clearly higher nuclear charge. Presumably, here the 5d states are already situated in the conduction band. As mentioned before, 5d and conduction band states may mix very efficiently, in strong contrast to the well-shielded 4f states. Therefore, the excitation actually occurs indeed into the conduction band, a situation like that depicted in Figure 8.26 (left). Hence, the emission occurs not via a $5d \rightarrow 4f$ emission but from a so-called impurity trapped exciton state. This is a similar situation as described for storage phosphors, but here with a very flat trap state. Such anomalous emissions can be proven by measuring the photoconductivity upon excitation. Finally, thermal quenching is already relevant at room temperature for the ytterbium compound while for the europium compound up to 200 °C no significant thermal quenching was noticed [364].

9 Magnetism

In this chapter, I will discuss selected magnetic properties of the rare-earth elements and their compounds. Due to the vast spectrum of meanwhile acquired property details and their applications, this selection has to be incomplete but aims to trigger interest and enable the reader for further self-studies on these topics. Therefore, I will first summarize relevant basics and then treat the properties of selected compounds as well as relevant applications.

9.1 Basic Principles

Magnetism describes the reaction of any medium upon exposure to an external magnetic field. For a proper understanding, some relevant parameters have to be defined and understood as described in a typical textbook [365]. An external magnetic field is a vector $\mathbf{H}$ typically caused in a coil by the flow of an electric current. It is usually given as amperes per meter, i. e., A/m, and thus defined by its origin. It causes a *magnetic induction* or *flux density* $\mathbf{B}$, which is related to the magnetic field via

$$\mathbf{B} = \mu_0 \mathbf{H} \tag{9.1}$$

with the permeability of the vacuum[80]

$$\mu_0 = 1.25663706212(19) \cdot 10^{-6} \frac{\text{Vs}}{\text{Am}} \approx 4\pi \cdot 10^{-7} \frac{\text{Vs}}{\text{Am}}, \tag{9.2}$$

if the coil is empty. As soon as matter is positioned inside the coil, it experiences a different flux density modified by its individual *magnetization* $\mathbf{M}$ according to

$$\mathbf{B} = \mu_0 (\mathbf{H} + \mathbf{M}) \tag{9.3}$$

and usually also given in A/m like $\mathbf{H}$. This different flux density is caused by the presence of electrons within the matter, which themselves are induced to local currents by the external field causing magnetic fields, and thus contribute to the flux density. Assuming that the matter under consideration is magnetically isotropic, $\mathbf{M}$ and $\mathbf{H}$ can only be parallel or antiparallel, and accordingly follow the relationships:

$$\mathbf{M} = \chi \mathbf{H} \tag{9.4}$$

---

[80] The actual value may be found on a NIST website (https://physics.nist.gov/cgi-bin/cuu/Value?mu0) based on the CODATA recommended values; the given value was actual in December 2023.

https://doi.org/10.1515/9783110680829-009

or

$$\mu_0\mathbf{M} = \chi\mathbf{B}_0 \tag{9.5}$$

if one considers $\mathbf{B}_0 = \mu_0\mathbf{H}$ as *an external magnetic induction* and includes the dimensionless magnetic *volume susceptibility* χ, which is zero in vacuum. Combining the previous Equations (9.3) and (9.4), which yields

$$\mathbf{B} = \mu_0(1 + \chi)\mathbf{H} = \mu_0\mu_r\mathbf{H} \tag{9.6}$$

after introduction of the dimensionless *relative permeability* according to

$$\mu_r = 1 + \chi, \tag{9.7}$$

which amounts to one in vacuum. Any the magnetic field disturbing bound and paired electrons within matter cause magnetizations with opposite direction with respect to the external field, thus weakening the measured flux density in accordance with a negative contribution to χ. Such *diamagnetic* contributions have to be considered for every material; materials with a negative χ with typical values below -10^{-4} and $\mu_r < 1$ are called diamagnetic materials. Superconductivity reduces μ_r to zero, and thus an ideal diamagnet is achieved. For practical reasons instead of the volume susceptibility χ, normally the *molar susceptibility*

$$\chi_{\text{mol}} = \frac{\chi}{\rho}M_r \tag{9.8}$$

and sometimes the *mass susceptibility*

$$\chi_\rho = \frac{\chi}{\rho} \tag{9.9}$$

employing the material's density ρ and molar mass M_r are used.

Since in literature different unit systems are in use, I add this paragraph on the most common unit systems, the International System of Units, SI and the Gaussian cgs (centimeter-gram-second) system. For instance, in Equation (9.2) μ_0 is given in SI units. In Gaussian units $\mu_0 = 1$, H is given there in *Oersted* and **B** in *Gauss*. Table 9.1 lists selected electromagnetism units in the SI and Gaussian systems including their conversion factors.

If unpaired or conducting, i. e., itinerant, electrons are present in a material, local currents are induced, which cause local magnetic fields. These act as local magnets and are oriented along the direction of the external field, contrary to diamagnetic materials. Such *paramagnetic* contributions contribute positively to χ, and accordingly materials with positive χ and $\mu_r > 1$ are called paramagnetic materials.

Table 9.1: Selected electromagnetism units in the SI and Gaussian system [366].

quantity	symbol	SI	Gaussian	conversion factor
magnetic **H** field	**H**	$\dfrac{A}{m}$	Oe	$\dfrac{H_G}{H_{SI}} = \sqrt{4\pi\mu_0} = \dfrac{4\pi \cdot 10^{-3}\,\text{Oe}}{1\,\text{A/m}}$
magnetic **B** field	**B**	$T = \dfrac{kg}{s^2 A}$	G	$\dfrac{B_G}{B_{SI}} = \sqrt{\dfrac{4\pi}{\mu_0}} = \dfrac{10^4\,\text{G}}{1\,\text{T}}$
vacuum permeability	μ_0	$\dfrac{V\,s}{A\,m} = \dfrac{kg\,m}{A\,s^2}$	1	$\dfrac{\mu_{0,G}}{\mu_{0,SI}} \approx \dfrac{1}{4\pi \cdot 10^{-7}}\dfrac{A\,m}{V\,s}$

Due to their electronic structure especially inner and outer transition metal atoms often feature such unpaired electrons localized in the respective 4f or nd shell frequently. In contrast to main-group elements, the number of unpaired electrons can be higher and these unpaired electrons are normally stable; thus, they are highly suited for magnetic applications. Among these, lanthanide elements may show the highest numbers of unpaired electrons since here seven 4f orbitals offer space for up to seven lone electrons while the five d orbitals only provide at maximum five lone electrons. Such atoms exhibit permanent magnetic moments and actually behave like atomic magnets. The volume susceptibility of this *Curie paramagnetism*[81] typically lies in the regime between $+10^{-5}$ and $+10^{-2}$ for inner and outer transition metal atoms with unpaired electrons.

Certainly, as soon as such materials show electronic conductivity, this will also cause rather weak paramagnetic contributions of itinerant electrons called *Pauli paramagnetism.*[82] Under certain circumstances, otherwise diamagnetic ions may show a weak and temperature independent paramagnetism by mixing of the diamagnetic ground state with other magnetic states. This is called *Van Vleck paramagnetism*[83]. Paramagnetic behavior is only a short term magnetization of a material, which requires an active external magnetic field. After turning off the external field, any magnetization vanishes quickly.

An external magnetic field may also induce magnetic ordering, which remains in the material below the respective ordering temperature, such as *ferromagnetism, antiferromagnetism* or *ferrimagnetism*. Normally, for these magnetically ordered materials the simple relationships between **H**, **B** and **M** become field dependent. The relative permeability μ_r may adopt values up to 10^6 for ferromagnets. Moreover, if materials behave magnetically nonisotropic, these field vectors are no longer necessarily parallel.

---

81 *Pierre Curie*, French physicist, Nobel Prize in Physics 1903 (*1859 †1906).
82 *Wolfgang Pauli*, Austrian physicist, Nobel Prize in Physics 1945 (*1900 †1958).
83 *John Hasbrouck Van Vleck*, American physicist (*1899 †1980).

9.2 Paramagnetism

If exposed to an external magnetic field, the atomic magnets in a material strive to align parallel to the external magnetic field vector. This competes with thermal motion, which randomizes any ordering. Accordingly, the macroscopic magnetization increases with decreasing temperature. This magnetization is the sum of all projections of the magnetic moments of the atoms present in the sample under consideration with respect to the external field. This causes the aforementioned positive contribution to the magnetic susceptibility. Hence, the *Curie law*[81] describes the reciprocal dependence of the molar susceptibility from the absolute temperature

$$\chi_{mol} = \frac{C}{T} \tag{9.10}$$

with the *Curie constant C* yielding a line intersecting the origin. Consequently, respective plots display χ_{mol}^{-1} versus T. With decreasing temperature and increasing field strength, the macroscopic magnetization will increasingly approach a saturation magnetization M_s. Therefore, Equation (9.10) is only applicable for weak fields and not too low temperatures. Moreover, the simple Curie law only holds for sufficiently separated atomic magnets, which do not interact with each other. To consider interactions within paramagnetic materials, Weiss[84] generalized this law by adding the *Weiss constant* or *Curie temperature* Θ_p to the denominator according to the *Curie–Weiss law*

$$\chi_{mol} = \frac{C}{T - \Theta_p} \tag{9.11}$$

where the line intersects the abscissa at $T = \Theta_p$. Given the case that ferromagnetic interactions, where the atomic moments align along the same direction, increase the susceptibility at low temperatures, a positive Θ_p will be found. In case of dominating antiferromagnetic interactions at low temperatures, where the atomic magnets align with opposite orientation, a negative Θ_p will be found. The overall discussion of Curie constants and temperatures is only reliable if atomic moments with a defined electronic configuration and sufficiently thermally isolated ground states contribute.

Looking at transition metal atoms as atomic magnets, these experience interactions with surrounding atoms and ligands. This latter interaction includes ligand-field effects, dubbed $\mathcal{H}_{lf}$, which according to our previous discussion in Chapter 2.1 is in the context of this chapter not relevant for rare-earth ions. This is in strong contrast to that of outer transition metals, which will be treated in Chapter 9.2.3. The main reason for this lies in the electronic structure, where the 4f valence electrons are well shielded by the filled $5s^2p^6$ shell against neighboring atoms. Moreover, the spin-orbit coupling $\mathcal{H}_{so}$ is less relevant for rare-earth ions than the electron-electron interactions $\mathcal{H}_{ee}$. Due to the relatively

---

84 *Pierre-Ernest Weiss*, French physicist (*1865 †1940).

low spin-orbit coupling for rare-earth ions, the Russel–Saunders coupling scheme can be employed to derive the ground state term symbols. Hence, rare-earth ions can in first and often very good approximation be treated as individual atomic magnets based on their respective ground states as listed in Tables 7.1 and 9.2. Langevin[85] derived via Boltzmann statistics the theoretical expression for the Curie constant

$$C = \mu_0 \frac{N_A \mu^2}{3k_B} \tag{9.12}$$

with Avogadro's constant N_A, the squared magnetic moment μ^2 and the Boltzmann constant k_B. Employing Equations (9.11) and (9.12) yield

$$\chi_{mol} = \mu_0 \frac{N_A n_{eff}^2 \mu_B^2}{3k_B} \frac{1}{T - \Theta_p}. \tag{9.13}$$

Here, the effective number of Bohr magnetons

$$\mu = n_{eff} \cdot \mu_B \tag{9.14}$$

is obtained, which will allow for comparison of experimentally obtained values with theoretically expected ones for inner and outer transition metal ions. For the determination of these via μ^2, we need the ground state term symbols of the contributing molecular magnets. The term symbols deliver the quantum numbers L, S and J. Since the coupling strength of electron spins and their angular moments is different in inner, the rare-earth elements, and outer transition metals, the d-block elements, I will focus first on the former, then shortly also on the latter.

9.2.1 Inner Transition Metals

According to theory for the rare-earth atoms, also dubbed inner transition metals, the aforementioned squared magnetic moment μ^2 of Equation (9.12) is obtained via

$$\mu^2 = J(J + 1)g_J^2 \mu_B^2 \tag{9.15}$$

with the Bohr magneton

$$\mu_B = 9.274 \cdot 10^{-24} \, \text{Am}^2 \tag{9.16}$$

and the g_J or Landé factor[86]

---

85 *Paul Langevin*, French physicist (*1872 †1946).
86 *Alfred Landé*, German-American physicist (*1888 †1976).

$$g_J = 1 + \frac{J(J+1) + S(S+1) - L(L+1)}{2J(J+1)} \tag{9.17}$$

considering the different contributions of spin and orbit angular moments to the magnetic moment [365]. The expected multiples of Bohr magnetons accordingly amount to

$$n_B = g_J \sqrt{J(J+1)}, \tag{9.18}$$

which is finally contrasted with experimentally obtained n_{eff} values in Table 9.2. For most of the 4f configurations, both fit very well. Only for the configurations close to a half-filled shell, $4f^5$ and $4f^6$, excited states are in reasonable proximity to be populated thermally also at low temperatures; here, a significant temperature dependence of n_{eff} is common. Figure 7.8 illustrates the situation for Eu^{3+} and the configuration $4f^6$, Figure 8.10 the situation for Sm^{3+}.

Table 9.2: Electronic 4f configurations and the chemically most relevant ions combined with the respective ground terms, their expected and typical experimental values for the number of Bohr magnetons as well as the saturation magnetization Bohr magnetons [365].

$4f^n$ configuration	Ln^{2+}	Ln^{3+}	Ln^{4+}	ground term	n_B	n_{eff}	n_{max}
$4f^0$		La^{3+}	Ce^{4+}	$^1S_0$	0		
$4f^1$		Ce^{3+}	Pr^{4+}	$^2F_{5/2}$	2.54	2.3–2.5	2.14
$4f^2$		Pr^{3+}		$^3H_4$	3.58	3.4–3.6	3.20
$4f^3$		Nd^{3+}		$^4I_{9/2}$	3.62	3.4–3.5	3.27
$4f^4$		Pm^{3+}		$^5I_4$	2.68	2.9	2.40
$4f^5$		Sm^{3+}		$^6H_{5/2}$	0.85	1.6	0.71
$4f^6$	Sm^{2+}	Eu^{3+}		$^7F_0$	0	3.5	0
$4f^7$	Eu^{2+}	Gd^{3+}	Tb^{4+}	$^8S_{7/2}$	7.94	7.7–7.9	7
$4f^8$		Tb^{3+}		$^7F_6$	9.72	9.7–9.8	9
$4f^9$		Dy^{3+}		$^6H_{15/2}$	10.65	10.2–10.6	10
$4f^{10}$		Ho^{3+}		$^5I_8$	10.61	10.3–10.5	10
$4f^{11}$		Er^{3+}		$^4I_{15/2}$	9.58	9.4–9.5	9
$4f^{12}$		Tm^{3+}		$^3H_6$	7.56	7.5	7
$4f^{13}$		Yb^{3+}		$^2F_{7/2}$	4.54	4.5	4
$4f^{14}$	Yb^{2+}	Lu^{3+}		$^1S_0$	0		

For the case of perfect ordering at low temperatures and a sufficiently strong magnetic field, the maximum molar saturation magnetization is given by

$$M_{s,mol} = N_A \cdot g_J J \mu_B = N_A \cdot n_{max} \mu_B \tag{9.19}$$

with

$$n_{max} = g_J J \tag{9.20}$$

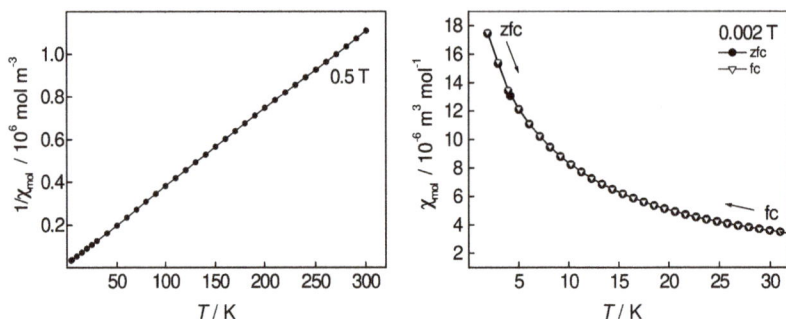

Figure 9.1: Magnetic data of $Gd_3(SiON_3)O$ showing the inverse magnetic susceptibility versus the absolute temperature recorded at a flux density of 0.5 T (left) and the susceptibility recorded at low temperatures and a flux density of 0.002 T in zfc (zero-field cooling) and fc (field-cooling) mode (right), data: [227, 367].

and also listed in Table 9.2 [365]. A typical representation is given in Figure 9.1 for $Gd_3(SiON_3)O$. Here, the analysis according to the Curie–Weiss law yields an experimental $n_{eff} = 7.68(5)$ per trivalent gadolinium ion very close to the theoretical value $n_B = 7.94$. The Curie temperature was determined to $\Theta_p \approx -7\,K$ indicating weak antiferromagnetic interactions. Accordingly, a further measurement with a relatively weak external field was conducted to scrutinize a possible ordering at low temperatures in *zero-field cooling* (zfc) and *field-cooling* mode (Figure 9.1, right). During the former, the sample was cooled without an external magnetic field, in the latter the sample was cooled under application of an external field. In the case of magnetic ordering, both measurements would deliver different results as in the latter the spins are already oriented during cooling and yield a higher magnetization while in the former the spins are oriented randomly in the beginning. If—like here—both curves coincide, no ordering is deduced.

This book focuses on 4f elements. Looking at 5f elements, things become significantly more complicated as here $\mathcal{H}_{ee} \approx \mathcal{H}_{so} \approx \mathcal{H}_{lf}$ and the expected magnetic moments are difficult to estimate.

9.2.2 Van Vleck Paramagnetism

The discussion so far assumes that only the magnetic ground state contributes to the magnetic behavior. But already at low temperatures excited states with differing magnetic moments may also be thermally populated if they are sufficiently close in energy, i. e., within $k_B T$. A view on Figures 7.8 and 8.10 (Chapters 7.8 and 8.4.4) discloses that regarding the trivalent rare-earth ions only Eu^{3+} and Sm^{3+} should show temperature independent paramagnetism. This effect is influenced by state broadenings caused by ligand-field effects, for instance. Accordingly, a temperature independent contribution to a magnetic susceptibility has to be considered in such cases, dubbed *Van Vleck paramagnetism*.[83] Thereby, Equation (9.11) is extended by a temperature-independent term χ_0 according to

$$\chi_{mol} = \frac{C}{T - \Theta_p} + \chi_0 \tag{9.21}$$

The coupling of the contributing terms is described and quantified by a coupling parameter λ, which relates to the energies of the Russel–Saunders terms of

$$E_J = \frac{\lambda}{2}\left(J(J + 1) - L(L + 1) - S(S + 1)\right) \tag{9.22}$$

with a difference between adjacent terms of

$$E_J - E_{J-1} = \lambda J. \tag{9.23}$$

Based on van Vleck's theory with Equation (9.17) and via

$$\chi_J = \frac{N_A g_J^2 \mu_B^2}{3k_B T}(J(J + 1)) + N_A \alpha_J \tag{9.24}$$

with the Avogadro constant N_A and a coupling parameter α_J relating the energies of the J term to its neighbors $J \pm 1$. Finally, the paramagnetic susceptibility of Eu^{3+} was derived as

$$\chi_{Eu^{3+}} = N_A \mu_B^2 \frac{A}{3\lambda Z} \tag{9.25}$$

with temperature dependent factors A and Z [368, 369]; A and Z are functions of the absolute temperature and λ only. The coupling parameter λ should be the larger, the weaker the mixing of adjacent J terms is. The borosulfate $Eu_2[B_2(SO_4)_6]$, mentioned earlier in this book as comprising an anion of weak coordination strength, shows a behavior in excellent agreement with this theory as shown in Figure 9.2. Table 9.3 lists selected val-

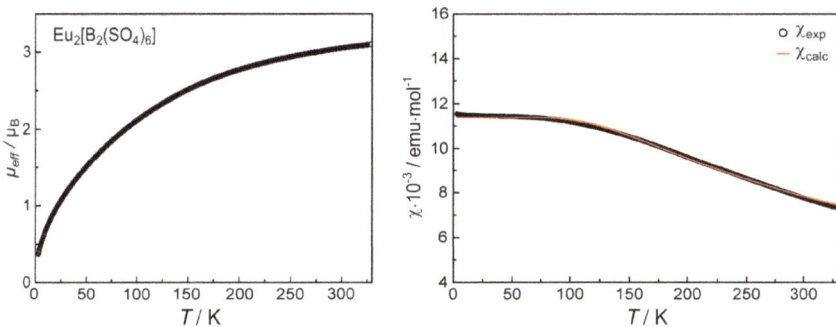

Figure 9.2: Magnetic data of $Eu_2[B_2(SO_4)_6]$ showing the effective magnetic moment via a zfc measurement (left) and the magnetic susceptibility versus the absolute temperature recorded at a flux density of 10 kOe fitted with a red curve based on van Vleck's theory (right), data: [292].

Table 9.3: Electronic 4f configurations and the chemically most relevant ions combined with the respective ground terms, their expected and typical experimental values for the number of Bohr magnetons as well as the saturation magnetization Bohr magnetons, data: [292, 369].

	Eu_2O_3	$Eu(BO_3)$	EuF_3	$Eu_2[B_2(SO_4)_6]$
n_{eff}	3.34	3.43	3.29	3.05
λ / K	460	471	490	589

ues for λ of oxidic compounds. These suggest that indeed with decreasing coordination strength a larger λ is obtained as then the europium ions experience less ligand-field and nephelauxetic effects.

The temperature dependent effective magnetic moment is calculated at not too low temperatures—typically at 300 K—via

$$\mu_{eff} = \sqrt{\frac{3k_B T}{\mu_0 N_A \mu_B^2} \chi_{mol}} \tag{9.26}$$

as this ignores the temperature independent contributions from Equation (9.24).

9.2.3 Outer Transition Metals

Earlier in this chapter, it was stated that d block elements, also dubbed outer transition metals, behave differently regarding ligand-field strength $\mathcal{H}_{lf}$ than inner transition metals. The obvious reason for this is that the relevant valence electrons in the d shell are not shielded against the ligands, which therefore directly interact with the d orbitals. Hence, $\mathcal{H}_{lf}$ competes with the electron-electron interactions $\mathcal{H}_{ee}$ while the spin-orbit coupling $\mathcal{H}_{so}$ is less relevant here, especially for the 3d elements, and the Russel–Saunders coupling scheme can be employed to derive the ground state term symbols. Due to the very weak spin-orbit coupling, the magnetism of such outer transition metals can mainly be treated as pure spin paramagnetism.

In the case of weak ligand-fields, $\mathcal{H}_{ee} > \mathcal{H}_{lf}$, the atoms behave magnetically like the free atoms—like the inner transition metals. Assuming pure spin paramagnetism ($L = 0$ and $J = S$), the g factor simplifies to 2 (Equation (9.17)). Then the squared magnetic moment μ^2 of Equation (9.12) is obtained via

$$\mu^2 = g_J S(S + 1)\mu_B^2 = 2S(S + 1)\mu_B^2 \tag{9.27}$$

with the Bohr magneton μ_B (Equation (9.16)). This expression agrees very well with n_{eff} values (Equation (9.14)) found for configurations $3d^1$ through $3d^5$ and well with the remaining ones due to increasing spin-orbit coupling leading to slightly higher figures. The highest magnetization amounts analogously Equation (9.19) to

$$n_{max} = 2S. \tag{9.28}$$

Thus, the highest possible magnetizations are expected for $3d^5$ configurations like Mn^{2+} with $n_{max} = 5$, which is only half the value expected for the best rare-earth configurations $4f^9$ and $4f^{10}$ according to Table 9.2.

In case of strong ligand-fields, $\mathcal{H}_{ee} < \mathcal{H}_{lf}$, spin-pairing may occur. This results in low-spin configurations and lower magnetic moments. Stronger ligand-fields are observed in case of 3d elements with strong ligands, and in case of 4d and 5d elements in almost all cases as the latter orbitals are comparably diffuse and, therefore, yield strong interaction with any ligand, and consequently, large ligand-field splittings. Moreover, the more diffuse the orbitals under consideration are, the weaker the interelectronic repulsion becomes further fostering spin-pairing.

9.3 Magnetic Ordering

Below sufficiently low temperatures, the atomic magnets may interact with their neighbors via *exchange interactions*. These may occur by direct interaction of the respective orbitals comprising unpaired electrons or by indirect interaction either via bridging ligands called *superexchange* (Figure 9.3). Alternatively, exchange interactions are achieved via conduction electrons, which are those close to the Fermi level in metals. With exception of the latter, which will be discussed in Chapter 9.3.2, such exchange interactions are restricted to small distances, typically up to the next but one neighbors. Below a *critical ordering temperature* T_c adjacent spins strive to align parallel or antiparallel if possible generating a long-range ordered magnetic structure. The exchange interaction may be ferromagnetic, ferrimagnetic or antiferromagnetic. Among these, the practically most important are ferromagnetism and ferrimagnetism. Experimentally, this phase change may be elucidated not only via magnetic measurements but also via neutron diffraction or heat capacity measurements. In Figure 9.3, the antiferromagnetic coupling of two gadolinium atoms occurs via an occupied p orbital of a bridging nitrogen atom, a situation found in the *rocksalt* structure of GdN displayed in Figure 6.20. For the quantitative description of such exchange interactions models have been developed.

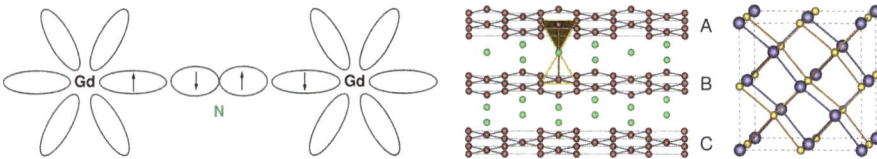

Figure 9.3: Indirect antiferromagnetic coupling of two gadolinium atoms via a nitrogen atom (left) and representation of the $MgCu_2$ type structure; illustration of the deficient close packed layers of dark red Cu atoms A, B and C with green Cu atoms in octahedral voids yielding yellow Cu_4 tetrahedra (center); the diamond-like arrangement of the yellow barycenters of these tetrahedra interpenetrates with the diamond-like network of blue Mg atoms (right).

The *Heisenberg model*[87] can be employed if the considered material is nonmetallic, i. e., the magnetism is well localized, and if the Russel–Saunders coupling scheme is applicable for description of the electronic states. The herein discussed rare-earth elements as well as the elements included in relevant examples like main-group and 3d elements fulfill at least the latter condition. Furthermore, the valence shell configurations of participating inner and outer transition metals should be close to half-occupied like Eu^{2+} or Gd^{3+}, which is the case for the discussion in Chapter 9.5 on GdN and the europium chalcogenides. For a more thorough derivation and discussion of the Heisenberg model, refer to more specialized literature. The relevant details regarding the purpose in this book are the Heisenberg exchange parameters J_1 and J_2, which will be obtained from a Heisenberg Hamiltonian

$$\mathcal{H} = -\frac{1}{2} \sum_i \mathbf{S}_i \left(J_1 \sum_j^n \mathbf{S}_j + J_2 \sum_j^{nn} \mathbf{S}_j \right) \tag{9.29}$$

with the normalized spin vectors $\mathbf{S}_i$ and $\mathbf{S}_j$, considering all atoms i and their neighbors j, and where the summations include the nearest (n) and next nearest neighbors (nn) [370]. Accordingly, the parameters J_1 and J_2 describe the exchange interactions as ferromagnetic if positive and antiferromagnetic if negative—in other words, a high-spin and a low-spin configuration.

> **i**   **Important crystal structure types no. 4 – MgCu$_2$:** The crystal structure of the $MgCu_2$ type, a *Laves phase*,[88] is illustrated in Figure 9.3. It consists of deficient closely packed layers of copper atoms (dark red), in which every fourth atom is removed; these so-called Kagomé nets are packed in a cubic close packed manner indicated as A, B and C in Figure 9.3. In a quarter of the generated octahedral voids, further copper atoms (bright green) are situated. Thus, Cu_4 tetrahedra are generated. The barycenters of these tetrahedra form a diamond-like network, which interpenetrates with a diamond-like network of Mg atoms. Hereby, the magnesium atoms are surrounded tetrahedrally by four magnesium and twelve copper atoms; the copper atoms are coordinated by six copper and six magnesium atoms. This structure type is usually adopted by intermetallic compounds where the atomic radii of both elements are sufficiently different. For the basics of this structure related to others, see also Appendix D.

9.3.1 Ferromagnetism and Antiferromagnetism

Even without an external magnetic field, exchange interactions between adjacent magnetic dipoles yield domains of as parallel as possible aligned dipoles in ferromagnets, so-called *magnetic domains*—in German literature also *Weiss domains*[84] is used. The size of these naturally forming domains is limited because each of these domains generates their own strong magnetic field and—given there is only a single domain—spreads

---

87 *Werner Heisenberg*, German physicist (*1901 †1976).

88 *Fritz Laves*, German mineralogist and crystallographer (*1906 †1978).

outside itself. This is energetically not favored due to magnetostatic reasons; accordingly, large domains split into smaller ones with differing magnetic orientation until a resulting magnetic field outside the material under consideration vanishes. Such domains have a typical size in the order of 10^{-6} to 10^{-4} μm [371].

Below the *Curie temperature* T_C adjacent magnetic domains seek a parallel alignment with the external field. In ferromagnetically ordering rare-earth compounds, the saturation magnetization M_s is closely related to the respective $4f^n$ configuration, regardless of the material being metallic, semiconducting intermetallic or electronically isolating. Hereby, a decreasing total spin S correlates with a decreasing T_C. For instance, the magnetic data of selected RIr_2 listed in Table 9.4 show this dependence. For iridium, no relevant magnetic moment can be estimated; they all crystallize in the $MgCu_2$ type structure and can therefore be compared easily.

Table 9.4: Ordering temperatures T_C/K of selected rare-earth iridium compounds RIr_2 with respect to the total spin S of the rare-earth atoms; data: [372].

R	Sm	Eu	Gd	Tb	Dy	Ho	Er
$4f^n$	5	6	7	8	9	10	11
S	2.5	3	3.5	3	2.5	2	1.5
T_C/K	36	70	88	43	22	12	3

Ferromagnetic ordering is identified by a strong increase of the magnetic susceptibility around T_C as shown in Figure 9.4 on the example of $Eu_2[Si_5N_8]$, which turns ferromagnetic around 13 K. In the field cooling mode, the steeply increasing ordering can be well seen, and the zero field cooling curve shows strong ordering with increasing temperature until 13 K. The inverse magnetic susceptibility develops like a normal paramagnet above 50 K and turns close to zero at 13 K. The magnetization isotherm approaches the expected saturation magnetization of approx. $7\,\mu_B$ at high magnetic fields and a temperature below $T_C = 13$ K; above T_C the magnetization is much weaker as here

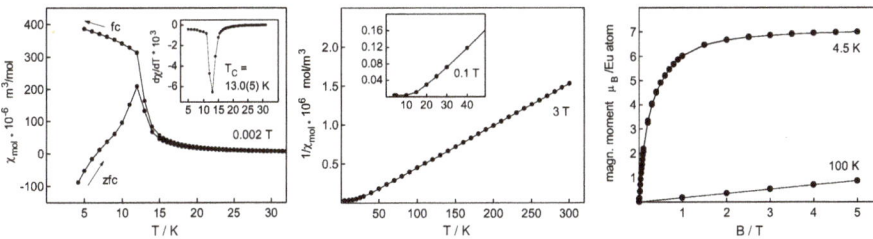

Figure 9.4: Magnetic data of $Eu_2[Si_5N_8]$ showing the magnetic susceptibility vs. the absolute temperature T in zfc (zero-field cooling) and fc (field-cooling) mode with an inset illustrating the derivative to determine T_C (left), the inverse magnetic susceptibility vs. T recorded at a flux density of 0.1 and 3 T (middle) and the field dependent magnetization below and above T_C (right); data: [227, 373].

no long-range ordering can be achieved and the structure of small domains is restored immediately.

After an initial magnetization in ferromagnets magnetic ordering remains even when the external magnetic field is switched off and as long as $T < T_C$. This is illustrated by *hysteresis loops* like those shown in Figure 9.5, where the response of the magnetic induction **B** of the ferromagnet to a local magnetic field **H** is displayed. The initial magnetization curve is shown as a broken line with increasing field strength until the *saturation magnetization* M_s is achieved (point S). Here, all domains are aligned parallel to the external field. Moreover, domain boundaries move or are even destroyed. After switching off the external magnetic field, the aforementioned magnetostatic reasons reduce the parallel alignment to the *remanent magnetization*, and the *remanence* (R) persists. Crystal defects and grain boundaries prevent further demagnetization and stabilize a permanent magnetization. To actually demagnetize the ferromagnet, an external field corresponding to the *coercive field* H_c or *coercivity* (−C) has to be applied in the opposite direction. By further increasing the external field in the opposite direction again saturation is achieved (−S) after switching off the field again a remanence (−R) persists; a further magnetization then closes the loop ending again at point S. The area the hysteresis loop encloses corresponds to an energy loss per volume, i. e., the product of flux density and magnetic field according to Table 9.1. This energy matches the energy needed to perform a single loop. Thus, the left loop in Figure 9.5 represents that of a *soft ferromagnet*, the right that of a *hard ferromagnet*. Soft magnets are rather easily magnetized, demagnetized and magnetically switched. Hard magnets normally need high fields for magnetization, are harder to switch or to demagnetize and are therefore used as permanent magnets. Moreover, the *maximum energy product* BH_{max} is equivalent to the area of the largest rectangle that can be inscribed under the hysteresis loop. This is obviously clearly larger for hard ferromagnets. It is proportional to the magnetostatic energy stored by the magnet. For these materials combining high Curie temperatures

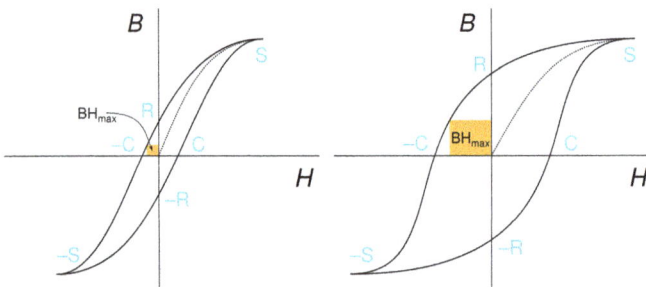

Figure 9.5: Hysteresis loops of a soft (left) and a hard ferromagnet (right) with the magnetic field **H** vs. the magnetic induction **B** and the relevant points saturation magnetization (S, −S), remanence (R, −R), coercivity (C) as well as the initial magnetization curve (broken line) and the maximum energy product BH_{max}.

of 3d metals and the large and stable magnetic moments of lanthanide elements proves advantageous like in $SmCo_5$, Sm_2Co_{17}, Sm_2Fe_{17} or $Nd_2Fe_{14}B$.

Most ferromagnets are metallic, but there are also semiconductors or isolators like the aforementioned $Eu_2[Si_5N_8]$ or the chalcogenides EuO and EuS, which will be treated in Chapter 9.5 since the heavier ones order antiferromagnetically. The pure existence of semiconducting or isolating ferromagnets was heavily disputed up to the 1960s.

9.3.2 Magnetism of Metallic 4f Systems

The magnetism of outer transition metals is usually described by the effect of *spin polarization*. Thereby, the electronic states in the band structure of the considered material deviate in energy regarding the spin orientation of the contained electrons. Accordingly, the band structure splits into a spin-up and a spin-down set of bands below the respective ordering temperature T_c. In this context also, the band gap should decrease at T_c.

For broad bands, such as s or p bands the spin-polarized split bands of the itinerant electrons are energetically shifted by applying an external magnetic field with slightly higher antiparallel (low-spin) and somewhat lower parallel spin alignment (high-spin) as shown in Figure 9.6. Thus, below the Fermi level empty states in the parallel spin band result, which are filled by spin-flipped electrons from the antiparallel band. Hence, a small surplus quantity of parallel spins is generated yielding paramagnetic behavior known as Pauli paramagnetism of itinerant electrons, present for all metals. Broad bands mean high delocalization and, therefore, this *jellium* or *electron gas* approach

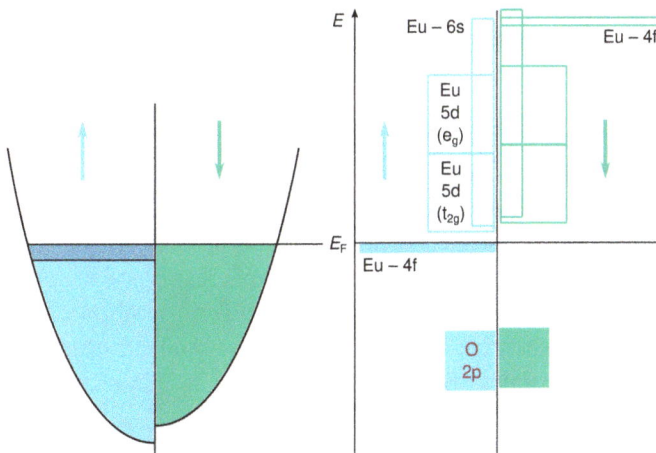

Figure 9.6: The darker blue regime illustrates the surplus spins yielding a blue majority spin half-band with parallel spin alignment and a green minority spin half-band (left); schematic electronic structure of EuO with spin-polarized half-bands, valence bands show a colored filling while conduction bands only show a colored edge (right).

works very well for broad bands. Within this concept, magnetic ordering of transition metals cannot be explained.

To understand magnetic ordering of 3d elements under these circumstances, it has to be considered that 3d bands are significantly narrower than s or p bands, i. e., the electrons therein are significantly better localized. A closer look on the band structure of iron reveals that in a nonmagnetic calculation antibonding states around the Fermi level would be occupied. In such cases, nature usually strives for a distortion of the structure. In iron, the electronic structure instead of the crystal structure can distort yielding spin-polarized half-bands, and thus minimizing antibonding interactions around the Fermi level if shifted with respect to each other. Figure 9.6 shows this understanding on the example of EuO, treated below. Then, in the same manner as described for the electron gas above, some itinerant electrons swap their spins populating the available states in the stabilized half-band below and depopulate states of the destabilized half-band above the Fermi level. Due to this local spin polarization, the surplus spin electrons are shielded worse from the nuclear charges of the adjacent atoms—since the number of electrons with opposite spin decreased upon spin polarization and, accordingly, this half-band is energetically stabilized, which is in line with the driving force of this distortion. Finally, this leads to spontaneous local magnetic ordering within magnetic domains as described in Chapter 9.3.1. Eventually, these magnetic domains based on the spin-polarized bands yield bulk magnetic ordering.

Transitioning to the lanthanides, the 4f bands are even narrower than those of 3d elements, thus even better localized. Accordingly, contrary to the outer transition metals, the magnetism of metallic 4f systems not primarily stems from itinerant conduction electrons but from the highly localized 4f states. The magnetic ordering in rare-earth metals as well as their metallic derivatives occurs normally via exchange interactions based on itinerant electrons. So, the magnetism of outer transition metals is *based on* itinerant electrons while that of inner transition metals is *mediated by* itinerant electrons.

The magnetic behavior of conducting rare-earth materials is best understood by the *RKKY model* based on groundbreaking contributions of *Ruderman*,[89] *Kasuya*,[90] *Kittel*[91] and *Yosida*[92] [374–376]. Due to their specific electronic structure lanthanide atoms, where the 4f states are well shielded by a filled $5s^2p^6$ shell, exhibit pronounced magnetic moments based on their 4f configuration even in metallic systems. These local moments polarize the spin densities of the conduction electrons around the Fermi level via a direct exchange. Accordingly, the spin density of parallel spins is increased and that of antiparallel spins is decreased around a magnetically active lanthanide atom. This polarization

89 *Malvin Ruderman*, American physicist and astrophysicist (*1927).

90 *Tadao Kasuya*, Japanese physicist (*1927).

91 *Charles Kittel*, American physicist (*1916 †2019).

92 *Kei Yosida*, Japanese physicist (*1922).

affects the neighboring lanthanide atoms and leads to an indirect coupling. Since this polarization is not restricted to the close neighborhood, but of considerable scope and, moreover, oscillating, it induces ferromagnetic or antiferromagnetic coupling with adjacent lanthanide atoms depending on the distances between the magnetic centers. The oscillation wavelength depends from the wavelength of the itinerant electrons around the Fermi level.

9.4 Magnetic Behavior of the Lanthanide metals

The magnetic data of the lanthanide metals are summarized in Table 9.5. Most of the data were taken from a review in the famous Handbook on the Physics and Chemistry of Rare-Earths, which nicely summarizes the crucial data [377]. Many details of the magnetic properties are—in parts—under severe discussion as the precise description of 4f electrons by means of theoretical calculations is still challenging. So, I will focus in this chapter on the basic magnetic properties relevant for most of their applications and the deeper understanding of lanthanide elements as such.

Table 9.5: Metals with the core and number of itinerant electrons, $4f^n$ configuration, the expected maximum saturation magnetization in Bohr magnetons n_{max} (Equation (9.20)) from the 4f configuration, the observed saturation moment n_{obs}, ordering temperatures T_N and T_C in Kelvins and (anisotropic) Curie temperatures Θ_p / K.

metal	$4f^n$	ground term	n_{max}	n_{obs}	T_N	T_C	$\Theta_{p,\|c}$	$\Theta_{p,\perp c}$
$La^{3+} \cdot 3\,e^-$	$4f^0$	$^1S_0$			–	–		
$Ce^{3+} \cdot 3\,e^-/Ce^{4+} \cdot 4\,e^-$	$4f^1/4f^0$	$^2F_{5/2}/^1S_0$			14	–		
$Pr^{3+} \cdot 3\,e^-$	$4f^2$	$^3H_4$	3.20	2.7	0.03	–		
$Nd^{3+} \cdot 3\,e^-$	$4f^3$	$^4I_{9/2}$	3.27	2.2	19, 8	–		
$Sm^{3+} \cdot 3\,e^-$	$4f^5$	$^6H_{5/2}$	0.71	0.5	106, 14	–		
$Eu^{2+} \cdot 2\,e^-$	$4f^7$	$^8S_{7/2}$	7	5.1	91	–	18	18
$Gd^{3+} \cdot 3\,e^-$	$4f^7$	$^8S_{7/2}$	7	7.63	–	293	317	317
$Tb^{3+} \cdot 3\,e^-$	$4f^8$	$^7F_6$	9	9.33	230	220	195	239
$Dy^{3+} \cdot 3\,e^-$	$4f^9$	$^6H_{15/2}$	10	10.2	178	86	121	169
$Ho^{3+} \cdot 3\,e^-$	$4f^{10}$	$^5I_8$	10	10.3	133	19	73	88
$Er^{3+} \cdot 3\,e^-$	$4f^{11}$	$^4I_{15/2}$	9	9.0	84, 52	18	62	33
$Tm^{3+} \cdot 3\,e^-$	$4f^{12}$	$^3H_6$	7	7.12	56	32	41	-17
$Yb^{2+} \cdot 2\,e^-/Yb^{3+} \cdot 3\,e^-$	$4f^{14}/4f^{13}$	$^1S_0/^2F_{7/2}$			–	–		
$Lu^{3+} \cdot 3\,e^-$	$4f^{14}$	$^1S_0$			–	–		

The discussion will address six main groups of the rare-earth metals. Further, consider also Chapter 3.3 discussing the temperature dependent phases of the rare-earth metals. The first series starts with *terbium, dysprosium and holmium*. Since the exchange interaction is believed to be mediated by the conduction electrons it seems reasonable that also a certain portion of these itinerant electrons become polarized. Accordingly, the observed magnetization n_{obs} can be larger than the expected n_{max} based purely on the unpaired 4f electrons. For instance, for *terbium* a slightly higher value of 9.33 was observed. Terbium orders antiferromagnetically at 230 K with a spiral structure of the spins and then ferromagnetically at 220 K. The antiferromagnetic ordering can also be turned into a ferromagnetic one by applying a stronger field of 1 T. Terbium metal exhibits a magnetic anisotropy, which can be seen well by looking at the different paramagnetic Curie temperatures of 195 and 239 K with parallel and perpendicular orientation with respect to the crystallographic c-axis. This confirms the predominant ferromagnetic interactions, which are conveniently achieved along the easy direction parallel the *b* axis. Above T_N terbium shows a simple Curie–Weiss behavior with a magnetic moment close to the theoretically expected value (Table 9.2). Also, *dysprosium* adopts a spiral spin structure in the regime of antiferromagnetic ordering between 178 and 86 K, where it eventually orders ferromagnetically. Apparently, here the antiferromagnetic ordering is more resilient. The easy axis here is the *a* axis. The phase transition at T_C is apparently first order including a discontinuous development of the lattice parameters leading also to a structural phase transition as outlined in Chapter 3.3. As observed for terbium, also here a certain portion of conduction electrons become polarized enhancing the total magnetization above the pure value expected for the Dy^{3+} core. *Holmium* displays a similar behavior like the aforementioned metals. Below 133 K, it orders antiferromagnetically with a spiral spin structure similar to the other two, switching to a more fan-like arrangement of the spins and at 19 K it finally turns ferromagnetic. In all three cases, the spins in the ferromagnetic phase are confined more or less to the basal plane with the *b* axis being the easy one. The difference between $\Theta_{p,\|c}$ and $\Theta_{p,\perp c}$ is the smallest in this series indicative for a decreasing uniaxial magnetic anisotropy.

The second block of similarly behaving *erbium and thulium* starts with erbium for which an antiferromagnetic ordering was detected below 84 K. This ordering coincides with an incommensurate sinusoidally modulated arrangement of the magnetic moments along the *c* axis; the wave vector changes smoothly toward the lower ordering temperature. Below 52 K, the previous ordering is superimposed by a second one, which is spiral parallel to the basal plane. At 18 K also erbium orders ferromagnetically, the easy axis is the *c* axis here. In summary, erbium features the most complicated and puzzling magnetic properties of the four so far considered metals. In *thulium* below 56 K, the same antiferromagnetic ordering along the easy *c* axis is found, but in contrast to the former no further ordering in the basal plane follows below. Instead, the modulation changes yielding a ferrimagnetic transition around 32 K where four layers of moments align parallel and three subsequent layers antiparallel the *c* axis. Under the influence of higher fields, this ferrimagnetic phase turns ferromagnetic.

As third group follow the magnetic properties of *gadolinium and europium*. Due to its maximum number of unpaired electrons, *gadolinium* features the highest Curie temperature of the lanthanide elements. In contrast to the aforementioned metals, no orbital angular moments contribute, the magnetism is apparently purely driven by spin. Below 293 K, it orders ferromagnetically; also, gadolinium exhibits weak magnetic anisotropy, with the c axis being easy, despite the absence of orbital angular moments. Approaching low temperatures, the easy axis slightly tilts toward the basal plane. Really pure gadolinium is a rather soft ferromagnet—for a definition, see Figure 9.5. Above 400 K, it shows Curie–Weiss behavior and an effective magnetic moment of 7.98 Bohr magnetons per atom, very close to the theoretical value (Table 9.2). As discussed previously, *europium* strives for a half-filled 4f shell, and hence also in the metallic state this is achieved. Therefore, its core is isoelectronic with that of gadolinium. The main difference between both are the atomic radii where europium atoms are significantly larger than gadolinium atoms and, of course, the lower charge of the core (Figure 3.1). Consequently, according to Chapter 8.3 the 5d bands contribute stronger in the europium case. Due to the larger size and smaller charge the density of states is lower at the Fermi level for europium. This reduces the indirect exchange interaction with neighboring europium atoms, and thus also the ordering temperature. Below 91 K, a first-order transition occurs yielding antiferromagnetic ordering with a spiral spin structure along the a axis of the body-centered cubic unit cell. Moreover, also the Curie temperature is clearly lower and no ferromagnetic ordering is achieved without tremendously strong fields. Due to europium's high oxygen affinity during any magnetic measurements of europium compounds a possible contamination with EuO (Chapter 9.5) has to be considered carefully, which alters the magnetic properties dramatically around its T_C = 69 K. At lower temperatures, EuH_2 (T_C = 18 K) might produce strange results.

The fourth series covers the light elements *praseodymium, neodymium and samarium*. As outlined in Chapter 3.3, *praseodymium and neodymium* adopt the *hc* structure in the temperature range interesting here. Accordingly, there are atoms on c (A) and h (B and C) layers, respectively. Hence, the local electronic structures differ substantially like the splitting of the J sublevels. Moreover, for understanding the magnetic properties also thermally populated sublevels are relevant leading to the clearly lower oberserved saturation moments n_{obs} of 2.7 (praseodymium) and 2.2 (neodymium) than expected for the 4f configurations (Table 9.5). Magnetic ordering in *praseodymium* at low temperatures has been disputed for decades, but at temperatures far below 1 K antiferromagnetic ordering is assumed. For *neodymium* an antiferromagnetic ordering on the hexagonal sites below 19 K and on the cubic sites below 8 K was reported. *Samarium* also adopts a structure, *hhc*, where c and h layers alternate (Chapter 3.3). The atoms on hexagonal sites order antiferromagnetically below 106 K with the moments being aligned along the c axis. Thereby, two ferromagnetic layers with a spin-up configuration alternate with two spin-down layers. Below 14 K also the atoms on the cubic sites feature an antiferromagnetic ordering in the same manner of pairwise ferromagnetic layers. Similar to Eu^{3+}, the magnetic susceptibilities are influenced by admixtures of excited states, such as $^6H_{7/2}$, and bear resemblance to Van Vleck paramagnetism (Chapter 9.2.2).

Cerium and ytterbium form the next group of similar elements. The magnetic behavior of *cerium* was already briefly discussed in Chapter 3.4. Similar to cerium, *ytterbium* also oscillates between two electron configurations, a diamagnetic and a paramagnetic. Therefore, it is not surprising that a structural phase transition with wide hysteresis is also observed in ytterbium. Here, between 100 and 360 K the structure glides with increasing temperature from the hexagonal close to the cubic close packing (Chapter 3.3). In the case of cerium, the tetravalent state is diamagnetic, while in ytterbium, it is the divalent state. During this phase transition, apparently, a portion of the ytterbium atoms shifts from the divalent to the paramagnetic trivalent state. Hence, the cubic close packing shows a temperature dependent paramagnetism. So, cerium is not only the spectroscopic, but also the magnetic twin of ytterbium.

The final group covers the elements *lanthanum and lutetium* as well as the rare-earth elements *scandium and yttrium*. All of them lack a partially filled 4f shell. So their magnetism shows contributions of itinerant electrons and diamagnetic susceptibilities of the core electrons. Yttrium, lanthanum and lutetium show typical diamagnetic behavior. Therefore, their susceptibilities are very sensitive to impurities like lanthanide atoms. For *scandium* the largest susceptibility within this group was determined, followed by yttrium and the remaining two. Scandium exhibits paramagnetism in agreement with the Curie-Weiss law with an effective moment n_{eff} of 1.65 μ_B per atom; this aligns with a quite localized single 3d electron, in accordance with the discussion in Chapter 2.1 on the special situation of orbitals without radial nodes. The paramagnetic Curie temperature of −850 K suggests antiferromagnetic interactions.

9.5 The Europium Chalcogenides, Antiferromagnetism and Ferrimagnetism

Magnetic ordering where adjacent magnetic moments of the same magnitude do not align parallel but antiparallel occurs in antiferromagnets. If the magnitudes of the opposing magnetic moments are different, a net magnetic moment results; such ordering is named *ferrimagnetic*. The ordering temperature is named here *Néel temperature* T_N,[93] below of which the magnetic moment vanishes completely or almost.

A prominent example are the europium chalcogenides EuX (X = Se, Te), which in contrast to their oxygen and sulfur counterparts show antiferromagnetic and ferrimagnetic ordering while the former order ferromagnetically [378, 381]. Accordingly, the europium chalcogenides are a suited series of compounds to discuss the underlying ordering mechanism. Moreover, all four representatives adopt a simple crystal structure, the rocksalt type. Magnetic measurements confirm that divalent europium is present in line with a composition $Eu^{2+}X^{2-}$. Gadolinium nitride, GdN, can be also be considered in this discussion as both rare-earth ions, Gd^{3+} and Eu^{2+}, are isoelectronic and GdN is

---

93 *Louis Néel*, French physicist (*1904 †2000).

isostructural with EuX. It should be noted that it is challenging to obtain pure GdN without contaminations of oxygen, which strongly influences the electronic structure and certainly also the magnetic properties.

On the example of EuO, Figure 9.6 depicts both sets of spin-polarized half-bands. Locally, an exchange splitting of 0.6 eV into spin-polarized half-bands occurs with the blue bands representing a parallel spin alignment with respect to neighboring europium ions—the majority spins—and the green bands representing antiparallel minority spins. The 4f states of europium experience the most prominent exchange splitting of ca. 10 eV. In the blue set, the 4f half-band, highly localized and, therefore, showing a very small band dispersion, form the top of the valence band at the Fermi level E_F. The bottom of the conduction band comprises 5d and 6s bands of Eu^{2+}; the latter features a very broad dispersion, the former splits according to the ligand-field splitting in an octahedral environment yielding a t_{2g} and a e_g band. An important result of the exchange splitting of the half-bands is that upon long-range magnetic ordering the band gap decreases. In the case of EuO, it then at least approaches a semiconductor-to-metal transition; for GdN, there is an ongoing discussion whether or not it turns metallic below T_C. Figure 9.7 illustrates the schematic electronic structure of the europium chalcogenides and gadolinium nitride. Optical reflection measurements confirm that all chalcogenides are semiconductors above the respective ordering temperature T_c [378]. Crucial for the understanding of the magnetic behavior are only the bands around the Fermi level. Hence, in Figure 9.7 for further comparison only the blue majority spin half-bands are shown. Because GdN shows a quite comparable magnetic behavior with EuO it is included here.

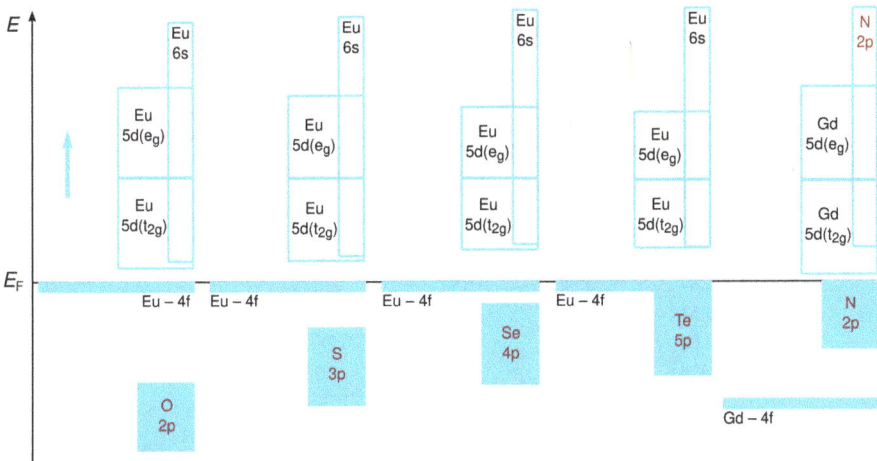

Figure 9.7: Schematic representation of the relevant spin-polarized bands in EuO, EuS, EuSe, EuTe and GdN (from left) only showing the majority spin states indicated in blue; valence bands show a colored filling while conduction bands only show a colored edge.

EuO orders ferromagnetically at $T_C = 69\,\mathrm{K}$. Its magnetic moment amounts to $n_{eff} = 6.94$, close to the theoretically expected value of divalent europium of $n_B = 7.94$ (Table 9.2). EuS exhibits essentially the same behavior, just at lower temperatures of $T_C = 17\,\mathrm{K}$. A direct exchange interaction of the $4f^7$ shell of neighboring europium ions is not reasonable due to the high localization and excellent screening by the filled $5s^2p^6$ shell; instead, the ferromagnetic coupling occurs via the energetically close 5d bands, which extend over longer ranges in the structure and enable for sure such a cation-cation superexchange between adjacent europium ions as soon as an electron is temporarily excited from the local 4f states to the 5d band, which then is polarized by the magnetic moment. In the row oxygen, sulfur, selenium and tellurium the electronegativity decreases and the size of the X^{2-} anions increases, in line with increasing lattice parameters (Table 9.6). Therefore, the ligand-field splitting—based on electrostatic interactions—decreases along this row, the nephelauxetic effect increases slightly, though. In summary, the oxide features the smallest band gap, which then increases until the telluride. With the increasing band gap, the ferromagnetic coupling path via the 5d states is weakened in the same series. The ferromagnetic ordering in GdN is believed to work in a similar way although here the 4f states are localized at much lower energies well below the Fermi level; here, the $4f^7$ valence electrons again spin-polarize the itinerant electrons in the 5d band, which then spin-polarize the neighboring gadolinium ions. Moreover, an antiferromagnetic superexchange via 2p bands seemingly does not play a role as the gap between these and the 4f states is too large [382]. As mentioned earlier in Chapters 8.3 and 8.4.1, the higher charge of Gd^{3+} compared with Eu^{2+} yields a strikingly different electronic structure.

Table 9.6: Band gap at room temperature, Curie temperature Θ_p, ordering temperatures T_C and T_N in Kelvins, magnetic moment n_{eff} in Bohr magnetons, unit cell lattice parameters and averaged experimental Heisenberg exchange coupling constants for nearest, J_1, and next nearest neighbors, J_2, in meV; data: [365, 370, 378–380].

	GdN	EuO	EuS	EuSe	EuTe
band gap / eV	0.98	1.12	1.65	1.80	2.00
Θ_p/K	81	79	18	9	−4
T_C/K	58	69	17		
T_N/K				4.6, 2.8, 1.8	9
n_{eff}	6.84	6.94	6.97		
J_1	+0.54	+1.54	+0.56	+0.33	+0.10
J_2	0.00	−0.28	−0.33	−0.26	−0.58
a / pm	498	514	597	620	660

Due to the decreasing electronegativity, the chalcogenide p bands and the 4f states of europium converge within the row EuO, EuS, EuSe and EuTe. Thus, in EuSe and EuTe significant mixing of these 4f states with 4p and 5p states of selenide and telluride occurs [370, 380, 383]. This mixing enables an increasing antiferromagnetic coupling of neighboring Eu^{2+} ions via a superexchange mediated by the chalcogenide anions. This is nicely reflected in the relative contribution of the coupling constants J_1 and J_2 listed in Table 9.6.

The Heisenberg exchange constant J_1 relates to the ferromagnetic interaction of adjacent europium ions via the 5d band as discussed above while J_2 relates to the antiferromagnetic superexchange via the chalcogenides' p states [384]. In the oxide and sulphide J_1 dominates, for the telluride J_2 prevails over J_1. Following these data, for EuO and EuS ferromagnetic and for EuTe antiferromagnetic ordering is expected. Indeed, for EuTe antiferromagnetic ordering below 9 K was found. Neutron scattering was employed to determine the magnetic structure. These investigations showed the presence of ferromagnetically ordered layers perpendicular the threefold axis in the cubic unit cell, which are pairwise antiferromagnetically coupled.

Regarding the selenide, the Heisenberg constants do not show a clear preference for either ordering mechanism, and accordingly EuSe shows a complicated magnetic behavior [385]. It turns antiferromagnetic below 5 K (antiferromagnetic phase 1), then below approximately 3 K ferrimagnetic and below 2 K again antiferromagnetic (antiferromagnetic phase 2). With increasing external field the antiferromagnetic phases become ferrimagnetic and eventually even ferromagnetic. Such behavior, where the magnetic ordering is also field dependent, is called *metamagnetic*. In the case of EuSe, this behavior can be expected as the Curie temperature is positive, and ferromagnetic interactions dominate above T_N.

9.6 Rare-Earth Ferromagnets—Often Really Hard and Permanent

Permanent magnets are employed in a manifold of applications such as wind generators, car motors, computer hard-disk drives, loudspeakers and microphones. Applications like electric vehicle traction in cars or trains and aerospace thrusters require stabilities of permanent magnets of up to 200 and 500 °C, respectively [386, 387]. Further applications of ferromagnets are *multiferroic materials* or as *magnetocaloric materials*, the latter will be introduced in the following chapter [388, 389]. Essentially, permanent magnets are hard ferromagnets. Preconditions for good permanent magnets are a high saturation magnetization, a high Curie temperature and a high remanence of the magnetization fostered by a high magnetic anisotropy.

The saturation magnetization M_s and the Curie temperature T_C of most permanent magnets are mainly determined by the magnetically active outer transition metal atoms. Such are especially late ones like iron or cobalt. Their magnetization derives from the spin S of the individual atoms. The interactions between these individual magnetic centers are strong enough to yield quite high ordering temperatures T_C allowing for employing such materials at high temperatures. Iron and cobalt order ferromagnetically at 1044 and 1388 K with saturation magnetizations of 2.2 and 1.7 μ_B, respectively.

The third precondition, the high *magnetic* or *magnetocrystalline anisotropy* is mainly provided by the rare-earth sublattice. Due to their relativistic spin-orbit coupling, the anisotropy is clearly higher for heavy elements such as rare-earth elements. Simply spoken, the anisotropic electrostatic crystal field modifies the orbital motion of the electrons. Via the relativistic spin-orbit coupling, the electrostatic interaction between the rare-earth atoms and the local ligand field yields an isotropy energy, which inhibits easy spin conversion. Accordingly, good permanent magnetic materials feature uniaxial crystal structures like tetragonal, rhombohedral or hexagonal systems with a preferential or principal axis along which the magnetization is easier to achieve compared to other *hard* directions, i. e., the *easy axis*. This is the case for all materials described in this chapter. The *anisotropy energy* E_{anis} per volume scales with the *anisotropy coefficient* K_1 via

$$E_{anis} = K_1 \cdot (\sin \theta)^2 \tag{9.30}$$

with the angle θ quantifying the deviation from the easy axis. K_1 depends from temperature and the composition of the considered material. According to

$$H_c = \alpha \frac{2K_1}{\mu_0 M_s} - \beta M_s \tag{9.31}$$

a high *coercivity field* H_c results, which is congruent with a higher remanence; here, the two empirical figures α and β may be determined from the temperature dependence of the materials parameters [390]. For a competitive hard ferromagnet to be suited as permanent magnet, the dimensionless *magnetic hardness parameter*

$$\kappa = \sqrt{\frac{K_1}{\mu_0 M_s^2}} \tag{9.32}$$

should exceed one. This empirical rule of thumb gives the estimation that a ferromagnetic material withstands self-demagnetization in any possible shape. Finally, the maximum energy product BH_{max} within the hysteresis loop (Figure 9.5) is proportional to the storable *magnetostatic energy* by the magnet per volume. It depends strongly from the shape and the synthesis of a specific magnetic material. An overview of the respective data for relevant materials is given in Table 9.7. *Magnetite*, i. e., Fe_3O_4, was the first permanent magnet in mankind and was described already around 600 BC by Thales of Miletus. It is therefore added for comparison purposes to this table. For further comparison, $SrFe_{12}O_{19}$, one of the most competitive materials not containing rare-earth elements is given.

Looking at the data in Table 9.7, the synergy of 3d outer and 4f inner transition metals is apparent. Thus, it is not really surprising that these mixed magnets dominate the market. Regarding the choice of rare-earth elements both samarium and neodymium

Table 9.7: Magnetic properties of selected commercial permanent magnets; data: [365, 386, 387].

composition	T_C / K	$\mu_0 M_s$ (r. t.) / T	K_1 (r. t.) / MJm^{-3}	κ	$\mu_0 M_r$ / T	$\mu_0 H_c$ / T	BH$_{max}$ / kJm^{-3}
SmCo$_5$	1003	1.07	17.2	4.4	0.88	2.1	150
Sm$_2$Co$_{17}$	1190	1.20	3.3	1.9	1.08	1.4	220
Nd$_2$Fe$_{14}$B	588	1.61	4.3	1.5	1.28	1.3	516
Sm$_2$Fe$_{17}$N$_3$	749	1.54	8.6	2.1	1.54	1.3	472
Fe$_3$O$_4$	858	0.60	−0.01	—	0.15	0.025	0.75
SrFe$_{12}$O$_{19}$	733	0.46	0.35	1.3	0.42	0.35	275

have an odd number of 4f electrons, which guarantees a magnetic ground state. Further, both contribute a high spin and a high orbit contribution to the magnetic moment.

There are two main strategies to develop high-performance permanent magnets. On one hand, the structure and the exact composition of the material determines the saturation magnetization, the Curie temperature as well as the magnetic anisotropy. On the other hand, extrinsic properties such as coercivity H_c and the energy product BH$_{max}$ may be tuned by nanostructuring or the design of the grain boundaries [386, 387]. For all of the following examples, the early rare-earth elements couple ferromagnetically with iron and cobalt, respectively, and the late rare-earth elements couple antiferromagnetically. These are accordingly ferrimagnets and show only low saturation magnetizations [391]. Generally, you might wonder how the interaction between outer and inner transition metal sublattices can be understood as the 4f states are well localized. In the case of iron or cobalt, the outer transition metals form 3d and 4s bands where a direct interaction via itinerant electrons occurs, where magnetization occurs according the mechanism discussed in Chapter 9.3.2. There we also saw that 4f electrons interact with others via spin-polarized bands. They employ itinerant electrons as mediator, and here both 3d and 4f elements meet. Thus, the outer transition metals are responsible for the magnetization as such, influence the Curie temperature while the inner transition metals enhance indirectly this magnetization and especially stabilize the magnetic ordering and the coercivity via their anisotropy [392].

9.6.1 SmCo$_5$

Samarium-cobalt alloys were the first highly competitive rare-earth permanent magnets, and they are still important today. Their story started in 1967 with SmCo$_5$ published by *Strnat*.[94] Samarium-cobalt SmCo$_5$ showed a really convincing magnetocrystalline anisotropy yielding an excellent K_1 coefficient combined with a high Curie tem-

---

94 *Karl Josef Strnat*, Austrian physicist (*1929 †1992).

perature and a good saturation magnetization [393]. It is thermodynamically very stable and can thus be employed up to higher temperatures. Accordingly, it can be synthesized first as ingots starting from melts of the elements. These ingots are subsequently grinded and magnetically saturated by applying a magnetic field. A side-effect of the grinding is the introduction of grain boundaries enhancing the coercivity of the particles. For the production of magnets, the material is then compressed and sintered, again under influence of a magnetic field.

Besides the main representative, also the other isostructural rare-earth compounds were investigated. The RCo_5 feature Curie temperatures ranging between 650 and 1050 K. As mentioned before, the early rare-earth elements couple ferromagnetically with the transition metal sublattice while the late rare-earth elements yield a ferrimagnetic ordering. The magnetization of the ferromagnets declines with increasing temperature. Interestingly, the magnetization of the ferrimagnets increases with rising temperature, which makes the late rare-earth elements interesting as admixture to increase the magnetization of the ferromagnets at elevated temperatures. The highest anisotropy among all RCo_5 shows the samarium compound, which is therefore the most relevant.

According to Figure 9.8, the coordination of the samarium atoms is quite anisotropic here. Along the principal axis, direct interactions of the rare-earth atoms are feasible while perpendicular to the principal axis only cobalt atoms coordinate. So, most of the employed RCo_5 contain as main rare-earth component samarium, but with admixtures of further late rare-earth elements.

With the goal to reduce the material's price and to enhance the coercivity, many attempts were made to substitute cobalt at least partially by iron. This proved challenging due to the low solubility of iron in the RCo_5. In the end, this work succeeded by the discovery that doping with copper fosters the solubility for iron in $SmCo_5$; in the course of

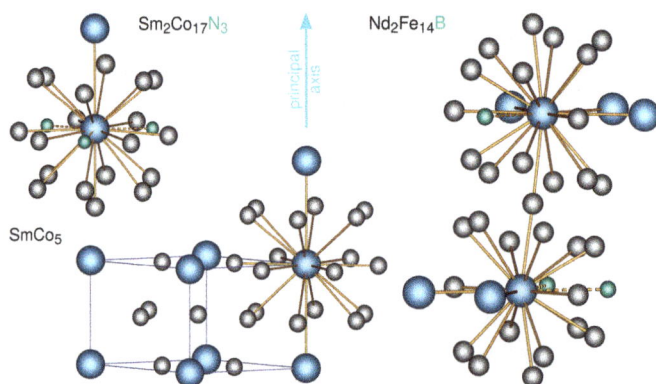

Figure 9.8: Coordination of the rare-earth atoms (blue) in $Sm_2Co_{17}N_3$, Sm_2Co_{17} (ignoring the green spheres in $Sm_2Co_{17}N_3$), $SmCo_5$ and $Nd_2Fe_{14}B$, the surrounding transition metal atoms are grey, the non-metal atoms green and the direction of the crystallographic principal axis is given; data: [200, 204, 394].

this development, finally another important cobalt based magnet, namely Sm_2Co_{17}, was found, which will be described in the following section [391, 395].

9.6.2 Sm_2Co_{17}

Because of the high Curie temperatures, the magnets based on Sm_2Co_{17} and $SmCo_5$ are the best for high-temperature applications. For example, the aforementioned application in aerospace thrusters requires a stability up to at least 500 °C. The first syntheses and the crystal structure for Sm_2Co_{17} were reported in 1966 [396, 397]. Also here, the samarium atoms are anisotropically coordinated as shown in Figure 9.8—just omit the green nitrogen atoms in $Sm_2Co_{17}N_3$. Contrarily to $SmCo_5$ the coordination is more hemispherical, though. Hence, it is not surprising that the anisotropy coefficient K_1 is smaller here.

Most important for the application as a permanent magnet is that the easy magnetization axis coincides with the principal axis. This is not the case for many rare-earth compounds R_2Co_{17}—only for Sm_2Co_{17}, Er_2Co_{17} and Tm_2Co_{17}. The adjacent atoms along the principal axis predominantly deliver negative, those perpendicular positive contributions to K_1; it depends from the actual combination of R and transition metal, which contribution dominates in the end. Moreover, only the light rare-earth elements like samarium yield ferromagnetic ordering, while in the case of the heavier like erbium and thulium ferrimagnetism occurs. Hence, Sm_2Co_{17} is the material of choice here. The other R_2Co_{17} feature Curie temperatures ranging between 1070 and 1220 K while the respective iron compounds show too low ordering temperatures below 500 K for a practical application. By introduction of nitrogen to form $Sm_2Fe_{17}N_3$, the Curie temperatures of the iron compound became interesting, though [391].

9.6.3 $Sm_2Fe_{17}N_3$

In 1990, Coey et al.[95] reported that the compounds R_2Fe_{17} absorbed nitrogen upon treatment with gaseous ammonia or nitrogen [398, 399]. The amount of nitrogen may vary between zero and six atoms per formula unit. It provides really superior magnetic properties such as a higher anisotropy parameter and an energy product BH_{max} of up to $375\,kJm^{-3}$. But unfortunately its limited thermal stability prevents a broader application as it cannot be sintered appropriately into large magnets. *Skomski*[96] cooperated intensely with Coey on the extraordinary improvement of the energy products of the

---

95 *John Michael David Coey*, Northern-Irish experimental physicist (*1945).
96 *Ralph Skomski*, German theoretical physicist (*1961 †2022).

boride and nitride magnets by combining nanoscaled hard and soft magnetic materials to achieve up to $1\,\mathrm{MJm}^{-3}$ [400].

In Chapter 6.4.2, the crystal structures of Sm_2Co_{17} and $Sm_2Fe_{17}N_3$ are described. It is evident that variable nitrogen contents can be achieved because the incorporation of nitrogen as nitride occurs on interstitial sites in Sm_2Co_{17}, the structure is maintained. The electronegative nitrogen atoms add to the coordination of the samarium atoms solely in the a-b plane. As this is perpendicular to the principal axis as depicted in Figure 9.8, this apparently enhances K_1 significantly. Moreover, the lattice expands especially in the a-b plane by a few percent upon uptake of nitrogen. This causes larger interatomic distances and, accordingly, a higher saturation magnetization, which is known as *magnetovolume effect*.

9.6.4 $Nd_2Fe_{14}B$

We learned from the previous example that incorporation of a nonmetal rises the Curie temperature. Thus, also the iron compound became competitive. Apparently, the nonmetal mediates a stronger interaction, a stronger superexchange between neighboring transition metal atoms. Then magnetic ordering becomes more stable, and the Curie temperature rises.

In a similar approach as before starting from a composition R_2Fe_{17}, we end up here with $Nd_2Fe_{14}B$ after the introduction of boron. $Nd_2Fe_{14}B$ was discovered in the 1980s and shows the highest magnetization of all rare-earth compounds $R_2Fe_{14}B$. The magnetization of the neodymium compound amounts to almost the threefold value of the dysprosium, approximately the double value of the holmium compound, which is due to the ferrimagnetic behavior there. $Nd_2Fe_{14}B$ outperforms also the previously described magnets regarding magnetization; see Table 9.7. On the other hand, the dysprosium and holmium compounds feature the highest anisotropy energies, i. e., a higher anisotropy parameter, suggesting a higher remanence. Since dysprosium is the more abundant element out of the latter two, this was chosen for high-remanence ferromagnets based on the neodymium compound [401]. The addition of dysprosium or terbium ensures a better thermal stability of the permanent magnetism allowing for operating temperatures of up to 480 K, i. e., roughly 200 °C. Because of the clearly higher price of the heavier rare-earth elements, there is a motivation to reduce the amount of dysprosium as much as possible. Hereby, it was advantageous that the maximum effect for the high application temperature can already be achieved by concentrating the dysprosium admixtures on the surface of the particles.

Compared with the previous rare-earth magnets the boride magnets feature considerable lower Curie temperatures between 400 and 700 K. The Curie temperature of $Nd_2Fe_{14}B$ lies at 585 K, luckily a sufficient value. In solid solutions with cobalt an even higher T_C of up to 800 K was found. But with increasing cobalt content the coercivity declined and the price of the transition metal rises.

The crystal structure was described in Chapter 6.4 and depicted in Figure 6.19. Important is the highly anisotropic surrounding of the rare-earth atoms as displayed in Figure 9.8. In contrast to the other examples, here the direct contacts to adjacent rare-earth atoms occur within the a-b plane as do the interaction to boron. But nevertheless, also the coordination of the neodymium atoms is highly anisotropic. The ternary compound $Nd_2Fe_{14}B$ is less high-temperature stable than the samarium-cobalt intermetallic phases described before. So, for high-temperature applications the latter will be preferred despite the better magnetic performance than even dysprosium doped $Nd_2Fe_{14}B$ [203, 391]. At room temperature, applications such as hard-disc drives the neodymium-iron-boron magnets are the best choice.

9.7 Magnetocaloric Materials

Among the most relevant challenges of environmental and climate change are air-conditioning and heating of buildings. These were responsible for more than half of the building energy consumption according to the United Nations annual report 2018—and the need especially for cooling will rapidly grow further [402]. Efficient electricity driven refrigeration or heating is therefore a challenge for chemistry and physics. Actually, conventional approaches suffer from the use of refrigerants like CFCs and the emission of climate-damaging gases.[97]

Materials showing the magnetocaloric effect might be helpful in achieving more efficient devices for refrigeration and heat pumps, potentially saving up to 30 % of the energy [403, 404]. Among such materials are many that contain rare-earth elements. Here, a ferromagnet is employed close to its ordering temperature, the Curie temperature T_C. The ferromagnet is magnetized and upon relaxation after turning off the external field heat is consumed, *magnetic cooling* or *magnetic refrigeration* occurs [389]. More precisely, upon *adiabatic magnetization* the magnetic moments of a paramagnetic substance are aligned along the magnetic field. This provides heat to the system. After turning off the magnetic field, the individual magnetic moments relax, and the magnetization declines. This *adiabatic demagnetization* consumes heat. In the case of a ferromagnet, below the transition temperature, the magnetic moments are permanently aligned. And above the Curie temperature the bulk magnetization vanishes, usually showing a more or less pronounced hysteresis. Simply spoken, the magnetization and demagnetization of a ferromagnetic or paramagnetic substance may transport heat if the heated magnetized and cooled demagnetized substances are contacted with outdoor and indoor reservoirs, respectively. Besides the environmentally-benign applications also liq-

---

[97] Environmentally hazardous chlorofluorocarbons (CFCs), such as CH_2ClF, contribute to the depletion of atmospheric ozone and exhibit strong absorption in the atmospheric window, intensifying the greenhouse effect.

uefaction of gases or cooling toward extremely low temperatures are applications of such magnetocaloric materials.

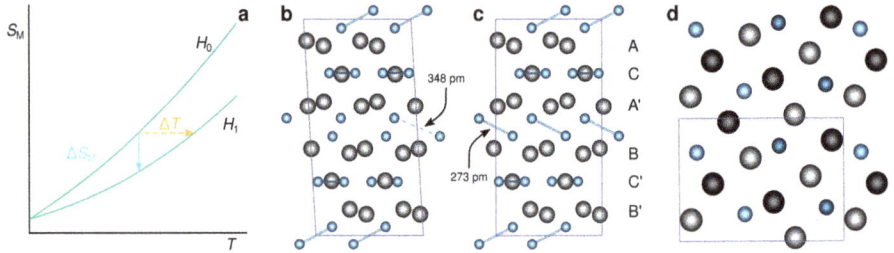

Figure 9.9: (a) Thermodynamic principles of the magnetocaloric effect and the structures of (b) monoclinic and (c) orthorhombic $Gd_5(Si_2Ge_2)$, the distances between the relevant Si/Ge atoms are indicated; (d) layers A (light) and B (dark) in Eu_2In; rare-earth atoms grey, main group atoms blue; data: [405, 406].

Thermodynamically spoken, the magnetization of a substance reduces its magnetic entropy, S_M. The demagnetization increases the magnetic entropy. This is displayed in Figure 9.9. Under a magnetic field H_1, the substance has the entropy S_1, and under the smaller, normally zero magnetic field H_0 the entropy S_0. Moreover, the entropy increases with increasing temperature. Hence, changing the magnetic field from H_0 to H_1 causes an entropy change and a temperature change. According to the second law of thermodynamics, entropy and temperature are related via the heat capacity c_p,

$$dS_{p,H} = \frac{c_{p,H}}{T}\,dT \tag{9.33}$$

with respect to a magnetic field H. A further derivation leads to an expression for the change of magnetic entropy

$$\Delta S_M = S_M(T,H_1) - S_M(T,H_0) = \mu_0 \int_{H_0}^{H_1} \left(\frac{\partial M_{T,H}}{\partial T}\right)_H dH \tag{9.34}$$

with zero field H_0, the magnetic field H_1 and the magnetization M as well as the permeability of vacuum μ_0. Hence, the magnetocaloric effect is then large, when the heat capacity strongly depends on the magnetic field. And this is the case close to the ordering temperature for ferromagnets or with decreasing temperature for paramagnetic substances [407]. Further, ΔS_M grows with large magnetic moments of the atoms within a substance. Therefore, elements like gadolinium, terbium, dysprosium, holmium, erbium, thulium, europium or samarium are interesting here according to Table 9.2. According to Table 9.8, in gadolinium metal a change of the magnetic field by two tesla yields an adiabatic temperature change of 4.7 K around its ferromagnetic Curie temperature of 293 K [408].

Table 9.8: Parameters of selected magnetocaloric substances containing rare-earth elements for a change of the magnetic field by two tesla and the latent heat ΔL; data: [408–411].

	ΔS_M / J kg^{-1} K^{-1}	ΔS_M / mJ cm^{-3} K^{-1}	ΔT / K	T_C / K	ΔL / kJ kg^{-1}
Gd	−5.2	−41	4.7	293	
Gd$_5$(Si$_2$Ge$_2$)	−27	−202	7	270	5.1
Eu$_2$In	−26	−169	5.0	55	1.4
TmZn	−19.6	−175	3.3	8.4	0.2

If accompanied by a first-order phase transition, also the latent heat of the phase transition contributes yielding a *giant magnetocaloric effect*, thus highly interesting for applications. Moreover, the entropy change may be enhanced by different lattice and electronic entropies of the phases. The latter scales with the density of states around the Fermi level; this is large near a quarter or three quarters filled f or d bands. A magnetocaloric effect is named "giant" if it exceeds that of metallic gadolinium [408].

9.7.1 Gd$_5$(Si$_2$Ge$_2$)

The first prominent example featuring a giant magnetocaloric effect, containing a rare-earth element was the intermetallic compound Gd$_5$(Si$_2$Ge$_2$) [412]. Pure Gd$_5$Si$_4$ shows ferromagnetic ordering at 335 K, while Gd$_5$Ge$_4$ orders antiferromagnetically at a much lower temperature. Halfway through the solid solution a distinct phase Gd$_5$(Si$_2$Ge$_2$) was identified. It orders ferromagnetically at a lower temperature than the pure silicon compound, close to room temperature, which is advantageous for an application in buildings. Paramagnetic Gd$_5$(Si$_2$Ge$_2$) is a monoclinically distorted variant of the orthorhombic structure of Gd$_5$Si$_4$ depicted in Figure 9.9. In the Gd$_5$Si$_4$ structure, a gadolinium lattice consisting of two close packed layers, A and B, and a further layer C with half rare-earth atom density hosts Si$_2$ dumbbells.

In Gd$_5$(Si$_2$Ge$_2$) a coupled first-order *magnetostructural* phase transition occurs. To classify a phase transition as a first-order transition, the exchange of latent heat is compulsory. Therefore, their value is also given in Table 9.8. Upon cooling below the Curie temperature, Gd$_5$(Si$_2$Ge$_2$) becomes ferromagnetic, and simultaneously its structure converts into the orthorhombic Gd$_5$Si$_4$ structure. Magnetically, the ferromagnetic ordering reduces the magnetic entropy of Gd$_5$(Si$_2$Ge$_2$). Structurally, a relative shift of the rare-earth and the main group element atoms M occurs between the layers A' and B yielding a volume increase of approximately 1 %. More importantly, as shown in Figure 9.9 selected M–M bonds form during the phase transition by reducing the Si/Ge–Si/Ge distances from 348 to 273 pm. This apparently decreases the lattice entropy of Gd$_5$(Si$_2$Ge$_2$) upon cooling. Thus, both transitions lead to a reduced entropy. A further contribution of a changed electronic entropy was considered to be small [403, 413].

9.7.2 Eu$_2$In

Contrarily to Gd$_5$(Si$_2$Ge$_2$), for the magnetocaloric material Eu$_2$In a coupled first-order *magnetoelastic* phase transition was found. In such a phase transition, the symmetry of the structure does not change. Instead, the phase transition is mainly driven by a reconstruction of the electronic structure around the Fermi level, which then also causes huge differences in the magnetic behavior. Eventually, the same changes of magnetic entropy add up with the overall entropy change to yield a giant magnetocaloric effect. Normally, for such magnetic effects driven by changes of the electronic structure, transition metals like in TmZn or rare-earth doped manganites R$_{1-x}$M$_x$[MnO$_3$] (R = La, Nd, Pr; M = Ca, Sr, Ba) are needed [411, 414, 415]. These provide the itinerant electrons, which convey the interaction of the strong magnetic moments of 4f elements according to the RKKY mechanism. In Eu$_2$In, this role has to be adopted by the main-group element indium.

The structure of the intermetallic phase Eu$_2$In can be understood by flat close packed layers with ordered distribution of europium and indium as shown in Figure 9.9. These layers are packed in an alternating fashion with the two europium atoms on both layers are crystallographic distinct; the layers feature very similar interatomic distances. The distances between the metal atoms do not show clearly anisotropic behavior, so it is indeed a three-dimensional structure. The underlying structure type is that of Co$_2$Si.

Upon cooling below the Curie temperature of 55 K, a phase transition occurs where the unit cell volume shrinks discontinuously by only 0.1 %, in line with an extraordinarily small hysteresis of the phase transition. This tiny change does not imply any valence change of the divalent europium atoms as Eu^{3+} would be significantly smaller. As it is a magnetoelastic transition, the structure remains the same. The determined magnetic moments of 8.5 μ_B indicate the presence of divalent europium, isoelectronic with Gd^{3+}. The magnetization at very low temperatures of 5 K of n_{max} = 14.4 μ_B suggests a collinear arrangement of the magnetic moments of both crystallographically different europium atoms. For comparison with the basic values for Eu^{2+}, consult Table 9.2. The slightly higher values than typical for divalent europium were ascribed to minor contributions from populated 5d states strongly hybridized with indium's 5p states. This intense interaction is believed to be responsible for the apparently electronically driven unique behavior of Eu$_2$In as a magnetocaloric material [410].

Epilogue

Our journey through the exciting seventeen elements named rare-earth elements ends here. Hopefully, it triggered some more thorough studies to find out more about their liaison with further elements. Alike the world, also the periodic table is colorful and versatile. Remember always:

> Discoveries? If we seek them we won't find them—we find them if we are not seeking them. Stay open for the unexpected!

I thank my group members for patience and for critically reading this book prior to submission, especially Vivien Wessels, Erich Turgunbajew, Philip Ettlinger and Alexander Weiß. For great ideas, I thank Marion Eisele and Martin Mangei, and for his extraordinary patience and inspiration my husband, Peter Schwab.

A modified representation of The Periodic Table, which was developed by Otto Theodor Benfey.

https://doi.org/10.1515/9783110680829-010

A Ionic Radii

Table A.1 lists the ionic radii of relevant rare-earth ions as well as further ions, which provide sites for doping with rare-earth elements discussed in this book. Together with the ionic radii of hydride (122 pm), fluoride (131 pm), chloride (181 pm), bromide (196 pm), iodide (220 pm), oxide (138 pm) and nitride (146 pm), most of the bond distances in the herein given compounds can be estimated [50].

Table A.1: Shannon's ionic radii of the R^{3+} with respect to coordination numbers in pm [50].

R	R^{3+}						R^{2+}	R^{4+}
	[6]	[7]	[8]	[9]	[10]	[12]	[7]	[8]
scandium	75		87					
yttrium	90	96	102	108				
lanthanum	103	110	116	122	127	136		
cerium	101	107	114	120	125	134		97
praseodymium	99		113	118				96
neodymium	98		111	116		127		
samarium	96	102	108	113		124	122	
europium	95	101	107	112			120	
gadolinium	94	100	105	111				
terbium	92	98	104	109				88
dysprosium	91	97	103	108				
holmium	90		102	107	112			
erbium	89	95	100	106				
thulium	88		99	105			109	
ytterbium	87	93	98	104			108	
lutetium	86		98	103				
calcium							106	
strontium							121	
barium							138	
zirconium								84

https://doi.org/10.1515/9783110680829-011

B Physical Properties

Table B.1 lists relevant physical parameters of the rare-earth elements.

Table B.1: Melting and boiling points, Pauling electronegativities, electronegativities based on configurational energies [20] scaled to Pauling's electronegativities, first through fourth ionization energies I, electron affinites A, chemical hardnesses η^0 of the elements (from first ionization energies and electron affinities) and η^{3+} of the R^{3+} estimated from the third and fourth ionization energies; data: [20, 29].

	m. p. °C	b. p. °C	X_P	$X_{RZH, P}$	I^{1st} eV	I^{2nd} eV	I^{3rd} eV	I^{4th} eV	A eV	η^0 eV	η^{3+} eV
Sc	1539	2832	1.20	1.2	6.56	12.8	24.76	73.49	0.19	3.2	24.4
Y	1523	3337	1.11	1.1	6.22	12.24	20.52	60.6	0.31	3.0	20.0
La	920	3454	1.08	1.0	5.58	11.06	19.18	49.95	0.55	2.5	15.4
Ce	798	3468	1.07	1.2	5.54	10.85	20.2	36.76	0.57	2.5	8.3
Pr	931	3017	1.07	1.1	5.47	10.55	21.62	38.98	0.96	2.3	8.7
Nd	1010	3027	1.07	1.2	5.53	10.73	22.1	40.41	1.92	1.8	9.2
Pm	1080	2730	1.07	1.2	5.58	10.9	22.3	41.1	0.13	2.7	9.4
Sm	1072	1804	1.07	1.4	5.64	11.07	23.4	41.4	0.16	2.7	9.0
Eu	822	1439	1.01	1.6	5.67	11.24	24.92	42.7	0.12	2.8	8.9
Gd	1311	3000	1.11	2.3	6.15	12.09	20.63	44.0	0.14	3.0	11.7
Tb	1360	2480	1.10	1.3	5.86	11.52	21.91	39.79	1.17	2.3	8.9
Dy	1409	2335	1.10	1.4	5.94	11.67	22.8	41.47	0.35	2.8	9.3
Ho	1470	2720	1.10	1.4	6.02	11.8	22.84	42.5	0.34	2.8	9.8
Er	1522	2510	1.11	1.3	6.11	11.93	22.74	42.7	0.31	2.9	10.0
Tm	1545	1725	1.11	1.5	6.18	12.05	23.68	42.7	1.03	2.6	9.5
Yb	824	1193	1.06	1.7	6.25	12.18	25.05	43.56	-0.02	3.1	9.3
Lu	1656	3315	1.14	1.1	5.43	13.9	20.96	45.25	0.24	2.6	12.1

https://doi.org/10.1515/9783110680829-012

C Chemical Shift Model Parameters

Tables C.1 and C.2 list parameters relevant for the chemical shift model. The correction parameter α (n) considering the lanthanoid contraction was determined to 0.095 eV/pm (n = 2) and 0.098 eV/pm (n = 3). All data were taken from [252], but own general U-value was used.

Table C.1: The basic energies of a valence electron $E_0(R^{n+})$, the interelectronic repulsion terms E_{rep} (n) (all energies in eV, radii differences in pm, data: [252]) and the calculated data for Figure 8.3, $\Delta_r = r_{Eu^{3+}} - r_{R^{3+}}$ based on Shannon's radii for an eightfold coordination [50].

R	$E_0(R^{2+})$	$E_{rep}(R^{2+})$	Δ_r	$E_0(R^{3+})$	$E_{rep}(R^{3+})$	$E_0^{fd}(R^{2+})$	$E_0^{fd}(R^{3+})$
La	−18.17	−0.12	−9.4			−0.89	
Ce	−19.72	−0.48	−7.7	−36.59	−0.17	0.41	6.17
Pr	−21.09	−0.54	−6.0	−38.37	−0.61	1.59	7.68
Nd	−22.33	0.22	−4.3	−39.90	−0.71	1.90	8.95
Pm	−23.33	0.96	−2.7	−41.39	0.19	1.94	9.26
Sm	−24.39	0.79	−1.3	−42.63	1.07	3.00	9.40
Eu	−25.25	0.33	0.0	−43.87	0.90	4.20	10.46
Gd	−25.90	5.57	1.3	−44.85	0.40	0.60	11.77
Tb	−26.54	4.63	2.6	−45.79	6.42	1.85	7.81
Dy	−27.08	4.19	3.9	−46.71	5.49	2.55	9.33
Ho	−27.81	4.97	5.1	−47.52	5.09	2.49	10.02
Er	−28.57	5.83	6.2	−48.48	6.00	2.35	9.92
Tm	−29.24	5.56	7.2	−49.38	6.96	3.08	9.84
Yb	−29.91	4.88	8.1	−50.19	6.63	4.38	10.93
Lu			8.9	−51.07	5.81		12.24

Table C.2: The exchange parameters in eV quantifying the splitting of the high- and low-spin branches of the 5d states employed for Figure 8.26 [252].

R	Eu	Gd	Tb	Dy	Ho	Er	Tm	Yb	Lu
$\Delta E^{ex}(R^{2+})$	0.00	0.84	0.60	0.40	0.33	0.28	0.21	0.19	
$\Delta E^{ex}(R^{3+})$		0.00	1.00	0.88	0.42	0.32	0.28	0.22	0.17

https://doi.org/10.1515/9783110680829-013

D Structure Types Derived From Close Packings

Herein, important structure types derived from closest packings are illustrated based on the filling of *octahedral* and *tetrahedral voids* (OVs and TVs), respectively. Orange-filled boxes represent filled voids while blank ones show unoccupied voids. Packing atoms and layers are colored cyan. You may recognize that the same or at least very similar filling schemes of tetrahedral, octahedral or related voids occur in both close packings—thus allowing for a systematic view on important structure types. Structures derived from cubic close packings are listed in the left column, structures derived from hexagonal close packings in the right column. It does of course not replace a concise description of the respective crystal structure.

① structure types with filled OVs

NaCl (rocksalt)

NiAs

CdCl$_2$, NdOF

CdI$_2$, Mg(OH)$_2$ (brucite)

TiO$_2$ (anatase)

TiO$_2$ (brookite)

TiO$_2$ (rutile), linear chains
α-PbO$_2$, zig-zag chains

https://doi.org/10.1515/9783110680829-014

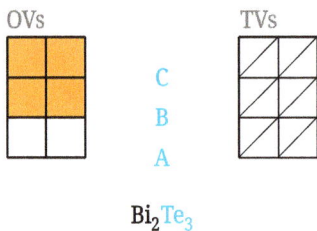

Bi_2Te_3

not known

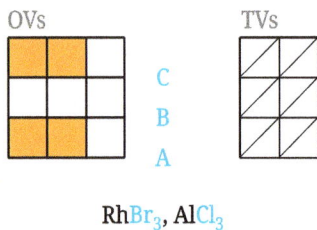

$RhBr_3$, $AlCl_3$

CaTiO$_3$ (perovskite)
ReO$_3$□

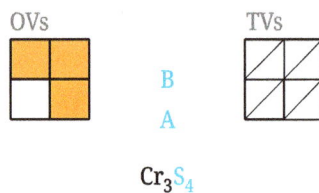

Cr_3S_4

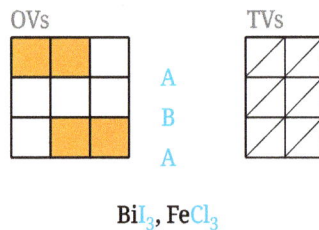

ZrI$_3$ (columnar structure)

BiI$_3$, FeCl$_3$

Al$_2$O$_3$ (corundum)
FeTiO$_3$ (ilmenite)

② adjacent face-sharing OVs may fuse to *tricapped triangular prisms* (TTP)

not possible

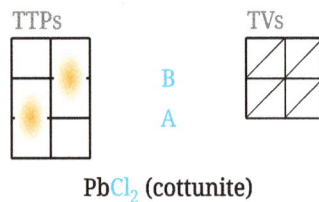

PbCl$_2$ (cottunite)

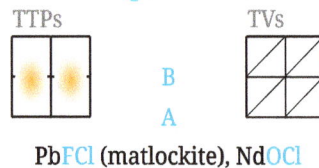

PbFCl (matlockite), NdOCl

③ structure types with filled TVs

CaF$_2$ (fluorite)   not known

ZnS (zincblende), CC (diamond)   ZnS (wurtzite)

④ structure types with simultaneously filled TVs and OVs

AlMnCu$_2$ (Heusler phase)   R$_2$O$_3$ (A type)

MgAl$_2$O$_4$ (spinel)   Al$_2$BeO$_4$ (chrysoberyl)
                                 Mg$_2$[SiO$_4$] (olivines)

⑤ further related crystal structures—here deficient closely packed layers are indicated in blue; in the latter kagomé nets are packed, in which every fourth atom is removed with respect to a closest packed layer

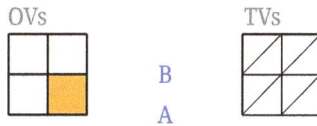

Mg$_2$CuCu$_3$ (MgCu$_2$-type)   Mg$_2$ZnZn$_3$ (MgZn$_2$-type)

E Exemplary Calculation of the Transition Integral's Behavior

Based on Table E.1, the symmetry races of contributing levels are employed to determine whether transitions between them are allowed as electric or magnetic dipole transitions. The product of any race with itself always contains the total-symmetric representation,

Table E.1: Character table of point group D_{3h}; given are the symmetry elements, the characters of the symmetry races, the assignment of relevant functions and rotational and translational movements.

D_{3h}	E	$2\,C_3$	$3\,C_2$	σ_h	$2\,S_3$	$3\sigma_v$		functions
a_1'	1	1	1	1	1	1		$x^2 + y^2, z^2$
a_2'	1	1	−1	1	1	−1	R_z	
e'	2	−1	0	2	−1	0	T_x, T_y	$(x^2 - y^2, xy), (x, y)$
a_1''	1	1	1	−1	−1	−1		
a_2''	1	1	−1	−1	−1	1	T_z	z
e''	2	−1	0	−2	1	0	R_x, R_y	(xz, yz)

so it is sufficient that the product of the electronic states' races yield the respective race of rotation (magnetic dipole) or translation (electric dipole). In our example, the total symmetric representation is a_1'. This representation ensures that the product of the originally contributing wavefunctions, for which the symmetry races represent the symmetric behavior, yields a positive integral according to Equations (7.9) and (7.11).

Let the wavefunctions ψ_1, ψ_2 and ψ_3 be represented by the symmetry races a_1' (e. g., a d_{z^2} orbital), e' (e. g., a p_x orbital) or a_2'' (e. g., a p_z orbital). Then the product of ψ_1 and ψ_2 behaves symmetrically like the product of their symmetry races $a_1' \otimes e'$. By multiplying the respective characters in Table E.1, this yields symmetry race e'. Since the two translations T_x and T_y transform like e', this transition is allowed as an electric dipole. As a second example, we consider the transition between ψ_2 and ψ_3. Here, the product yields e'' according to the symmetry behavior of the rotations R_x and R_y. Hence, this transition is allowed as a magnetic dipole transition while it is forbidden as an electric dipole transition.

https://doi.org/10.1515/9783110680829-015

F Excerpts of MAPLE Calculations

These calculations in this book were employed to determine or confirm coordination numbers of selected compounds.

F.1 Coordination and MAPLE of LaH$_3$ and GdH$_3$

LaH$_3$

```
ELEMENTARZELLE LaH3 - LaH3 28/ 7/2023
-----------------------------------------------------------------------
Raumgruppe : FM3M
a = 560.3000 b = 560.3000 c = 560.3000
Zahl der Formeleinheiten : 4
Eingegebene Atome
x y z IonRad Ladung Besetzung Suchrad.
La1 0.000000 0.000000 0.000000 116.00 + 3.0000 1.000000 500.0
H1 0.250000 0.250000 0.250000 114.00 - 1.0000 1.000000 500.0
H2 0.500000 0.500000 0.500000 114.00 - 1.0000 1.000000 500.0
----------------------
- Ergebnisse für La1 -
----------------------
ZT La1 0.0000 0.0000 0.0000 Abstand ECoN( 1) ECoN( 3)
1 H1 0.2500 0.2500 -0.2500 242.617 1.146 1.174
2 H1 -0.2500 -0.2500 0.2500 242.617 1.146 1.174
3 H1 0.2500 -0.2500 0.2500 242.617 1.146 1.174
4 H1 -0.2500 0.2500 -0.2500 242.617 1.146 1.174
5 H1 -0.2500 0.2500 0.2500 242.617 1.146 1.174
6 H1 0.2500 -0.2500 -0.2500 242.617 1.146 1.174
7 H1 -0.2500 -0.2500 -0.2500 242.617 1.146 1.174
8 H1 0.2500 0.2500 0.2500 242.617 1.146 1.174
9 H2 0.0000 0.0000 0.5000 280.150 0.351 0.372
10 H2 0.0000 0.0000 -0.5000 280.150 0.351 0.372
11 H2 0.0000 0.5000 0.0000 280.150 0.351 0.372
12 H2 0.0000 -0.5000 0.0000 280.150 0.351 0.372
13 H2 0.5000 0.0000 0.0000 280.150 0.351 0.372
14 H2 -0.5000 0.0000 0.0000 280.150 0.351 0.372
Nächster Ligand :
15 H1 0.2500 0.2500 0.7500 464.576 0.000 0.000
Letzter Ligand :
46 H2 -0.5000 -0.5000 -0.5000 485.234 0.000 0.000
```

https://doi.org/10.1515/9783110680829-016

```
Koordinationszahl : 14
---------------------
- Ergebnisse für H1 -
---------------------
ZT H1 0.2500 0.2500 0.2500 Abstand ECoN( 1) ECoN( 2)
1 La1 0.5000 0.5000 0.0000 242.617 1.000 1.000
2 La1 0.5000 0.0000 0.5000 242.617 1.000 1.000
3 La1 0.0000 0.5000 0.5000 242.617 1.000 1.000
4 La1 0.0000 0.0000 0.0000 242.617 1.000 1.000
Nächster Ligand :
5 La1 0.5000 0.5000 1.0000 464.576 0.000 0.000
Letzter Ligand :
16 La1 0.0000 0.0000 1.0000 464.576 0.000 0.000
Koordinationszahl : 4
---------------------
- Ergebnisse für H2 -
---------------------
ZT H2 0.5000 0.5000 0.5000 Abstand ECoN( 1) ECoN( 2)
1 La1 0.5000 0.5000 1.0000 280.150 1.000 1.000
2 La1 0.5000 0.5000 0.0000 280.150 1.000 1.000
3 La1 0.5000 1.0000 0.5000 280.150 1.000 1.000
4 La1 0.5000 0.0000 0.5000 280.150 1.000 1.000
5 La1 1.0000 0.5000 0.5000 280.150 1.000 1.000
6 La1 0.0000 0.5000 0.5000 280.150 1.000 1.000
Nächster Ligand :
7 La1 1.0000 1.0000 1.0000 485.234 0.000 0.000
Letzter Ligand :
14 La1 0.0000 0.0000 0.0000 485.234 0.000 0.000
Koordinationszahl : 6
-----------------------
- Ergebnisse für MAPLE -
-----------------------
Ladung Abstand Potential PMF MAPLE *MAPLE
La1 + 3.0000 242.62 -1.97412 7.18432 982.5481 109.1720
H1 - 1.0000 242.62 0.72653 0.88134 120.5345 120.5345
H2 - 1.0000 242.62 0.52106 0.63209 86.4471 86.4471
Madelungkonstante : 9.5791 +/- 0.000001
Coulombanteil der Gitterenergie : 1310.0642 +/- 0.000100 kcal/mol
Coulombanteil der Gitterenergie : 5483.9300 +/- 0.000417 kJ/mol
```

LaH₃ in the GdH₃ Type

```
ELEMENTARZELLE LaH3 - GdH3 28/ 7/2023
-----------------------------------------------------------------------
Raumgruppe : P-3C1
a = 703.6200 b = 703.6200 c = 731.3000
Zahl der Formeleinheiten : 6
Eingegebene Atome
x y z IonRad Ladung Besetzung Suchrad.
La1 0.666667 0.000000 0.250000 110.00 + 3.0000 1.000000 500.0
H2 0.333333 0.666667 0.167000 114.00 - 1.0000 1.000000 500.0
H1 0.356000 0.028000 0.096000 114.00 - 1.0000 1.000000 500.0
H3 0.000000 0.000000 0.250000 114.00 - 1.0000 1.000000 500.0
----------------------
- Ergebnisse für La1 -
----------------------
ZT La1 0.6667 0.0000 0.2500 Abstand ECoN( 1) ECoN( 3)
1 H3 1.0000 0.0000 0.2500 234.540 1.344 1.373
2 H2 0.6667 0.3333 0.3330 242.267 1.156 1.186
3 H2 0.3333 -0.3333 0.1670 242.267 1.156 1.186
4 H1 0.9720 0.3280 0.0960 250.039 0.967 0.998
5 H1 0.6440 -0.3280 0.4040 250.039 0.967 0.998
6 H1 0.6720 0.0280 0.5960 253.678 0.880 0.911
7 H1 0.6440 -0.0280 -0.0960 253.678 0.880 0.911
8 H1 0.3280 -0.0280 0.4040 255.265 0.843 0.874
9 H1 0.3560 0.0280 0.0960 255.265 0.843 0.874
10 H1 1.0280 0.3560 0.4040 276.373 0.413 0.437
11 H1 0.6720 -0.3560 0.0960 276.373 0.413 0.437
Nächster Ligand :
12 H2 0.6667 0.3333 -0.1670 384.714 0.000 0.000
Letzter Ligand :
40 H1 -0.0280 -0.6720 0.0960 494.014 0.000 0.000
Koordinationszahl : 11
----------------------
- Ergebnisse für H2 -
----------------------
ZT H2 0.3333 0.6667 0.1670 Abstand ECoN( 1) ECoN( 2)
1 La1 0.3333 0.3333 0.2500 242.267 1.000 1.000
2 La1 0.0000 0.6667 0.2500 242.267 1.000 1.000
3 La1 0.6667 1.0000 0.2500 242.267 1.000 1.000
Nächster Ligand :
4 La1 0.6667 0.6667 -0.2500 384.714 0.000 0.000
```

```
Letzter Ligand :
12 La1 0.3333 1.0000 0.7500 486.602 0.000 0.000
Koordinationszahl : 3
Abstand Zentralatom - Ebene(La1,La1,La1) : 60.698
---------------------
- Ergebnisse für H1 -
---------------------
ZT H1 0.3560 0.0280 0.0960 Abstand ECoN( 1) ECoN( 3)
1 La1 0.3333 0.3333 0.2500 250.039 1.142 1.150
2 La1 0.3333 0.0000 -0.2500 253.678 1.056 1.064
3 La1 0.6667 0.0000 0.2500 255.265 1.019 1.026
4 La1 0.0000 -0.3333 0.2500 276.373 0.559 0.566
Nächster Ligand :
5 La1 0.6667 0.6667 -0.2500 464.238 0.000 0.000
Letzter Ligand :
14 La1 0.3333 -0.6667 0.2500 494.014 0.000 0.000
Koordinationszahl : 4
---------------------
- Ergebnisse für H3 -
---------------------
ZT H3 0.0000 0.0000 0.2500 Abstand ECoN( 1) ECoN( 2)
1 La1 0.0000 -0.3333 0.2500 234.540 1.000 1.000
2 La1 -0.3333 0.0000 0.2500 234.540 1.000 1.000
3 La1 0.3333 0.3333 0.2500 234.540 1.000 1.000
Nächster Ligand :
4 La1 -0.3333 -0.3333 0.7500 434.406 0.000 0.000
Letzter Ligand :
12 La1 0.6667 0.0000 0.2500 469.080 0.000 0.000
Koordinationszahl : 3
Abstand Zentralatom - Ebene(La1,La1,La1) : 0.000
------------------------
- Ergebnisse für MAPLE -
------------------------
Ladung Abstand Potential PMF MAPLE *MAPLE
La1 + 3.0000 234.54 -1.91239 6.53493 951.8232 105.7581
H2 - 1.0000 242.27 0.66858 0.76155 110.9209 110.9209
H1 - 1.0000 227.81 0.65838 0.74993 109.2293 109.2293
H3 - 1.0000 234.54 0.80213 0.91367 133.0779 133.0779
Madelungkonstante : 8.8471 +/- 0.000001
Coulombanteil der Gitterenergie : 1288.5883 +/- 0.000100 kcal/mol
Coulombanteil der Gitterenergie : 5394.0319 +/- 0.000417 kJ/mol
```

GdH₃

```
ELEMENTARZELLE GdH3 28/ 7/2023
-----------------------------------------------------------------------
Raumgruppe : P-3C1
a = 646.6200 b = 646.6200 c = 671.7000
Zahl der Formeleinheiten : 6
Eingegebene Atome
x y z IonRad Ladung Besetzung Suchrad.
Gd1 0.666667 0.000000 0.250000 100.00 + 3.0000 1.000000 500.0
H2 0.333333 0.666667 0.167000 114.00 - 1.0000 1.000000 500.0
H1 0.356000 0.028000 0.096000 114.00 - 1.0000 1.000000 500.0
H3 0.000000 0.000000 0.250000 114.00 - 1.0000 1.000000 500.0
----------------------
- Ergebnisse für Gd1 -
----------------------
ZT Gd1 0.6667 0.0000 0.2500 Abstand ECoN( 1) ECoN( 3)
1 H3 1.0000 0.0000 0.2500 215.540 1.343 1.372
2 H2 0.6667 0.3333 0.3330 222.633 1.155 1.185
3 H2 0.3333 -0.3333 0.1670 222.633 1.155 1.185
4 H1 0.6440 -0.3280 0.4040 229.759 0.967 0.997
5 H1 0.9720 0.3280 0.0960 229.759 0.967 0.997
6 H1 0.6720 0.0280 0.5960 233.004 0.883 0.913
7 H1 0.6440 -0.0280 -0.0960 233.004 0.883 0.913
8 H1 0.3280 -0.0280 0.4040 234.562 0.843 0.874
9 H1 0.3560 0.0280 0.0960 234.562 0.843 0.874
10 H1 1.0280 0.3560 0.4040 253.962 0.412 0.437
11 H1 0.6720 -0.3560 0.0960 253.962 0.412 0.437
Nächster Ligand :
12 H2 0.3333 -0.3333 0.6670 353.430 0.000 0.000
Letzter Ligand :
40 H1 -0.0280 -0.6720 0.0960 453.981 0.000 0.000
Koordinationszahl : 11
----------------------
- Ergebnisse für H2 -
----------------------
ZT H2 0.3333 0.6667 0.1670 Abstand ECoN( 1) ECoN( 2)
1 Gd1 0.3333 0.3333 0.2500 222.633 1.000 1.000
2 Gd1 0.0000 0.6667 0.2500 222.633 1.000 1.000
3 Gd1 0.6667 1.0000 0.2500 222.633 1.000 1.000
Nächster Ligand :
4 Gd1 0.6667 0.6667 -0.2500 353.430 0.000 0.000
```

```
Letzter Ligand :
12 Gd1 0.3333 1.0000 0.7500 447.000 0.000 0.000
Koordinationszahl : 3
Abstand Zentralatom - Ebene(Gd1,Gd1,Gd1) : 55.751
---------------------
- Ergebnisse für H1 -
---------------------
ZT H1 0.3560 0.0280 0.0960 Abstand ECoN( 1) ECoN( 3)
1 Gd1 0.3333 0.3333 0.2500 229.759 1.142 1.149
2 Gd1 0.3333 0.0000 -0.2500 233.004 1.058 1.065
3 Gd1 0.6667 0.0000 0.2500 234.562 1.018 1.025
4 Gd1 0.0000 -0.3333 0.2500 253.962 0.559 0.565
Nächster Ligand :
5 Gd1 0.6667 0.6667 -0.2500 426.563 0.000 0.000
Letzter Ligand :
14 Gd1 0.3333 -0.6667 0.2500 453.981 0.000 0.000
Koordinationszahl : 4
---------------------
- Ergebnisse für H3 -
---------------------
ZT H3 0.0000 0.0000 0.2500 Abstand ECoN( 1) ECoN( 2)
1 Gd1 0.3333 0.3333 0.2500 215.540 1.000 1.000
2 Gd1 0.0000 -0.3333 0.2500 215.540 1.000 1.000
3 Gd1 -0.3333 0.0000 0.2500 215.540 1.000 1.000
Nächster Ligand :
4 Gd1 -0.3333 -0.3333 0.7500 399.065 0.000 0.000
Letzter Ligand :
12 Gd1 -0.6667 -0.6667 0.2500 431.080 0.000 0.000
Koordinationszahl : 3
Abstand Zentralatom - Ebene(Gd1,Gd1,Gd1) : 0.000
-----------------------
- Ergebnisse für MAPLE -
-----------------------
Ladung Abstand Potential PMF MAPLE *MAPLE
Gd1 + 3.0000 215.54 -2.08133 6.53268 1035.9092 115.1010
H2 - 1.0000 222.63 0.72738 0.76101 120.6758 120.6758
H1 - 1.0000 209.25 0.71661 0.74974 118.8888 118.8888
H3 - 1.0000 215.54 0.87264 0.91299 144.7753 144.7753
Madelungkonstante : 8.8438 +/- 0.000001
Coulombanteil der Gitterenergie : 1402.3959 +/- 0.000100 kcal/mol
Coulombanteil der Gitterenergie : 5870.4305 +/- 0.000417 kJ/mol
```

GdH$_3$ in the LaH$_3$ Type

```
ELEMENTARZELLE GdH3 - LaH3 28/ 7/2023
----------------------------------------------------------------------
Raumgruppe : FM3M
a = 532.3000 b = 532.3000 c = 532.3000
Zahl der Formeleinheiten : 4
Eingegebene Atome
x y z IonRad Ladung Besetzung Suchrad.
Gd1 0.000000 0.000000 0.000000 104.00 + 3.0000 1.000000 500.0
H1 0.250000 0.250000 0.250000 114.00 - 1.0000 1.000000 500.0
H2 0.500000 0.500000 0.500000 114.00 - 1.0000 1.000000 500.0
In der Elementarzelle sind 16 Atome enthalten.
----------------------
- Ergebnisse für Gd1 -
----------------------
ZT Gd1 0.0000 0.0000 0.0000 Abstand ECoN( 1) ECoN( 3)
1 H1 0.2500 0.2500 -0.2500 230.493 1.146 1.174
2 H1 -0.2500 -0.2500 0.2500 230.493 1.146 1.174
3 H1 0.2500 -0.2500 0.2500 230.493 1.146 1.174
4 H1 -0.2500 0.2500 -0.2500 230.493 1.146 1.174
5 H1 -0.2500 0.2500 0.2500 230.493 1.146 1.174
6 H1 0.2500 -0.2500 -0.2500 230.493 1.146 1.174
7 H1 -0.2500 -0.2500 -0.2500 230.493 1.146 1.174
8 H1 0.2500 0.2500 0.2500 230.493 1.146 1.174
9 H2 0.0000 0.0000 0.5000 266.150 0.351 0.372
10 H2 0.0000 0.0000 -0.5000 266.150 0.351 0.372
11 H2 0.0000 0.5000 0.0000 266.150 0.351 0.372
12 H2 0.0000 -0.5000 0.0000 266.150 0.351 0.372
13 H2 0.5000 0.0000 0.0000 266.150 0.351 0.372
14 H2 -0.5000 0.0000 0.0000 266.150 0.351 0.372
Nächster Ligand :
15 H1 0.2500 0.2500 0.7500 441.360 0.000 0.000
Letzter Ligand :
46 H2 -0.5000 -0.5000 -0.5000 460.985 0.000 0.000
Koordinationszahl : 14
----------------------
- Ergebnisse für H1 -
----------------------
ZT H1 0.2500 0.2500 0.2500 Abstand ECoN( 1) ECoN( 2)
1 Gd1 0.5000 0.5000 0.0000 230.493 1.000 1.000
2 Gd1 0.5000 0.0000 0.5000 230.493 1.000 1.000
```

```
3 Gd1 0.0000 0.5000 0.5000 230.493 1.000 1.000
4 Gd1 0.0000 0.0000 0.0000 230.493 1.000 1.000
Nächster Ligand :
5 Gd1 0.5000 0.5000 1.0000 441.360 0.000 0.000
Letzter Ligand :
16 Gd1 0.0000 0.0000 1.0000 441.360 0.000 0.000
Koordinationszahl : 4
---------------------
- Ergebnisse für H2 -
---------------------
ZT H2 0.5000 0.5000 0.5000 Abstand ECoN( 1) ECoN( 2)
1 Gd1 0.5000 0.5000 1.0000 266.150 1.000 1.000
2 Gd1 0.5000 0.5000 0.0000 266.150 1.000 1.000
3 Gd1 0.5000 1.0000 0.5000 266.150 1.000 1.000
4 Gd1 0.5000 0.0000 0.5000 266.150 1.000 1.000
5 Gd1 1.0000 0.5000 0.5000 266.150 1.000 1.000
6 Gd1 0.0000 0.5000 0.5000 266.150 1.000 1.000
Nächster Ligand :
7 Gd1 1.0000 1.0000 1.0000 460.985 0.000 0.000
Letzter Ligand :
14 Gd1 0.0000 0.0000 0.0000 460.985 0.000 0.000
Koordinationszahl : 6
------------------------
- Ergebnisse für MAPLE -
------------------------
Ladung Abstand Potential PMF MAPLE *MAPLE
Gd1 + 3.0000 230.49 -2.07796 7.18432 1034.2320 114.9147
H1 - 1.0000 230.49 0.76474 0.88134 126.8748 126.8748
H2 - 1.0000 230.49 0.54847 0.63209 90.9944 90.9944
Madelungkonstante : 9.5791 +/- 0.000001
Coulombanteil der Gitterenergie : 1378.9761 +/- 0.000100 kcal/mol
Coulombanteil der Gitterenergie : 5772.3952 +/- 0.000417 kJ/mol
```

F.2 Coordinations in EuBr$_2$

```
ELEMENTARZELLE SrBr2 20/ 5/2023
-----------------------------------------------------------------------
Raumgruppe : P4/N
a = 1156.3000 b = 1156.3000 c = 709.1000
Eingegebene Atome
x y z IonRad Ladung Besetzung Suchrad.
```

```
Eu1 0.250000 0.250000 0.355500 117.00 + 2.0000 1.000000 500.0
Eu2 0.103000 0.587100 0.748900 117.00 + 2.0000 1.000000 500.0
Br1 0.161000 0.042400 0.602600 196.00 - 1.0000 1.000000 500.0
Br2 0.039500 0.155400 0.125600 196.00 - 1.0000 1.000000 500.0
Br3 0.250000 0.750000 0.500000 196.00 - 1.0000 1.000000 500.0
Br4 0.250000 0.750000 0.000000 196.00 - 1.0000 1.000000 500.0
---------------------
Ergebnisse für Eu1
---------------------

ZT Eu1 0.2500 0.2500 0.3555 Abstand ECoN( 1) ECoN( 2)
1 Br2 0.3446 0.0395 0.1256 312.707 1.017 1.017
2 Br2 0.0395 0.1554 0.1256 312.707 1.017 1.017
3 Br2 0.1554 0.4605 0.1256 312.707 1.017 1.017
4 Br2 0.4605 0.3446 0.1256 312.707 1.017 1.017
5 Br1 0.4576 0.1610 0.6026 314.508 0.982 0.983
6 Br1 0.1610 0.0424 0.6026 314.508 0.982 0.983
7 Br1 0.0424 0.3390 0.6026 314.508 0.982 0.983
8 Br1 0.3390 0.4576 0.6026 314.508 0.982 0.983
Letzter Ligand :
8 Br1 0.3390 0.4576 0.6026 314.508 0.982 0.983
Koordinationszahl : 8
---------------------
Ergebnisse für Eu2
---------------------

ZT Eu2 0.1030 0.5871 0.7489 Abstand ECoN( 1) ECoN( 3)
1 Br3 0.2500 0.7500 0.5000 309.067 1.119 1.125
2 Br4 0.2500 0.7500 1.0000 309.960 1.102 1.108
3 Br2 0.1554 0.4605 1.1256 310.568 1.090 1.096
4 Br1 -0.0424 0.6610 0.3974 312.559 1.052 1.058
5 Br1 0.0424 0.3390 0.6026 313.004 1.044 1.049
6 Br2 -0.1554 0.5395 0.8744 316.580 0.976 0.981
7 Br1 0.3390 0.4576 0.6026 328.103 0.763 0.769
8 Br2 -0.0395 0.8446 0.8744 351.743 0.395 0.400
Nächster Ligand :
9 Br2 0.1554 0.4605 0.1256 469.520 0.000 0.000
Letzter Ligand :
10 Br1 -0.0424 0.6610 1.3974 497.023 0.000 0.000
Koordinationszahl : 8
```

F.3 Coordinations in B Type Sm$_2$O$_3$

```
ELEMENTARZELLE B-Typ Sm2O3 10/ 8/2023
-------------------------------------------------------------------------
Raumgruppe : C2/M
a = 1418.3800 b = 362.7400 c = 885.7700
α = 90.0000 β = 100.0000 γ = 90.0000
Zahl der Formeleinheiten : 6
Eingegebene Atome
x y z IonRad Ladung Besetzung Suchrad.
Sm1 0.135700 0.500000 0.489700 110.00 + 3.0000 1.000000 500.0
Sm2 0.180000 0.500000 0.136200 110.00 + 3.0000 1.000000 500.0
Sm3 0.467200 0.500000 0.186500 110.00 + 3.0000 1.000000 500.0
O1 0.131200 0.000000 0.286700 140.00 - 2.0000 1.000000 500.0
O2 0.325400 0.500000 0.029600 140.00 - 2.0000 1.000000 500.0
O3 0.298600 0.500000 0.369400 140.00 - 2.0000 1.000000 500.0
O4 0.472200 0.000000 0.345900 140.00 - 2.0000 1.000000 500.0
O5 0.000000 0.500000 0.000000 140.00 - 2.0000 1.000000 500.0
---------------------
Ergebnisse für Sm1
---------------------
ZT Sm1 0.1357 0.5000 0.4897 Abstand ECoN( 1) ECoN( 3)
1 O4 0.0278 0.5000 0.6541 228.839 1.232 1.263
2 O3 0.2014 1.0000 0.6306 230.469 1.190 1.222
3 O3 0.2014 0.0000 0.6306 230.469 1.190 1.222
4 O4 -0.0278 0.5000 0.3459 244.429 0.839 0.871
5 O1 0.1312 1.0000 0.2867 254.695 0.604 0.634
6 O1 0.1312 0.0000 0.2867 254.695 0.604 0.634
7 O3 0.2986 0.5000 0.3694 270.724 0.310 0.333
Nächster Ligand :
8 O1 0.3688 0.5000 0.7133 354.680 0.000 0.000
Letzter Ligand :
19 O1 -0.1312 0.0000 0.7133 491.400 0.000 0.000
Koordinationszahl : 7
---------------------
Ergebnisse für Sm2
---------------------
ZT Sm2 0.1800 0.5000 0.1362 Abstand ECoN( 1) ECoN( 3)
1 O2 0.1746 1.0000 -0.0296 232.661 1.183 1.191
2 O2 0.1746 0.0000 -0.0296 232.661 1.183 1.191
3 O2 0.3254 0.5000 0.0296 241.268 0.966 0.974
4 O1 0.1312 1.0000 0.2867 242.202 0.943 0.951
```

```
5 O1 0.1312 0.0000 0.2867 242.202 0.943 0.951
6 O3 0.2986 0.5000 0.3694 242.689 0.931 0.939
7 O5 0.0000 0.5000 0.0000 262.754 0.484 0.491
Nächster Ligand :
8 O4 -0.0278 0.5000 0.3459 374.681 0.000 0.000
Letzter Ligand :
22 O1 0.3688 0.5000 -0.2867 496.859 0.000 0.000
Koordinationszahl : 7
---------------------
Ergebnisse für Sm3
---------------------
ZT Sm3 0.4672 0.5000 0.1865 Abstand ECoN( 1) ECoN( 3)
1 O2 0.3254 0.5000 0.0296 223.738 1.232 1.259
2 O4 0.4722 1.0000 0.3459 229.200 1.089 1.116
3 O4 0.4722 0.0000 0.3459 229.200 1.089 1.116
4 O1 0.6312 0.5000 0.2867 234.129 0.962 0.989
5 O5 0.5000 1.0000 0.0000 254.987 0.480 0.503
6 O5 0.5000 0.0000 0.0000 254.987 0.480 0.503
Nächster Ligand :
7 O3 0.2986 0.5000 0.3694 311.270 0.009 0.010
Letzter Ligand :
20 O3 0.2986 -0.5000 0.3694 477.984 0.000 0.000
Koordinationszahl : 6
```

General Index

absolute hardness 25
acceptor 105
acetylacetonato complexes 34
activator 105
adiabatic demagnetisation 201
adiabatic magnetisation 201
afterglow 143
Alexandrite effect 146
aluminates 86
angular node 13
anisotropy coefficient 196
anisotropy energy 196
antenna phosphors 105, 120, 128, 131, 132, 139, 149
anti-Stokes process 148
antiferromagnetism 176, 192
applications
– adiabatic demagnetisation 201
– adiabatic magnetisation 201
– aerospace thrusters 195
– alloy additive 43
– antenna phosphors 120, 139, 149
– anti-Stokes phosphor 148
– bio-imaging 145
– brand identification 150
– car motors 195
– catalysis 74, 75
– compact fluorescent lamp 128, 130, 138, 154, 160
– computer hard-disk drives 195
– contact process 75
– data storage 82
– detector for radiation 145
– downconversion 154, 157
– electric vehicle traction 195
– electron emitter 78
– energy saving lamps 138
– four-level laser 151
– hard-disc drive 201
– hot cathode 78
– hydrogen storage 54
– imaging plate detector 145, 163
– infrared sensor 82
– IPDS 145
– laser 147, 155
– lasers in medicine 152, 155
– lighters 43
– liquefaction of gases 202

– long lasting luminescence 163
– loudspeaker 195
– magnetic cooling 201
– magnetic refrigeration 201
– magnetocaloric materials 195, 201
– microphone 195
– multiferroic materials 195
– optical data storage 172
– optical memory 172
– oxygen pressure buffer 75
– oxygen storage 74
– permanent magnets 79, 195
– persistent phosphors 143, 163, 166
– photocatalysis 145
– photovoltaics 149
– plasma display panels (PDPs) 140, 153
– product identification 150
– product safety markers 165
– quantum cutting 154, 157
– safety markers 143, 166
– scintillation phosphor 128
– scintillation phosphors 161
– signal amplification 147
– solar cells 149
– solar luminescent concentrators 172
– solid state laser 146, 150–152, 155
– spectral hole burning 141
– storage phosphor 161, 163
– superconductors 77
– thermoluminescence 143
– thermometry 128
– three-way catalytic converters 75
– transparent display 149
– upconversion 134, 146–148, 156, 157
– upconversion laser 156
– water disinfection 165
– white light emitting diode 160, 161, 166, 167
– wind generators 195
– wind turbine 79
– Xe plasma lamp 140, 153
aqua complexes 31
assignment of bands 151
atomic properties
– diffuseness 12
– electron affinity 12
– electronegativity 12

https://doi.org/10.1515/9783110680829-017

Formula Index

https://doi.org/10.1515/9783110680829-018

Bibliography

[1]   J. Gadolin. Undersökning af en svart tung Stenart ifrån Ytterby Stenbrott i Roslagen. Proc. R. Acad. (Stockholm, new series) 15 (1794), 137–155.

[2]   M. Grohol. Study on the critical raw materials for the EU 2023 (2023).

[3]   U. S. G. Survey. Mineral commodity summaries 2022. Mineral Commodity Summaries (2022), 202. ISSN 2022. https://doi.org/10.3133/mcs2022.

[4]   T. Prohaska, J. Irrgeher, J. Benefield, J. K. Böhlke, L. A. Chesson, T. B. Coplen, T. Ding, P. J. H. Dunn, M. Gröning, N. E. Holden, H. A. J. Meijer, H. Moossen, A. Possolo, Y. Takahashi, J. Vogl, T. Walczyk, J. Wang, M. E. Wieser, S. Yoneda, X.-K. Zhu and J. Meija. Standard atomic weights of the elements 2021 (IUPAC Technical Report). Pure Appl. Chem. 94 (2022), 573–600. https://doi.org/10.1515/pac-2019-0603.

[5]   D. Holtstam and U. B. Andersson. The REE minerals of the bastnas-type deposits, South-Central Sweden. Can. Mineral. 45 (2007), 1073–1114. https://doi.org/10.2113/gscanmin.45.5.1073.

[6]   A. Pieczka, A. Szuszkiewicz, E. Szełęg, S. Ilnicki, K. Nejbert and K. Turniak. Samarskite-group minerals and alteration products: an example from the Julianna Pegmatitic System, Piława Górna, SW Poland. Can. Mineral. 52 (2014), 303–319. https://doi.org/10.3749/canmin.52.2.303.

[7]   F. Szabadvary. Chapter 73 The history of the discovery and separation of the rare earths. In Handbook on the Physics and Chemistry of Rare Earths, pp. 33–80. Elsevier (1988). https://doi.org/10.1016/s0168-1273(88)11005-2.

[8]   M. Klaproth. Chemical analysis of a Swedish fossil, in which a new Earth was discovered. Med. Phys. J. 12 (1804), 69–73.

[9]   W. D. Hisinger and J. Berzelius. XXVI. On cerium, a new metal found in a mineral substance of bastnas in Sweden, called tungsten. Philos. Mag. 20 (1804), 154–158. https://doi.org/10.1080/14786440408676616.

[10] C. Mosander. XXX. On the new metals, lanthanium and didymium, which are associated with cerium; and on erbium and terbium, new metals associated with yttria. The London, Edinburgh, and Dublin Philosophical Magazine and Journal of Science 23 (1843), 241–254. https://doi.org/10.1080/14786444308644728.

[11] J. A. Marinsky, L. E. Glendenin and C. D. Coryell. The chemical identification of radioisotopes of neodymium and of element 61. J. Am. Chem. Soc. 69 (1947), 2781–2785. https://doi.org/10.1021/ja01203a059.

[12] Anuraag. Electronegativity (2015). Accessed 2022-08-19.

[13] S. T. Liddle. International year of the periodic table: lanthanide and actinide chemistry. Angew. Chem. Int. Ed. 58 (2019), 5140–5141. https://doi.org/10.1002/anie.201901578.

[14] G. Steinhauser. Wohin mit dem f-Block? Nachr. Chem. 67 (2019), 8–11. https://doi.org/10.1002/nadc.20194086536.

[15] T. Benfey. The biography of a periodic spiral. Bull. Hist. Chem. 34 (2009), 141–145.

[16] M. Kaupp. The role of radial nodes of atomic orbitals for chemical bonding and the periodic table. J. Comput. Chem. 28 (2007), 320–325. https://doi.org/10.1002/jcc.20522.

[17] R. Dronskowski. Computational Chemistry of Solid State Materials. Wiley VCH, 1. edition (2005).

[18] L. C. Allen. Electronegativity is the average one-electron energy of the valence-shell electrons in ground-state free atoms. J. Am. Chem. Soc. 111 (1989), 9003–9014.

[19] L. C. Allen. Extension and completion of the periodic table. J. Am. Chem. Soc. 114 (1992), 1510–1511.

[20] M. Rahm, T. Zeng and R. Hoffmann. Electronegativity seen as the ground-state average valence electron binding energy. J. Am. Chem. Soc. 141 (2018), 342–351. https://doi.org/10.1021/jacs.8b10246.

[21] P. Dorenbos. f $\longrightarrow$ d transition energies of divalent lanthanides in inorganic compounds. J. Phys. Condens. Matter 15 (2003), 575–594.

[22] S. Alvarez. A cartography of the van der Waals territories. Dalton Trans. 42 (2013), 8617–8636. https://doi.org/10.1039/C3DT50599E.

https://doi.org/10.1515/9783110680829-019

[23]   F. H. Spedding, A. H. Daane and K. W. Herrmann. The crystal structures and lattice parameters of high-purity scandium, yttrium and the rare earth metals. Acta Crystallogr. 9 (1956), 559–563. https://doi.org/10.1107/s0365110x5600156x.

[24]   K. Gschneidner, R. Elliott and R. McDonald. Effects of alloying additions on the alpha—gamma transformation of cerium. J. Phys. Chem. Solids 23 (1962), 1191–1199. https://doi.org/10.1016/0022-3697(62)90166-x.

[25]   K. Takemura and K. Syassen. Pressure-volume relations and polymorphism of europium and ytterbium to 30 GPa. J. Phys. F, Met. Phys. 15 (1985), 543–559. https://doi.org/10.1088/0305-4608/15/3/010.

[26]   A. H. Daane, R. E. Rundle, H. G. Smith and F. H. Spedding. The crystal structure of samarium. Acta Crystallogr. 7 (1954), 532–535. https://doi.org/10.1107/s0365110x54001818.

[27]   F. Spedding and B. Beaudry. The effect of impurities, particularly hydrogen, on the lattice parameters of the "ABAB" rare earth metals. J. Less-Common Met. 25 (1971), 61–73. https://doi.org/10.1016/0022-5088(71)90066-x.

[28]   B. J. Beaudry and F. H. Spedding. The solubility of RH_{2-x} in Gd, Er, Tm, Lu and Y from ambient to 850 °C. Metall. Trans. B 6 (1975), 419–427. https://doi.org/10.1007/bf02913827.

[29]   W. M. Haynes. CRC Handbook of Chemistry and Physics, 95th Edition. Taylor and Francis Group (2014). ISBN 9781482208689.

[30]   B. J. Beaudry and K. A. J. Gschneidner. Preparation and Basic Properties of the Rare Earth Metals Chapter 2. North-Holland, Netherlands (1978).

[31]   D. C. Koskenmaki and K. A. Gschneidner. Chapter 4 Cerium. In Metals, pp. 337–377. Elsevier (1978). https://doi.org/10.1016/s0168-1273(78)01008-9.

[32]   B. Johansson and A. Rosengren. Generalized phase diagram for the rare-earth elements: Calculations and correlations of bulk properties. Phys. Rev. B 11 (1975), 2836–2857. https://doi.org/10.1103/physrevb.11.2836.

[33]   B. N. Harmon and A. J. Freeman. Spin-polarized energy-band structure, conduction-electron polarization, spin densities, and the neutron magnetic form factor of ferromagnetic gadolinium. Phys. Rev. B 10 (1974), 1979–1993. https://doi.org/10.1103/physrevb.10.1979.

[34]   A. Rosengren and B. Johansson. Alloy theory of the intermediate valence state: Application to europium metal. Phys. Rev. B 13 (1976), 1468–1472. https://doi.org/10.1103/physrevb.13.1468.

[35]   B. J. Jensen, F. J. Cherne and N. Velisavljevic. Dynamic experiments to study the alpha-epsilon phase transition in cerium. J. Appl. Phys. 127 (2020). https://doi.org/10.1063/1.5142508.

[36]   K. A. Munro, D. Daisenberger, S. G. MacLeod, S. McGuire, I. Loa, C. Popescu, P. Botella, D. Errandonea and M. I. McMahon. The high-pressure, high-temperature phase diagram of cerium. J. Phys. Condens. Matter 32 (2020), 335401. https://doi.org/10.1088/1361-648x/ab7f02.

[37]   A. V. Nikolaev and A. V. Tsvyashchenko. The puzzle of the gamma to alpha and other phase transitions in cerium. Phys. Usp. 55 (2012), 657–680. https://doi.org/10.3367/ufne.0182.201207b.0701.

[38]   W. H. Zachariasen. On the crystal structure of alpha'-cerium. J. Appl. Phys. 48 (1977), 1391–1394. https://doi.org/10.1063/1.323877.

[39]   W. H. Zachariasen. Crystal structures of the alpha-cerium phases. Proc. Natl. Acad. Sci. 75 (1978), 1066–1067. https://doi.org/10.1073/pnas.75.3.1066.

[40]   S. Endo, H. Sasaki and T. Mitsui. A new allotropic phase of cerium above 121 kbar. J. Phys. Soc. Jpn. 42 (1977), 882–885. https://doi.org/10.1143/jpsj.42.882.

[41]   X.-G. Zhu, Y. Liu, Y.-W. Zhao, Y.-C. Wang, Y. Zhang, C. Lu, Y. Duan, D.-H. Xie, W. Feng, D. Jian, Y.-H. Wang, S.-Y. Tan, Q. Liu, W. Zhang, Y. Liu, L.-Z. Luo, X.-B. Luo, Q.-Y. Chen, H.-F. Song and X.-C. Lai. Kondo scenario of the gamma-palpha phase transition in single crystalline cerium thin films. Quantum Mater. 5 (2020). https://doi.org/10.1038/s41535-020-0248-y.

[42]   I. Loa, E. I. Isaev, M. I. McMahon, D. Y. Kim, B. Johansson, A. Bosak and M. Krisch. Lattice dynamics and superconductivity in cerium at high pressure. Phys. Rev. Lett. 108 (2012), 045502. https://doi.org/10.1103/physrevlett.108.045502.

[43] W. Chen, D. V. Semenok, X. Huang, H. Shu, X. Li, D. Duan, T. Cui and A. R. Oganov. High-temperature superconducting phases in cerium superhydride with a Tc up to 115 K below a pressure of 1 Megabar. Phys. Rev. Lett. 127 (2021), 117001. https://doi.org/10.1103/physrevlett.127.117001.

[44] R. G. Parr and R. G. Pearson. Absolute hardness: companion parameter to absolute electronegativity. J. Am. Chem. Soc. 105 (1983), 7512–7516. https://doi.org/10.1021/ja00364a005.

[45] R. G. Parr and Z. Zhou. Absolute hardness: unifying concept for identifying shells and subshells in nuclei, atoms, molecules, and metallic clusters. Acc. Chem. Res. 26 (1993), 256–258. https://doi.org/10.1021/ar00029a005.

[46] L. Pauling. The Nature of the Chemical Bond and the Structure of Molecules and Crystals: an Introduction to Modern Structural Chemistry. The George Fisher Baker Non-Resident Lectureship in Chemistry at Cornell University. Cornell University Press, 3. edition (1960). ISBN 9780801403330.

[47] G. Borzone, R. Raggio and R. Ferro. Thermochemistry and reactivity of rare earth metals. Phys. Chem. Chem. Phys. 1 (1999), 1487–1500. https://doi.org/10.1039/a900312f.

[48] L. R. Morss. Thermochemical properties of yttrium, lanthanum, and the lanthanide elements and ions. Chem. Rev. 76 (1976), 827–841. https://doi.org/10.1021/cr60304a007.

[49] J. A. Rard. Chemistry and thermodynamics of europium and some of its simpler inorganic compounds and aqueous species. Chem. Rev. 85 (1985), 555–582. https://doi.org/10.1021/cr00070a003.

[50] R. D. Shannon. Revised effective ionic radii and systematic studies of interatomie distances in halides and chalcogenides. Acta Crystallogr. A 32 (1976), 751–767. https://doi.org/10.1107/S0567739476001551.

[51] R. Albrecht, T. Doert and M. Ruck. The hydrogarnets $Sr_3[RE(OH)_6]_2$ (RE = Sc, Y, Ho – Lu): Syntheses, crystal structures, and their thermal decomposition to ternary rare-earth metal oxides. Z. Anorg. Allg. Chem. 646 (2020), 1517–1524. https://doi.org/10.1002/zaac.202000031.

[52] R. Hoppe and H. Sabrowsky. Über Scandate, Yttrate, Indate und Thallate der Alkalimetalle. Z. Anorg. Allg. Chem. 357 (1968), 202–214. https://doi.org/10.1002/zaac.19683570408.

[53] C. Hoch, J. Bender, A. Wohlfarth and A. Simon. Die Suboxometallate A_9MO_4 (A = Rb, Cs; M = Al, Ga, In, Fe, Sc). Z. Anorg. Allg. Chem. 635 (2009), 1777–1782. https://doi.org/10.1002/zaac.200900193.

[54] L. Helm and A. E. Merbach. Inorganic and bioinorganic solvent exchange mechanisms. Chem. Rev. 105 (2005), 1923–1960. https://doi.org/10.1021/cr030726o.

[55] M. Duvail, R. Spezia and P. Vitorge. A dynamic model to explain hydration behaviour along the lanthanide series. ChemPhysChem 9 (2008), 693–696. https://doi.org/10.1002/cphc.200700803.

[56] A. Zalkin, J. D. Forrester and D. H. Templeton. Crystal structure of cerium magnesium nitrate hydrate. J. Chem. Phys. 39 (1963), 2881–2891. https://doi.org/10.1063/1.1734120.

[57] T. A. Beineke and J. Delgaudio. Crystal structure of ceric ammonium nitrate. Inorg. Chem. 7 (1968), 715–721. https://doi.org/10.1021/ic50062a020.

[58] G. Meyer, E. Manek and A. Reller. $(NH_4)_2[Pr(NO_3)_5(H_2O)_2] \cdot 2\,H_2O$ Kristallstruktur und thermisches Verhalten. Z. Anorg. Allg. Chem. 591 (1990), 77–86. https://doi.org/10.1002/zaac.19905910109.

[59] E. Manek and G. Meyer. Synthese und Struktur der ternären Ammoniumnitrate $(NH_4)_2[M(NO_3)_5]$ (M = Tb...Lu, Y). Z. Anorg. Allg. Chem. 619 (1993), 1237–1240. https://doi.org/10.1002/zaac.19936190715.

[60] Y.-W. Zhang, Z.-M. Wang, J.-T. Jia, C.-S. Liao and C.-H. Yan. CCDC 135821: Experimental Crystal Structure Determination (1999). https://doi.org/10.5517/CC4KBBB.

[61] D. Matković-Čalogović. Structure of potassium triaqua(ethylenediaminetetraacetato)neodymate(III) pentahydrate. Acta Crystallogr., Sect. C 44 (1988), 435–437. https://doi.org/10.1107/s0108270187011004.

[62] A. B. Ganaie and K. Iftikhar. Theoretical modeling (Sparkle RM1 and PM7) and crystal structures of the luminescent dinuclear Sm(III) and Eu(III) complexes of 6,6,7,7,8,8,8-heptafluoro-2,2-dimethyl-3,5-octanedione and 2,3-Bis(2-pyridyl)pyrazine: Determination of individual spectroscopic parameters for two unique Eu3+ sites. ACS Omega 6 (2021), 21207–21226. https://doi.org/10.1021/acsomega.0c05976.

[63]   R. E. Sievers and R. E. Rondeau. New superior paramagnetic shift reagents for nuclear magnetic resonance spectral clarification. J. Am. Chem. Soc. 93 (1971), 1522–1524. https://doi.org/10.1021/ja00735a049.

[64]   C. F. v. Weizsäcker. Zur Theorie der Kernmassen. Z. Phys. 96 (1935), 431–458. https://doi.org/10.1007/bf01337700.

[65]   E. Gapon and D. Iwanenko. Zur Bestimmung der Isotopenzahl. Naturwissenschaften 20 (1932), 792–793. https://doi.org/10.1007/bf01494007.

[66]   M. Attrep and P. Kuroda. Promethium in pitchblende. J. Inorg. Nucl. Chem. 30 (1968), 699–703. https://doi.org/10.1016/0022-1902(68)80427-0.

[67]   K. H. Wedepohl. The composition of the continental crust. Geochim. Cosmochim. Acta 59 (1995), 1217–1232. ISSN 0016-7037. https://doi.org/10.1016/0016-7037(95)00038-2.

[68]   L. A. Haskin and T. Paster. Chapter 21 Geochemistry and mineralogy of the rare earths. In Non-metallic Compounds – I, volume 3 of Handbook on the Physics and Chemistry of Rare Earths, pp. 1–80. Elsevier (1979). https://doi.org/10.1016/S0168-1273(79)03004-X.

[69]   Y. Ni, J. M. Hughes and A. N. Mariano. Crystal chemistry of the monazite and xenotime structures. Am. Mineral. 80 (1995), 21–26. ISSN 0003-004X. https://doi.org/10.2138/am-1995-1-203.

[70]   J. M. Hughes, M. Cameron and K. D. Crowley. Structural variations in natural F, OH, and Cl apatites. Am. Mineral. 74 (1989), 870–876. ISSN 0003-004X.

[71]   T. Yamanaka, N. Hirai and Y. Komatsu. Structure change of $Ca_{1-x}Sr_xTiO_3$ perovskite with composition and pressure. Am. Mineral. 87 (2002), 1183–1189. ISSN 0003-004X. https://doi.org/10.2138/am-2002-8-917.

[72]   Y. Ni, J. M. Hughes and A. N. Mariano. The atomic arrangement of bastnäesite-(Ce), $Ce(CO_3)F$, and structural elements of synchysite-(Ce), röntgenite-(Ce), and parisite-(Ce). Am. Mineral. 78 (1993), 415–418. ISSN 0003-004X.

[73]   W. Liben, Y. Ni, J. M. Hughes, P. Bayliss and J. W. Drexler. The atomic arrangement of synchysite-(Ce), $CeCaF(CO_3)_2$. Can. Mineral. 32 (1994), 865–871.

[74]   P. Bonazzi, D. Holtstam, L. Bindi, P. Nysten and G. Capitani. Multi-analytical approach to solve the puzzle of an allanite-subgroup mineral from Kesebol, Vaästra Götaland, Sweden. Am. Mineral. 94 (2009), 121–134. ISSN 0003-004X. https://doi.org/10.2138/am.2009.2998.

[75]   J. A. Speer and G. V. Gibbs. The crystal structure of synthetic titanite, $CaTiOSiO_4$, and the domain textures of natural titanites. Am. Mineral. 61 (1976), 238–247. ISSN 0003-004X.

[76]   R. M. Hazen and L. W. Finger. Crystal structure and compressibility of zircon at high pressure. Am. Mineral. 64 (1979), 196–201. ISSN 0003-004X.

[77]   R. W. G. Wyckoff. Fluorite structure. Cryst. Struct. 1 (1963), 239–444.

[78]   Y. A. Pakhomovsky, Y. P. Men'shikov, V. N. Yakovenchuk, G. Y. Ivanyuk, S. V. Krivovichev and P. C. Burns. Cerite-La, $(La,Ce,Ca)_9(Fe,Ca,Mg)(SiO_4)_3[SiO_3(OH)]_4(OH)_3$, a new mineral species from the khibina alkaline massif: occurrence and crystal structure. Can. Mineral. 40 (2002), 1177–1184. https://doi.org/10.2113/gscanmin.40.4.1177.

[79]   Y. Sugitani, Y. Suzuki and K. Nagashima. Polymorphism of samarskite and its relationship to other structurally related Nb–Ta oxides with the α-PbO_2 structure. Am. Mineral. 70 (1985), 856–866. ISSN 0003-004X.

[80]   K. Long, B. Van Gosen, N. Foley and D. Cordier. The Principal Rare Earth Elements Deposits of theUnited States—A Summary of Domestic Depositsand a Global Perspective. U.S. Geological Survey Scientific Investigations Report 5220 (2010).

[81]   C. Mariano, A. Cox and J. Hedrick. Economic evaluation of REE and Y mineral deposits. In Presentation at the 2010 Annual Meeting: Phoenix, Arizona, p. 33. Society for Mining, Metallurgy & Exploration [SME] (2010).

[82]   A. Walters, P. Lusty and A. Hill. Rare Earth Elements. British Geological Survey, Nottingham (2011).

[83]   C. K. Gupta and N. Krishnamurthy. Extractive Metallurgy of Rare Earths. CRC (2005). ISBN 9780415333405.

[84]   W. Jackson and G. Christiansen. International strategic minerals inventory summary report; rare-earth oxides. US Geological Survey (1993). https://doi.org/10.3133/cir930n.

[85]   J. E. Powell. Chapter 22 Separation chemistry. In Non-metallic Compounds – I, pp. 81–109. Elsevier (1979). https://doi.org/10.1016/s0168-1273(79)03005-1.

[86]   B. J. Tickner, G. J. Stasiuk, S. B. Duckett and G. Angelovski. The use of yttrium in medical imaging and therapy: historical background and future perspectives. Chem. Soc. Rev. 49 (2020), 6169–6185. https://doi.org/10.1039/c9cs00840c.

[87]   H. Cheng and S. C. Nixon. Laboratory studies into the use of the scandium-46-EDTA complex as a tracer for groundwater. Technical Report INIS-mf-11357, Water Research Centre, Medmenham, International Atomic Energy Agency (IAEA) (1988).

[88]   L. Wang, X. Huang, Y. Yu, Y. Xiao, Z. Long and D. Cui. Eliminating ammonia emissions during rare earth separation through control of equilibrium acidity in a HEH(EHP)-Cl system. Green Chem. 15 (2013), 1889. https://doi.org/10.1039/c3gc40470f.

[89]   M. Regadío, T. V. Hoogerstraete, D. Banerjee and K. Binnemans. Split-anion solvent extraction of light rare earths from concentrated chloride aqueous solutions to nitrate organic ionic liquids. RSC Adv. 8 (2018), 34754–34763. https://doi.org/10.1039/c8ra06055j.

[90]   B. Dewulf, V. Cool, Z. Li and K. Binnemans. Effect of polar molecular organic solvents on non-aqueous solvent extraction of rare-earth elements. Sep. Purif. Technol. 294 (2022), 121197. https://doi.org/10.1016/j.seppur.2022.121197.

[91]   W. D. Bonificio and D. R. Clarke. Rare-Earth separation using bacteria. Environ. Sci. Technol. Lett. 3 (2016), 180–184. https://doi.org/10.1021/acs.estlett.6b00064.

[92]   K. Binnemans, P. T. Jones, B. Blanpain, T. V. Gerven, Y. Yang, A. Walton and M. Buchert. Recycling of rare earths: a critical review. J. Clean. Prod. 51 (2013), 1–22. https://doi.org/10.1016/j.jclepro.2012.12.037.

[93]   P. Boelens, Z. Lei, B. Drobot, M. Rudolph, Z. Li, M. Franzreb, K. Eckert and F. Lederer. High-gradient magnetic separation of compact fluorescent lamp phosphors: Elucidation of the removal dynamics in a rotary permanent magnet separator. Minerals 11 (2021), 1116. https://doi.org/10.3390/min11101116.

[94]   C. Ma, J. R. Beckett and G. R. Rossman. Allendeite $Sc_4Zr_3O_{12}$ and hexamolybdenum (Mo,Ru,Fe), two new minerals from an ultrarefractory inclusion from the Allende meteorite. Am. Mineral. 99 (2014), 654–666. https://doi.org/10.2138/am.2014.4667.

[95]   H. Yang, C. Li, R. A. Jenkins, R. T. Downs and G. Costin. Kolbeckite, $ScPO_4 \cdot 2H_2O$, isomorphous with metavariscite. Acta Crystallogr., Sect. C 63 (2007), i91–i92. https://doi.org/10.1107/s0108270107038036.

[96]   C. Keller. Über ternäre Oxide des Niobs und Tantals vom Typ ABO_4. Z. Anorg. Allg. Chem. 318 (1962), 89–106. https://doi.org/10.1002/zaac.19623180108.

[97]   W. Wang and C. Y. Cheng. Separation and purification of scandium by solvent extraction and related technologies: a review. J. Chem. Technol. Biotechnol. 86 (2011), 1237–1246. https://doi.org/10.1002/jctb.2655.

[98]   J. N. Huiberts, R. Griessen, J. H. Rector, R. J. Wijngaarden, J. P. Dekker, D. G. de Groot and N. J. Koeman. Yttrium and lanthanum hydride films with switchable optical properties. Nature 380 (1996), 231–234. https://doi.org/10.1038/380231a0.

[99]   Z. M. Geballe, H. Liu, A. K. Mishra, M. Ahart, M. Somayazulu, Y. Meng, M. Baldini and R. J. Hemley. Synthesis and stability of lanthanum superhydrides. Angew. Chem. Int. Ed. 57 (2018), 688–692. https://doi.org/10.1002/anie.201709970.

[100] G. Libowitz and A. Maeland. Chapter 26 Hydrides. In Non-metallic Compounds – I, volume 3 of Handbook on the Physics and Chemistry of Rare Earths, pp. 299–336. Elsevier (1979). https://doi.org/10.1016/S0168-1273(79)03009-9.

[101] M. E. Kost, N. T. Kuznetsov and A. L. Shilov. New hydride phases in the systems R-Mg-H (R = Y, La, Ce). Dokl., Chem. (Engl. Transl.); (United States) 292:3 (1987).

[102] M. Ellner, H. Reule and E. Mittemeijer. The structure of the trihydride GdH$_3$. J. Alloys Compd. 309 (2000), 127–131. https://doi.org/10.1016/s0925-8388(00)01055-0.

[103] R. Hoppe. Madelung constants. Angew. Chem. Int. Ed. 5 (1966), 95–106.

[104] R. Hoppe. Coordination number – An inorganic chameleon. Angew. Chem. Int. Ed. 9 (1970), 25–34.

[105] R. Hübenthal. MAPLE, Program for the Calculation of the Madelung Part of Lattice Energy. Software (1993).

[106] D. Colognesi, G. Barrera, A. Ramirez-Cuesta and M. Zoppi. Hydrogen self-dynamics in orthorhombic alkaline earth hydrides through incoherent inelastic neutron scattering. J. Alloys Compd. 427 (2007), 18–24. https://doi.org/10.1016/j.jallcom.2006.03.031.

[107] V. V. Burnasheva, V. A. Yarty's, N. Fadeeva, S. Solov'ev and K. N. Semenenko. The crystal structure of the deuteride LaNi 5 D 6.0. Sov. Phys. Dokl. 23 (1978), 97.

[108] H. Eick. Chapter 124 Lanthanide and actinide halides. In Lanthanides/Actinides: Chemistry, volume 18 of Handbook on the Physics and Chemistry of Rare Earths, pp. 365–411. Elsevier (1994). https://doi.org/10.1016/S0168-1273(05)80047-9.

[109] J. M. Haschke. Chapter 32 Halides. In Handbook on the Physics and Chemistry of Rare Earths, pp. 89–151. Elsevier (1979). https://doi.org/10.1016/s0168-1273(79)04005-8.

[110] R. Konings and A. Kovács. Thermodynamic properties of the lanthanide(III) halides. In Handbook on the Physics and Chemistry of Rare Earths, pp. 147–247. Elsevier (2003). https://doi.org/10.1016/s0168-1273(02)33003-4.

[111] G. Meyer and M. S. Wickleder. Simple and complex halides. In Handbook on the Physics and Chemistry of Rare Earths Volume 28, pp. 53–129. Elsevier (2000). https://doi.org/10.1016/s0168-1273(00)28005-7.

[112] G. Brauer. Handbuch der präparativen anorganischen Chemie. F. Enke (1975). ISBN 3432023286.

[113] J. M. Haschke. Specialty chemicals. An identity problem. J. Chem. Educ. 52 (1975), 157. https://doi.org/10.1021/ed052p157.

[114] S. L. Holt (editor). Inorganic Syntheses. Wiley (1984). https://doi.org/10.1002/9780470132531.

[115] J. M. Haschke. The phase equilibria, vaporization behavior, and thermodynamic properties of europium tribromide. J. Chem. Thermodyn. 5 (1973), 283–290. https://doi.org/10.1016/s0021-9614(73)80088-6.

[116] J. B. Reed, B. S. Hopkins, L. F. Audrieth, P. W. Selwood, R. Ward and J. J. Dejong. Anhydrous rare Earth chlorides. In Inorganic Syntheses, pp. 28–33. John Wiley & Sons, Inc. (2007). https://doi.org/10.1002/9780470132326.ch11.

[117] A. Zalkin, D. H. Templeton and T. E. Hopkins. The atomic parameters in the lanthanum trifluoride structure. Inorg. Chem. 5 (1966), 1466–1468. https://doi.org/10.1021/ic50042a047.

[118] A. Cheetham and N. Norman. The structures of yttrium and bismuth trifluorides by neutron. Acta Chem. Scand. A 28 (1974).

[119] K. Rotereau, P. Daniel, A. Desert and J. Y. Gesland. The high-temperature phase transition in samarium fluoride,: structural and vibrational investigation. J. Phys. Condens. Matter 10 (1998), 1431–1446. https://doi.org/10.1088/0953-8984/10/6/026.

[120] R. Løusch, C. Hebecker and Z. Ranft. Röntgenographische Untersuchungen an neuen ternären Fluoriden vom Typ TlIII MF$_6$ (M = Ga, In, Sc) sowie an Einkristallen von ScF$_3$. Z. Anorg. Allg. Chem. 491 (1982), 199–202. https://doi.org/10.1002/zaac.19824910125.

[121] B. Morosin. Crystal structures of anhydrous rare-earth chlorides. J. Chem. Phys. 49 (1968), 3007–3012. https://doi.org/10.1063/1.1670543.

[122] S. Lidin, T. Popp, M. Somer and H. G. von Schnering. Generalized edshammar polyhedra for the description of a family of solid-state structures. Angew. Chem. Int. Ed. 31 (1992), 924–927. https://doi.org/10.1002/anie.199209241.

[123] B. Hyde and S. Andersson. Inorganic Crystal Structures. Wiley (1989). ISBN 9780471628972.

[124] W. H. Zachariasen. Crystal chemical studies of the 5f-series of elements. I. New structure types. Acta Crystallogr. 1 (1948), 265–268. https://doi.org/10.1107/s0365110x48000703.

[125] S. Troyanov. Crystal structure of Ti(AlCl$_4$)$_2$ and refinement of the crystal structure of AlCl$_3$. ChemInform 23 (1992), no–no.

[126] W. Klemm and E. Krose. Die Kristallstrukturen von ScCl$_3$, TiCl$_3$ und VCl$_3$. Z. Anorg. Allg. Chem. 253 (1947), 218–225. https://doi.org/10.1002/zaac.19472530313.

[127] Z. Mazej. Room temperature syntheses of lanthanoid tetrafluorides LnF$_4$, Ln = Ce, Pr, Tb. J. Fluorine Chem. 118 (2002), 127–129. https://doi.org/10.1016/s0022-1139(02)00223-3.

[128] L. B. Asprey and B. B. Cunningham. Unusual oxidation states of some actinide and lanthanide elements. In Progress in Inorganic Chemistry, pp. 267–302. John Wiley & Sons, Inc. (2007). https://doi.org/10.1002/9780470166031.ch6.

[129] R. Schmidt and B. Müller. Einkristalluntersuchungen an Au[AuF$_4$]$_2$ und CeF$_4$, zwei unerwarteten Nebenprodukten. Z. Anorg. Allg. Chem. 625 (1999), 605–608.

[130] H. Bärnighausen, H. Pätow and H. P. Beck. Kristallchemische Studien an Seltenerd-Dihalogeniden. Die Kristallstruktur von Ytterbium(II)-chlorid, YbCl$_2$. Z. Anorg. Allg. Chem. 403 (1974), 45–55. https://doi.org/10.1002/zaac.19744030106.

[131] V. F. Goryushkin, I. S. Astakhova, A. P. Poshevneva and S. A. Zalymova. On crystalline holmium dichloride. Zh. Fiz. Khim. 34 (1989), 2469–2472. ISSN 0044-457X.

[132] V. F. Goryushkin and A. I. Poshevneva. Thermodynamic characteristics of holmium dichloride. Zh. Fiz. Khim. 68 (1994), 172–173. ISSN 0044-4537.

[133] J. D. Corbett. Conproportionation routes to reduced lanthanide halides. In Topics in f-Element Chemistry, pp. 159–173. Springer Netherlands (1991). https://doi.org/10.1007/978-94-011-3758-4_6.

[134] H. Mattausch, J. B. Hendricks, R. Eger, J. D. Corbett and A. Simon. Reduced halides of yttrium with strong metal-metal bonding: yttrium monochloride, monobromide, sesquichloride, and sesquibromide. Inorg. Chem. 19 (1980), 2128–2132. https://doi.org/10.1021/ic50209a057.

[135] K. R. Poeppelmeier and J. D. Corbett. Metal-metal bonding in reduced scandium halides. Synthesis and characterization of heptascandium decachloride Sc$_7$Cl$_{10}$. A novel metal-chain structure. Inorg. Chem. 16 (1977), 1107–1111. https://doi.org/10.1021/ic50171a026.

[136] K. R. Poeppelmeier and J. D. Corbett. Metal-metal bonding in reduced scandium halides. Synthesis and crystal structure of scandium monochloride. Inorg. Chem. 16 (1977), 294–297. https://doi.org/10.1021/ic50168a013.

[137] W. Klemm and H. Bommer. Zur Kenntnis der Metalle der seltenen Erden. Z. Anorg. Allg. Chem. 231 (1937), 138–171. https://doi.org/10.1002/zaac.19372310115.

[138] M. Ryazanov, L. Kienle, A. Simon and H. Mattausch. New synthesis route to and physical properties of lanthanum monoiodide. Inorg. Chem. 45 (2006), 2068–2074. https://doi.org/10.1021/ic051834r.

[139] G. Meyer. The reduction of rare-earth metal halides with unlike metals – Wöhlers metallothermic reduction. Z. Anorg. Allg. Chem. 633 (2007), 2537–2552. https://doi.org/10.1002/zaac.200700386.

[140] R. Masse and A. Simon. Electrochemical preparation of metal cluster compounds. Mater. Res. Bull. 16 (1981), 1007–1011. https://doi.org/10.1016/0025-5408(81)90143-4.

[141] J. Smith L. Holt. Inorganic Syntheses (1939–2018). https://doi.org/10.1002/series2146.

[142] J. P. Sanchez, J. M. Friedt, H. Bärnighausen and A. J. V. Duyneveldt. Structural, magnetic, and electronic properties of europium dihalides EuX$_2$ (X = Cl, Br, I). Inorg. Chem. 24 (1985), 408–415. https://doi.org/10.1021/ic00197a031.

[143] H. Bärnighausen and N. Schulz. Die Kristallstruktur der monoklinen Form von Europium(II)jodid EuJ$_2$. Acta Crystallogr. B 25 (1969), 1104–1110. https://doi.org/10.1107/s0567740869003591.

[144] D. Partin and M. O'Keeffe. The structures and crystal chemistry of magnesium chloride and cadmium chloride. J. Solid State Chem. 95 (1991), 176–183. https://doi.org/10.1016/0022-4596(91)90387-w.

[145] L. Asprey and F. Kruse. Divalent thulium. Thulium di-iodide. J. Inorg. Nucl. Chem. 13 (1960), 32–35. https://doi.org/10.1016/0022-1902(60)80232-1.

[146] A. N. Christensen. Crystal growth and characterization of the transition metal silicides MoSi$_2$ and WSi$_2$. J. Cryst. Growth 129 (1993), 266–268. https://doi.org/10.1016/0022-0248(93)90456-7.

[147] C. Felser, K. Ahn, R. Kremer, R. Seshadri and A. Simon. Giant negative magnetoresistance in GdI_2: Prediction and realization. J. Solid State Chem. 147 (1999), 19–25. https://doi.org/10.1006/jssc.1999.8274.

[148] G. Meyer. The synthesis and structures of complex rare-earth halides. Prog. Solid State Chem.. 14 (1982), 141–219. https://doi.org/10.1016/0079-6786(82)90005-x.

[149] G. Meyer and T. Staffel. Die Tieftemperatur-Synthese von Oxidhalogeniden, YOX (X = Cl, Br, I), als Quelle der Verunreinigung von Yttriumtrihalogeniden, YX_3, bei der Gewinnung nach der Ammoniumhalogenid-Methode. Die Analogie von YOCl und YSCl. Z. Anorg. Allg. Chem. 532 (1986), 31–36. https://doi.org/10.1002/zaac.19865320106.

[150] R. Farra, F. Girgsdies, W. Frandsen, M. Hashagen, R. Schlögl and D. Teschner. Synthesis and catalytic performance of CeOCl in deacon reaction. Catal. Lett. 143 (2013), 1012–1017. https://doi.org/10.1007/s10562-013-1085-4.

[151] Y.-P. Lan, H. Y. Sohn, A. Murali, J. Li and C. Chen. The formation and growth of CeOCl crystals in a molten KCl-LiCl flux. Appl. Phys. A 124 (2018). https://doi.org/10.1007/s00339-018-2122-3.

[152] F. H. Kruse, L. B. Asprey and B. Morosin. The crystal structure of the lanthanide oxyiodides, SmOI, TmOI and YbOI. Acta Crystallogr. 14 (1961), 541–542. https://doi.org/10.1107/s0365110x61001704.

[153] H. Bärnighausen, G. Brauer and N. Schultz. Darstellung und Kristallstruktur der Samarium-, Europium- und Ytterbium-oxidbromide LnOBr und Ln_3O_4Br. Z. Anorg. Allg. Chem. 338 (1965), 250–265. https://doi.org/10.1002/zaac.19653380505.

[154] F. Weige and V. Wishnevsky. Die Dampfphasenhydrolyse von Lanthaniden(III)-chloriden, 3 Wärmetönung und Gibbs-Energie der Reaktion MCl_3 (f) + H_2O(g) $\longleftarrow$ MOCl(f) + 2HCl(g) (M = Er, Tm). Chem. Ber. 103 (1970), 193–199. https://doi.org/10.1002/cber.19701030126.

[155] E. Garcia, J. D. Corbett, J. E. Ford and W. J. Vary. Low-temperature routes to new structures for yttrium, holmium, erbium, and thulium oxychlorides. Inorg. Chem. 24 (1985), 494–498. https://doi.org/10.1021/ic00198a013.

[156] A. Taoudi, J. Laval and B. Frit. Synthesis and crystal structure of three new rare earth oxyfluorides related to baddeleyite [LnOF Ln=Tm,Yb,Lu]. Mater. Res. Bull. 29 (1994), 1137–1147. https://doi.org/10.1016/0025-5408(94)90183-x.

[157] J. B. Burns, N. A. Stump and J. R. Peterson. Synthesis and polarized Raman studies of rhombohedral LuOCl single crystals. J. Raman Spectrosc. 26 (1995), 39–41. https://doi.org/10.1002/jrs.1250260108.

[158] L. Beaury, J. Derouet, J. Hölsä, M. Lastusaari and J. Rodriguez-Carvajal. Neutron powder diffraction studies of stoichiometric NdOF between 1.5 and 300 K. Solid State Sci. 4 (2002), 1039–1043. https://doi.org/10.1016/s1293-2558(02)01361-4.

[159] G. Brandt and R. Diehl. Preparation, powder data and crystal structure of YbOCl. Mater. Res. Bull. 9 (1974), 411–419. https://doi.org/10.1016/0025-5408(74)90208-6.

[160] G. Meyer and T. Schleid. Oxidchloride unter reduzierenden Bedingungen: Einkristalle von NdOCl und GdOCl. Z. Anorg. Allg. Chem. 533 (1986), 181–185. https://doi.org/10.1002/zaac.19865330222.

[161] S. Zimmermann and G. Meyer. Lutetium(III) oxide iodide. Acta Crystallogr., Sect. E: Struct. Rep. Online 63 (2007), i193. https://doi.org/10.1107/s160053680705283x.

[162] L. Jongen and G. Meyer. Scandium(III) oxide bromide, ScOBr. Acta Crystallogr., Sect. E: Struct. Rep. Online 61 (2005), i153–i154. https://doi.org/10.1107/s1600536805019914.

[163] S. Zimmermann and G. Meyer. A missing rare-earth oxide halide structure now observed for scandium oxide iodide, ScOI. Z. Anorg. Allg. Chem. 634 (2008), 2217–2220. https://doi.org/10.1002/zaac.200800164.

[164] D. Zagorac, J. Zagorac, M. Fonović, M. Pejić and J. C. Schön. Computational discovery of new modifications in scandium oxychloride (ScOCl) using a multi-methodological approach. Z. Anorg. Allg. Chem. 648 (2022). https://doi.org/10.1002/zaac.202200198.

[165] L. Eyring. Chapter 27 The binary rare earth oxides. In Non-metallic Compounds – I, volume 3 of Handbook on the Physics and Chemistry of Rare Earths, pp. 337–399. Elsevier (1979). https://doi.org/10.1016/S0168-1273(79)03010-5.

[166] R. Haire and L. Eyring. Chapter 125 Comparisons of the binary oxides. In Lanthanides/Actinides: Chemistry, volume 18 of Handbook on the Physics and Chemistry of Rare Earths, pp. 413–505. Elsevier (1994). https://doi.org/10.1016/S0168-1273(05)80048-0.

[167] R. J. M. Konings, O. Beneš, A. Kovács, D. Manara, D. Sedmidubský, L. Gorokhov, V. S. Iorish, V. Yungman, E. Shenyavskaya and E. Osina. The thermodynamic properties of the f-elements and their compounds. Part 2. The lanthanide and actinide oxides. J. Phys. Chem. Ref. Data 43 (2014), 013101. https://doi.org/10.1063/1.4825256.

[168] G. Schiller. Die Kristallstrukturen von Ce_2O_3 (A-Form), $LiCeO_2$ und CeF_3-Ein Beitrag zur Kristallchemie des dreiwertigen Cers. Ph.D. thesis, Universität Karlsruhe (1985).

[169] P. Aldebert and J. Traverse. Etude par diffraction neutronique des structures de haute temperature de La_2O_3 et Nd_2O_3. Mater. Res. Bull. 14 (1979), 303–323. https://doi.org/10.1016/0025-5408(79)90095-3.

[170] B. J. Kennedy and M. Avdeev. The structure of B-type Sm_2O_3. A powder neutron diffraction study using enriched 154Sm. Solid State Sci. 13 (2011), 1701–1703. https://doi.org/10.1016/j.solidstatesciences.2011.06.020.

[171] A. Saiki, N. Ishizawa, N. Mizutani and M. Kato. Structural change of C-rare earth sesquioxides Yb_2O_3 and Er_2O_3 as a function of temperature. J. Ceram. Soc. Jpn. 93 (1985), 649–654. ISSN 0372-7718.

[172] D. T. Cromer. The crystal structure of monoclinic Sm_2O_3. J. Phys. Chem. 61 (1957), 753–755. https://doi.org/10.1021/j150552a011.

[173] J. Zhang, R. V. Dreele and L. Eyring. Structures in the oxygen-deficient fluorite-related R_nO_{2n-2} homologous series: $Pr_{12}O_{22}$. J. Solid State Chem. 122 (1996), 53–58. https://doi.org/10.1006/jssc.1996.0081.

[174] J. Zhang, R. V. Dreele and L. Eyring. The structures of Tb_7O_{12} and $Tb_{11}O_{20}$. J. Solid State Chem. 104 (1993), 21–32. https://doi.org/10.1006/jssc.1993.1138.

[175] T. Montini, M. Melchionna, M. Monai and P. Fornasiero. Fundamentals and catalytic applications of CeO_2-based materials. Chem. Rev. 116 (2016), 5987–6041. https://doi.org/10.1021/acs.chemrev.5b00603.

[176] M. Zinkevich, D. Djurivic and F. Aldinger. Thermodynamic modelling of the cerium–oxygen system. Solid State Ion. 177 (2006), 989–1001. https://doi.org/10.1016/j.ssi.2006.02.044.

[177] V. Perrichon, A. Laachir, G. Bergeret, R. Fréty, L. Tournayan and O. Touret. Reduction of cerias with different textures by hydrogen and their reoxidation by oxygen. J. Chem. Soc. Faraday Trans. 90 (1994), 773–781. https://doi.org/10.1039/ft9949000773.

[178] C. Pascual and P. Duran. Subsolidus phase equilibria and ordering in the system ZrO_2-Y_2O_3. J. Am. Ceram. Soc. 66 (1983), 23–27. https://doi.org/10.1111/j.1151-2916.1983.tb09961.x.

[179] D. R. Modeshia, C. S. Wright, J. L. Payne, G. Sankar, S. G. Fiddy and R. I. Walton. Low-temperature redox properties of nanocrystalline cerium (IV) oxides revealed by in situ XANES. J. Phys. Chem. C 111 (2007), 14035–14039. https://doi.org/10.1021/jp075410p.

[180] Z. Yang, G. Luo, Z. Lu and K. Hermansson. Oxygen vacancy formation energy in Pd-doped ceria: A DFT+U study. J. Chem. Phys. 127 (2007), 074704. https://doi.org/10.1063/1.2752504.

[181] H. Bärnighausen. Untersuchungen am System EuO–Eu_2O_3. J. Prakt. Chem. 34 (1966), 1–14. https://doi.org/10.1002/prac.19660340101.

[182] R. C. Rau. The crystal structure of Eu_3O_4. Acta Crystallogr. 20 (1966), 716–723. https://doi.org/10.1107/s0365110x66001737.

[183] A. A. R. Fernandes, J. Santamaria, S. L. Bud'ko, O. Nakamura, J. Guimpel and I. K. Schuller. Effect of physical and chemical pressure on the superconductivity of high-temperature oxide superconductors. Phys. Rev. B 44 (1991), 7601–7606. https://doi.org/10.1103/physrevb.44.7601.

[184] S. Katano, J. Fernandez-Baca, S. Funahashi, N. Môri, Y. Ueda and K. Koga. Crystal structure and superconductivity of $La_{2-x}Ba_xCuO_4$ ($0.03 \leq x \leq 0.24$). Physica C 214 (1993), 64–72. https://doi.org/10.1016/0921-4534(93)90108-3.

[185] H. Bärnighausen and G. Brauer. Ein neues Europiumoxid Eu_3O_4 und die isotype Verbindung Eu_2SrO_4. Acta Crystallogr. 15 (1962), 1059. https://doi.org/10.1107/s0365110x62002807.

[186] R. Rau. X-ray Crystallographic Studies Of Europium Oxides and Hydroxides. Technical report, General Electric Co. Advanced Technology Services, Cincinnati (1962). https://doi.org/10.2172/4670398.

[187] K. Ahn, V. K. Pecharsky and K. A. Gschneidner. The magnetothermal behavior of mixed-valence Eu_3O_4. J. Appl. Phys. 106 (2009). https://doi.org/10.1063/1.3204662.

[188] J. G. Bednorz and K. A. Müller. Possible high T_c superconductivity in the Ba-La-Cu-O system. Z. Phys. B, Condens. Matter 64 (1986), 189–193. https://doi.org/10.1007/bf01303701.

[189] M. K. Wu, J. R. Ashburn, C. J. Torng, P. H. Hor, R. L. Meng, L. Gao, Z. J. Huang, Y. Q. Wang and C. W. Chu. Superconductivity at 93 K in a new mixed-phase Y-Ba-Cu-O compound system at ambient pressure. Phys. Rev. Lett. 58 (1987), 908–910. https://doi.org/10.1103/physrevlett.58.908.

[190] A. Molodyk, S. Samoilenkov, A. Markelov, P. Degtyarenko, S. Lee, V. Petrykin, M. Gaifullin, A. Mankevich, A. Vavilov, B. Sorbom, J. Cheng, S. Garberg, L. Kesler, Z. Hartwig, S. Gavrilkin, A. Tsvetkov, T. Okada, S. Awaji, D. Abraimov, A. Francis, G. Bradford, D. Larbalestier, C. Senatore, M. Bonura, A. E. Pantoja, S. C. Wimbush, N. M. Strickland and A. Vasiliev. Development and large volume production of extremely high current density $YBa_2Cu_3O_7$ superconducting wires for fusion. Sci. Rep. 11 (2021). https://doi.org/10.1038/s41598-021-81559-z.

[191] K. A. Gschneidner. Handbook on the Physics and Chemistry of Rare Earths. In Vol. 4, p. 602. North Holland (1979). ISBN 9780444852168.

[192] K. A. Gschneidner. Handbook on the Physics and Chemistry of Rare Earths. In Vol. 6, p. 574. Elsevier Science Ltd (1984). ISBN 9780444865922.

[193] K. A. Gschneidner. Handbook on the Physics and Chemistry of Rare Earths. In Vol. 15, p. 530. North Holland (1991). ISBN 9780444889669.

[194] L. Eyring. Handbook on the Physics and Chemistry of Rare Earths: Lanthanides/Actinides. In Vol. 17, p. 788. North Holland (1993). ISBN 9780444815026.

[195] K. A. Gschneidner and L. Eyring. Handbook on the Physics and Chemistry of Rare Earths. In Vol. 23, p. 664. North Holland (1996). ISBN 9780444825070.

[196] L. Eyring and J. M. Lemm. Handbook on the Physics and Chemistry of Rare Earths. In Vol. 25, p. 508. North Holland (1998). ISBN 9780444828712.

[197] L. Eyring. Handbook on the Physics and Chemistry of Rare Earths. In Vol. 32, p. 625. North Holland (2001). ISBN 9780444507624.

[198] K. A. Gschneidner and V. K. Pecharsky. Handbook on the Physics and Chemistry of Rare Earths. In Vol. 38, p. 500. North Holland (2008). ISBN 9780444521439.

[199] G. Ning and R. L. Flemming. Rietveld refinement of LaB_6: data from micro-XRD. J. Appl. Crystallogr. 38 (2005), 757–759. https://doi.org/10.1107/s0021889805023344.

[200] C. B. Shoemaker, D. P. Shoemaker and R. Fruchart. The structure of a new magnetic phase related to the sigma phase: iron neodymium boride $Nd_2Fe_{14}B$. Acta Crystallogr., Sect. C 40 (1984), 1665–1668. https://doi.org/10.1107/s0108270184009094.

[201] G. Schell, H. Winter, H. Rietschel and F. Gompf. Electronic structure and superconductivity in metal hexaborides. Phys. Rev. B 25 (1982), 1589–1599. https://doi.org/10.1103/physrevb.25.1589.

[202] I. Overland. The geopolitics of renewable energy: Debunking four emerging myths. Energy Res. Soc. Sci. 49 (2019), 36–40. https://doi.org/10.1016/j.erss.2018.10.018.

[203] O. Gutfleisch. Controlling the properties of high energy density permanent magnetic materials by different processing routes. J. Phys. D, Appl. Phys. 33 (2000), R157–R172. https://doi.org/10.1088/0022-3727/33/17/201.

[204] J. Liu, J. Brabers, A. Winkelman, A. Menovsky, F. de Boer and K. Buschow. Synthesis and magnetic properties of $R_2Co_{17}N_x$ type interstitial compounds. J. Alloys Compd. 200 (1993), L3–L6. https://doi.org/10.1016/0925-8388(93)90461-u.

[205] V. Babizhetskyy, O. Jepsen, R. K. Kremer, A. Simon, B. Ouladdiaf and A. Stolovits. Structure and bonding of superconducting LaC_2. J. Phys. Condens. Matter 26 (2013), 025701. https://doi.org/10.1088/0953-8984/26/2/025701.

[206] T. Mochiku, T. Nakane, H. Kito, H. Takeya, S. Harjo, T. Ishigaki, T. Kamiyama, T. Wada and K. Hirata. Crystal structure of yttrium sesquicarbide. Physica C 426–431 (2005), 421–425. https://doi.org/10.1016/j.physc.2005.02.065.

[207] A. Sousanis, P. Smet and D. Poelman. Samarium monosulfide (SmS): Reviewing properties and applications. Materials 10 (2017), 953. https://doi.org/10.3390/ma10080953.

[208] O. Massenet, J. M. D. Coey and F. Holtzberg. Phase transition and magnetism in Eu_3S_4. J. Phys., Colloq.. 37 (1976), C4-297–C4-299. https://doi.org/10.1051/jphyscol:1976452.

[209] A. A. Grizik, A. A. Eliseev, G. P. Borodulenko and V. A. Tolstova. Low-temperature form Ln_2S_3 (Ln-Eu, Sm, Gd). Zh. Neorg. Kh. 22 (1977), 558–559.

[210] H. Zimmer and K. Niedenzu. Annual Reports in Inorganic and General Syntheses-1975. Elsevier Science, Technology Books (2013). ISBN 9781483260136.

[211] R. Marchand, P. L'Haridon and Y. Laurent. Structure cristalline de $Eu_2(II)SiO_4$. J. Solid State Chem. 24 (1978), 71–76. https://doi.org/10.1016/s0022-4596(78)90184-6.

[212] J. M. S. Skakle, C. L. Dickson and F. P. Glasser. The crystal structures of $CeSiO_4$ and $Ca_2Ce_8(SiO_4)_6O_2$. Powder Diffr. 15 (2000), 234–238. https://doi.org/10.1017/s0885715600011143.

[213] Y. Smolin and S. Tkachev. Determination of the structure of gadolinium oxyorthosilicate (Gd_2SiO_5). Kristallografiya 14 (1969), 22.

[214] H. Okudera, A. Yoshiasa, Y. Masubuchi, M. Higuchi and S. Kikkawa. Determinations of crystallographic space group and atomic arrangements in oxide-ion-conducting $Nd_{9.33}(SiO4)6 O_2$. Z. Kristallogr. Cryst. Mater. 219 (2004), 27–31. https://doi.org/10.1524/zkri.219.1.27.25399.

[215] J. Felsche. The crystal structures of the dimorphic rare earth disilicate, $Pr_2Si_2O_7$. Z. Kristallogr. Cryst. Mater. 133 (1971), 364–385. https://doi.org/10.1524/zkri.1971.133.16.364.

[216] J. Felsche. A new silicate structure containing linear $[Si_3O_{10}]$ groups. Naturwissenschaften 59 (1972), 35–36. https://doi.org/10.1007/bf00594623.

[217] H. Müller-Bunz and T. Schleid. $La_2Si_2O_7$ im I–Typ: Gemäß $La_6[Si_4O_{13}][SiO_4]_2$ kein echtes Lanthandisilicat. Z. Anorg. Allg. Chem. 628 (2002), 564–569. https://doi.org/10.1002/1521-3749(200203)628:3<564::AID-ZAAC564>3.0.CO;2-T.

[218] F. Demartin, T. Pilati, V. Diella, P. Gentile and C. M. Gramaccioli. A crystal-chemical investigation of Alpine gadolinite. Can. Mineral. 31 (1993), 127–136. ISSN 0008-4476.

[219] C. Wang, X. Liu, M. E. Fleet, J. Li, S. Feng, R. Xu and Z. Jin. Helical chain observed under transmission electron microscope: Synthesis and structure refinement of lutetium disilicate $Lu_2Si_2O_7$. CrystEngComm 12 (2010), 1617. https://doi.org/10.1039/b919658g.

[220] L. Pauling. The principles determining the structure of complex ionic crystals. J. Am. Chem. Soc. 51 (1929), 1010–1026. https://doi.org/10.1021/ja01379a006.

[221] K. S. Bagdasarov, N. B. Bolotina and V. I. Kalinin. Photo-induced effects and real structure of yttrium-aluminium garnet crystals. Kristallografiya 36 (1991), 715–728. ISSN 0023-4761.

[222] M. Avdeev, S. Yakovlev, A. A. Yaremchenko and V. V. Kharton. Transitions between $P2_1$ and $P6_322$ modifications of $SrAl_2O_4$ by in situ high-temperature X-ray and neutron diffraction. J. Solid State Chem. 180 (2007), 3535–3544. ISSN 0022-4596.

[223] V. Efremov, N. CHYORNAYA, V. Trunov and V. Pisarenko. Crystal-structure of lanthanum-magnesium hexaaluminate. Kristallografiya 33 (1988), 38–42.

[224] N. Iyi, Z. Inoue and S. Kimura. The crystal structure and cation distribution of highly nonstoichiometric magnesium-doped potassium beta-alumina. J. Solid State Chem. 61 (1986), 236–244. https://doi.org/10.1016/0022-4596(86)90027-7.

[225] H. Huppertz and W. Schnick. Edge-sharing SiN_4-tetrahedra in the highly condensed nitridosilicate $BaSi_7N_{10}$. Chem. Eur. J. 3 (1997), 249–252. https://doi.org/10.1002/chem.19970030213.

[226] T. Schlieper, W. Milius and W. Schnick. Nitrido-silicate. II [1]. Hochtemperatur-Synthesen und Kristallstrukturen von $Sr_2Si_5N_8$ und $Ba_2Si_5N_8$. Z. Anorg. Allg. Chem. 621 (1995), 1380–1384. https://doi.org/10.1002/zaac.19956210817.

[227] H. A. Höppe. Optische, magnetische und strukturelle Eigenschaften von Nitridosilicaten, Oxonitridosilicaten und Carbidonitridosilicaten. Ph.D. thesis, LMU (Ludwig-Maximilians-Universität München) (2003).

[228] P. Pust, V. Weiler, C. Hecht, A. Tücks, A. S. Wochnik, A.-K. Henß, D. Wiechert, C. Scheu, P. J. Schmidt and W. Schnick. Narrow-band red-emitting Sr[LiAl$_3$N$_4$]:Eu^{2+} as a next-generation LED-phosphor material. Nat. Mater. 13 (2014), 891–896. https://doi.org/10.1038/nmat4012.

[229] H. A. Höppe. The synthesis, crystal structure and vibrational spectra of *alpha*-Sr(*PO3*)2 containing an unusual catena-polyphosphate helix. Solid State Sci. 7 (2005), 1209–1215. https://doi.org/10.1016/j.solidstatesciences.2005.06.014.

[230] K. Förg and H. A. Höppe. Synthesis, crystal structure, optical and thermal properties of lanthanide hydrogen-polyphosphates Ln[H(PO$_3$)$_4$] (Ln = Tb, Dy, Ho). Dalton Trans. 44 (2015), 19163–19174. https://doi.org/10.1039/c5dt02648b.

[231] H. A. Höppe and M. Daub. Synthesis, crystal structure and optical properties of the catena-metaphosphates Ce(PO$_3$)$_4$ and U(PO$_3$)$_4$. Z. Kristallogr. Cryst. Mater. 227 (2012), 535–539. https://doi.org/10.1524/zkri.2012.1489.

[232] B. Ewald, Y.-X. Huang and R. Kniep. Structural chemistry of borophosphates, metalloborophosphates, and related compounds. Z. Anorg. Allg. Chem. 633 (2007), 1517–1540. ISSN 1521-3749. https://doi.org/10.1002/zaac.200700232.

[233] H. Ehrenberg, S. Laubach, P. Schmidt, R. McSweeney, M. Knapp and K. Mishra. Investigation of crystal structure and associated electronic structure of Sr$_6$BP$_5$O$_{20}$. J. Solid State Chem. 179 (2006), 968–973. https://doi.org/10.1016/j.jssc.2005.12.033.

[234] P. Gross, A. Kirchhain and H. A. Höppe. The borosulfates K$_4$[BS$_4$O$_{15}$(OH)], Ba[B$_2$S$_3$O$_{13}$], and Gd$_2$[B$_2$S$_6$O$_{24}$]. Angew. Chem. Int. Ed. 55 (2016), 4353–4355. https://doi.org/10.1002/anie.201510612.

[235] S. G. Jantz, F. Pielnhofer, L. van Wüllen, R. Weihrich, M. J. Schäfer and H. A. Höppe. The first alkaline-Earth fluorooxoborate Ba[B$_4$O$_6$F$_2$]-characterisation and doping with Eu^{2+}. Chem. Eur. J. 24 (2017), 443–450. https://doi.org/10.1002/chem.201704324.

[236] H. A. Höppe, K. Kazmierczak, M. Daub, K. Förg, F. Fuchs and H. Hillebrecht. The first borosulfate K$_5$[B(SO$_4$)$_4$]. Angew. Chem. Int. Ed. 51 (2012), 6255–6257. https://doi.org/10.1002/anie.201109237.

[237] J. Bruns, H. A. Höppe, M. Daub, H. Hillebrecht and H. Huppertz. Borosulfates–synthesis and structural chemistry of silicate analogue compounds. Chem. Eur. J. 26 (2020), 7966–7980. https://doi.org/10.1002/chem.201905449.

[238] M. Fox. Optical Properties of Solids. Oxford Master Series in Physics. Oxford University Press Oxford (2010). ISBN 9780191576720.

[239] K. Binnemans. Interpretation of europium(III) spectra. Coord. Chem. Rev. 295 (2015), 1–45. ISSN 0010-8545. https://doi.org/10.1016/j.ccr.2015.02.015.

[240] W. C. Martin, R. Zalubas and L. Hagan. Atomic energy levels – the rare earth elements. (the spectra of lanthanum, cerium, praseodymium, neodymium, promethium, samarium, europium, gadolinium, terbium, dysprosium, holmium, erbium, thulium, ytterbium, and lutetium). [66 atoms and ions]. Technical report, National Bureau of Standards, Washington D.C. 20234 (1978).

[241] S. Kettle. Symmetry and Structure: Readable Group Theory for Chemists. John Wiley (2007). ISBN 9780470060391.

[242] T. Förster and T. Forster. Transfer mechanisms of electronic excitation energy. Radiat. Res. Suppl. 2 (1960), 326. https://doi.org/10.2307/3583604.

[243] F. Seitz. An interpretation of crystal luminescence. Trans. Faraday Soc. 35 (1939), 74. https://doi.org/10.1039/tf9393500074.

[244] D. Cooke, R. Muenchausen, B. Bennett, K. McClellan and A. Portis. Temperature-dependent luminescence of cerium-doped ytterbium oxyorthosilicate. J. Lumin. 79 (1998), 185–190. https://doi.org/10.1016/s0022-2313(98)00042-8.

[245] S. G. Jantz, R. Erdmann, S. Hariyani, J. Brgoch and H. A. Höppe. Sr$_6$(BO$_3$)$_3$(BN$_2$): An oxido–nitrido–borate phosphor featuring BN$_2$ dumbbells. Chem. Mater. 32 (2020), 8587–8594. https://doi.org/10.1021/acs.chemmater.0c02925.

[246] G. Dieke, H. Crosswhite and H. Crosswhite. Spectra and Energy Levels of Rare Earth Ions in Crystals. John Wiley & Sons, Inc., New York [u. a.] (1968). ISBN 0470213906.

[247] R. T. Wegh, A. Meijerink, R.-J. Lamminmäki and J. Hölsä. Extending Dieke's diagram. J. Lumin. 87–89 (2000), 1002–1004.

[248] P. Peijzel, A. Meijerink, R. Wegh, M. Reid and G. Burdick. A complete energy level diagram for all trivalent lanthanide ions. J. Solid State Chem. 178 (2005), 448–453. https://doi.org/10.1016/j.jssc.2004.07.046.

[249] G. H. Dieke and H. M. Crosswhite. The spectra of the doubly and triply ionized rare earths. Appl. Opt. 2 (1963), 675. https://doi.org/10.1364/ao.2.000675.

[250] P. Dorenbos. The 4f $\longrightarrow$ 4f^{n-1}5d transitions of the trivalent lanthanides in halogenides and chalcogenides. J. Lumin. 91 (2000), 91–106. https://doi.org/10.1016/s0022-2313(00)00197-6.

[251] P. Dorenbos. 5d-level energies of Ce^{3+} and the crystalline environment. IV. Aluminates and simple oxides. J. Lumin. 99 (2002), 283–299. https://doi.org/10.1016/s0022-2313(02)00347-2.

[252] P. Dorenbos. Improved parameters for the lanthanide 4f^q and 4f$^{(q-1)}$5d curves in HRBE and VRBE schemes that takes the nephelauxetic effect into account. J. Lumin. 222 (2020), 117164. https://doi.org/10.1016/j.jlumin.2020.117164.

[253] B. R. Judd. Optical absorption intensities of rare-earth ions. Phys. Rev. 127 (1962), 750–761. https://doi.org/10.1103/physrev.127.750.

[254] G. S. Ofelt. Intensities of crystal spectra of rare-earth ions. J. Chem. Phys. 37 (1962), 511–520. https://doi.org/10.1063/1.1701366.

[255] W. T. Carnall, P. R. Fields and B. G. Wybourne. Spectral intensities of the trivalent lanthanides and actinides in solution. I. Pr^{3+}, Nd^{3+}, Er^{3+}, Tm^{3+}, and Yb^{3+}. J. Chem. Phys. 42 (1965), 3797–3806. https://doi.org/10.1063/1.1695840.

[256] W. T. Carnall, H. Crosswhite and H. M. Crosswhite. Energy level structure and transition probabilities in the spectra of the trivalent lanthanides in LaF$_3$. Technical report, Argonne National Lab. (ANL), United States (1978). https://doi.org/10.2172/6417825.

[257] W. T. Carnall, P. R. Fields and K. Rajnak. Electronic energy levels of the trivalent lanthanide aquo ions. IV. Eu3+. J. Chem. Phys. 49 (1968), 4450–4455.

[258] P. Babu and C. Jayasankar. Optical spectroscopy of Eu^{3+} ions in lithium borate and lithium fluoroborate glasses. Physica B, Condens. Matter 279 (2000), 262–281. https://doi.org/10.1016/s0921-4526(99)00876-5.

[259] A. Ćirić, S. Stojadinović, M. Sekulić and M. D. Dramićanin. JOES: An application software for Judd-Ofelt analysis from Eu^{3+} emission spectra. J. Lumin. 205 (2019), 351–356. ISSN 0022-2313. https://doi.org/10.1016/j.jlumin.2018.09.048.

[260] A. Ćirić. JOES – Judd Ofelt from Emission Spectra Application Software. website (2019). https://doi.org/10.17632/k498ggvffd.5.

[261] C. Görller-Walrand and K. Binnemans. Chapter 167 Spectral intensities of f-f transitions. In Handbook on the Physics and Chemistry of Rare Earths, pp. 101–264. Elsevier (1998). https://doi.org/10.1016/s0168-1273(98)25006-9.

[262] M. P. Hehlen, M. G. Brik and K. W. Krämer. 50th anniversary of the Judd–Ofelt theory: An experimentalist's view of the formalism and its application. J. Lumin. 136 (2013), 221–239. https://doi.org/10.1016/j.jlumin.2012.10.035.

[263] A. Bronova, T. Bredow, R. Glaum, M. J. Riley and W. Urland. BonnMag: Computer program for ligand-field analysis of fn systems within the angular overlap model. J. Comput. Chem. 39 (2017), 176–186. https://doi.org/10.1002/jcc.25096.

[264] R. Glaum, W. Grunwald, N. Kannengießer and A. Bronova. Analysis of ligand field effects in europium(III) phosphates. Z. Anorg. Allg. Chem. 646 (2020), 184–192. https://doi.org/10.1002/zaac.202000019.

[265] C. E. Schäffer and C. K. Jørgensen. Proceedings International Symposium on the Chemistry of the Co-ordination Compounds. The nephelauxetic series of ligands corresponding to increasing tendency of partly covalent bonding. J. Inorg. Nucl. Chem. 8 (1958), 143–148. ISSN 0022-1902. https://doi.org/10.1016/0022-1902(58)80176-1.

[266] D. Newman. Ligand ordering parameters. Aust. J. Phys. 30 (1977), 315. https://doi.org/10.1071/ph770315.

[267] P. A. Tanner and Y. Y. Yeung. Nephelauxetic effects in the electronic spectra of Pr^{3+}. J. Phys. Chem. A 117 (2013), 10726–10735. https://doi.org/10.1021/jp408625s.

[268] C.-G. Ma, M. Brik, Q.-X. Li and Y. Tian. Systematic analysis of spectroscopic characteristics of the lanthanide and actinide ions with the 4f and 5f (N= 1...14) electronic configurations in a free state. J. Alloys Compd. 599 (2014), 93–101. https://doi.org/10.1016/j.jallcom.2014.02.044.

[269] P. Dorenbos. The nephelauxetic effect on the electron binding energy in the 4f ground state of lanthanides in compounds. J. Lumin. 214 (2019), 116536. https://doi.org/10.1016/j.jlumin.2019.116536.

[270] P. A. Tanner, Y. Y. Yeung and L. Ning. What factors affect the $^5D_0$ energy of Eu^{3+} – an investigation of nephelauxetic effects. J. Phys. Chem. A 117 (2013), 2771–2781. https://doi.org/10.1021/jp400247r.

[271] C. Morrison, D. Mason and C. Kikuchi. Modified slater integrals for an ion in a solid. Phys. Lett. A 24 (1967), 607–608. https://doi.org/10.1016/0375-9601(67)90642-1.

[272] P. Dorenbos. Lanthanide charge transfer energies and related luminescence, charge carrier trapping, and redox phenomena. J. Alloys Compd. 488 (2009), 568–573. ISSN 0925-8388. https://doi.org/10.1016/j.jallcom.2008.09.059. Proceedings of the 25th Rare Earth Research Conference, June 22-26, Tuscaloosa, Alabama, {USA}.

[273] I. Neefjes, J. J. Joos, Z. Barandiarán and L. Seijo. Mixed-valence lanthanide-activated phosphors: invariance of the intervalence charge transfer (IVCT) absorption onset across the series. J. Phys. Chem. C 124 (2020), 2619–2626. https://doi.org/10.1021/acs.jpcc.9b11084.

[274] O. Berkooz, M. Malamud and S. Shtrikman. Observation of electron hopping in $^{151}Eu_3S_4$ by Mössbauer spectroscopy. Solid State Commun. 6 (1968), 185–188. https://doi.org/10.1016/0038-1098(68)90029-x.

[275] G. Allen, M. Wood and J. Dyke. Spectroscopic properties of some mixed-valence transitional metal chalcogenides. J. Inorg. Nucl. Chem. 35 (1973), 2311–2318. https://doi.org/10.1016/0022-1902(73)80295-7.

[276] W. van Schaik, S. Lizzo, W. Smit and G. Blasse. Influence of impurities on the luminescence quantum efficiency of $(La,Ce,Tb)PO_4$. J. Electrochem. Soc. 140 (1993), 216–222. https://doi.org/10.1149/1.2056091.

[277] L. Seijo and Z. Barandiarán. Intervalence charge transfer luminescence: The anomalous luminescence of cerium-doped $Cs_2LiLuCl_6$ elpasolite. J. Chem. Phys. 141 (2014). https://doi.org/10.1063/1.4902384.

[278] J. J. Joos, L. Seijo and Z. Barandiarán. Direct evidence of intervalence charge-transfer states of Eu-doped luminescent materials. J. Phys. Chem. Lett. 10 (2019), 1581–1586. https://doi.org/10.1021/acs.jpclett.9b00342.

[279] F. T. Lange and H. Bärnighausen. Untersuchungen zum chemischen Transport der wasserfreien Europiumchloride $EuCl_2$, Eu_5Cl_{11}, Eu_4Cl_9, $Eu_{14}Cl_{33}$ und $EuCl_3$ mit $AlCl_3$. Z. Anorg. Allg. Chem. 619 (1993), 1747–1754. https://doi.org/10.1002/zaac.19936191021.

[280] C. Wickleder. KEu_2Cl_6 und $K_{1,6}Eu_{1,4}Cl_5$: Zwei neue gemischtvalente Europiumchloride. Z. Anorg. Allg. Chem. 628 (2002), 1815. https://doi.org/10.1002/1521-3749(200208)628:8<1815::aid-zaac1815>3.0.co;2-i.

[281] C. Wickleder. A new mixed valent europium chloride: $Na_5Eu_7Cl_{22}$. Z. Naturforsch. B 57 (2002), 901–907. https://doi.org/10.1515/znb-2002-0810.

[282] H. A. Höppe, K. Kazmierczak, C. Grumbt, L. Schindler, I. Schellenberg and R. Pöttgen. The oxonitridoborate $Eu_5(BO_{2.51(7)}N_{0.49(7)})_4$ and the mixed-valent borates $Sr_3Ln_2(BO_3)_4$ (Ln = Ho, Er). Eur. J. Inorg. Chem. 2013 (2013), 5443–5449. https://doi.org/10.1002/ejic.201300827.

[283] K. Kazmierczak and H. A. Höppe. Synthesis, crystal structure and optical spectra of europium borate fluoride $Eu_5(BO_3)_3F$. Eur. J. Inorg. Chem. 2010 (2010), 2678–2681. https://doi.org/10.1002/ejic.201000105.

[284] P. Dorenbos. The Eu^{3+} charge-transfer energy and the relation with the band gap of compounds. J. Lumin. 111 (2005), 89–104. https://doi.org/10.1016/j.jlumin.2004.07.003.

[285] C. K. Jørgensen. Orbitals in Atoms and Molecules. Academic Press (1962).

[286] P. Dorenbos. Ce^{3+} 5d-centroid shift and vacuum referred 4f-electron binding energies of all lanthanide impurities in 150 different compounds. J. Lumin. 135 (2013), 93–104. https://doi.org/10.1016/j.jlumin.2012.09.034.

[287] P. Dorenbos. Blasse's Pandora's box. Opt. Mater. 11 (2021), 100076. https://doi.org/10.1016/j.omx.2021.100076.

[288] G. Blasse. Energy transfer phenomena in the system (Y, Ce, Gd, Tb)F_3. Phys. Status Solidi A 73 (1982), 205–208. https://doi.org/10.1002/pssa.2210730126.

[289] P. Netzsch, M. Hämmer, E. Turgunbajew, T. P. van Swieten, A. Meijerink, H. A. Höppe and M. Suta. Beyond the energy gap law: The influence of selection rules and host compound effects on nonradiative transition rates in Boltzmann thermometers. Adv. Opt. Mater. 10 (2022), 2200059. https://doi.org/10.1002/adom.202200059.

[290] J. Capobianco, P. Kabro, F. Ermeneux, R. Moncorgé, M. Bettinelli and E. Cavalli. Optical spectroscopy, fluorescence dynamics and crystal-field analysis of Er^{3+} in YVO_4. Chem. Phys. 214 (1997), 329–340. https://doi.org/10.1016/s0301-0104(96)00318-7.

[291] M. Suta, Ž. Antić, V. Đorđević, S. Kuzman, M. D. Dramićanin and A. Meijerink. Making Nd^{3+} a sensitive luminescent thermometer for physiological temperatures—an account of pitfalls in Boltzmann thermometry. Nanomaterials 10 (2020), 543. https://doi.org/10.3390/nano10030543.

[292] P. Netzsch, M. Hämmer, P. Gross, H. Bariss, T. Block, L. Heletta, R. Pöttgen, J. Bruns, H. Huppertz and H. A. Höppe. $R_2[B_2(SO_4)_6]$ (R = Y, La–Nd, Sm, Eu, Tb–Lu): a silicate-analogous host structure with weak coordination behaviour. Dalton Trans. 48 (2019), 4387–4397. https://doi.org/10.1039/c9dt00445a.

[293] M. Hämmer, O. Janka, J. Bönnighausen, S. Klenner, R. Pöttgen and H. A. Höppe. On the phosphors $Na_5M(WO_4)_4$ (M = Y, La...Nd, Sm...Lu, Bi) – crystal structures, thermal decomposition, and optical and magnetic properties. Dalton Trans. 49 (2020), 8209–8225. https://doi.org/10.1039/d0dt00782j.

[294] J.-C. G. Bünzli and C. Piguet. Taking advantage of luminescent lanthanide ions. Chem. Soc. Rev. 34 (2005), 1048–1077. https://doi.org/10.1039/b406082m.

[295] G. Blasse and A. Bril. On the Eu^{3+} fluorescence in mixed metal oxides. III. Energy transfer in Eu^{3+}-activated tungstates and molybdates of the type Ln_2WO_6 and Ln_2MoO_6. J. Chem. Phys. 45 (1966), 2350–2355. https://doi.org/10.1063/1.1727945.

[296] G. Blasse. Luminescence of calcium halophosphate-Sb^{3+},Mn^{2+} at low temperatures. Chem. Phys. Lett. 104 (1984), 160–162. ISSN 0009-2614.

[297] H. A. Höppe, M. Daub and M. C. Bröhmer. Coactivation of α-$Sr[PO_3]_2$ and $SrM(P_2O_7)$ (M = Zn, Sr) with Eu^{2+} and Mn^{2+}. Chem. Mater. 19 (2007), 6358.

[298] J. Rubio, O. A. Muñoz F., C. Zaldo and S. Murrieta. Energy transfer in monocrystalline KCl codoped with europium and manganese. Solid State Commun. 65 (1988), 251–255. https://doi.org/10.1016/0038-1098(88)90780-6.

[299] M. Daub, A. J. Lehner and H. A. Höppe. Synthesis, crystal structure and optical properties of $Na_2R(PO_4)(WO_4)$ (R = Y, Tb–Lu). Dalton Trans. 41 (2012), 12121. https://doi.org/10.1039/c2dt31358h.

[300] I. P. Roof, M. D. Smith, S. Park and H.-C. zur Loye. $EuKNaTaO_5$: Crystal growth, structure and photoluminescence property. J. Am. Chem. Soc. 131 (2009), 4202–4203.

[301] C. K. Jørgensen and B. Judd. Hypersensitive pseudoquadrupole transitions in lanthanides. Mol. Phys. 8 (1964), 281–290. ISSN 0026-8976. https://doi.org/10.1080/00268976400100321.

[302] C. K. Jørgensen. Modern Aspects of Ligand Field Theory. North-Holland Pub. Co. (1971). ISBN 0720402182.

[303] D. Henrie, R. Fellows and G. Choppin. Hypersensitivity in the electronic transitions of lanthanide and actinide complexes. Coord. Chem. Rev. 18 (1976), 199–224. https://doi.org/10.1016/s0010-8545(00)82044-5.

[304] S. F. Mason, R. D. Peacock and B. Stewart. Ligand-polarization contributions to the intensity of hypersensitive trivalent lanthanide transitions. Mol. Phys. 30 (1975), 1829–1841. https://doi.org/10.1080/00268977500103321.

[305] S. F. Mason and B. Stewart. The anisotropic ligand polarization intensity mechanism in dihedral lanthanide(III) complexes. Mol. Phys. 55 (1985), 611–620. https://doi.org/10.1080/00268978500101581.

[306] F. Auzel. f–f oscillator strengths, hypersensitivity, branching ratios and quantum efficiencies discussed in the light of forgotten results. J. Alloys Compd. 380 (2004), 9–14. https://doi.org/10.1016/j.jallcom.2004.03.087.

[307] T. Jüstel, H. Nikol and C. Ronda. New developments in the field of luminescent materials for lighting and displays. Angew. Chem. Int. Ed. 37 (1998), 3084.

[308] H. A. Höppe. New developments of inorganic phosphors. Angew. Chem. Int. Ed. 48 (2009), 3572–3582. https://doi.org/10.1002/ange.200804005.

[309] T. Jüstel, H. Bechtel, W. Mayr and D. U. Wiechert. Blue emitting $BaMgAl_{10}O_{17}$:Eu with a blue body color. J. Lumin. 104 (2003), 137–143. ISSN 0022-2313. https://doi.org/10.1016/S0022-2313(03)00010-3.

[310] N. Riesen, K. Badek and H. Riesen. Data storage in a nanocrystalline mixture using room temperature frequency-selective and multilevel spectral hole-burning. ACS Photonics 8 (2021), 3078–3084. https://doi.org/10.1021/acsphotonics.1c01115.

[311] W. Yen, S. Shionoya and H. Yamamoto. Phosphor Handbook. The CRC Press laser and optical science and technology series. CRC Press/Taylor and Francis (2007). ISBN 9780849335648.

[312] P. Dorenbos. Mechanism of persistent luminescence in Eu^{2+} and Dy^{3+} codoped aluminate and silicate compounds. J. Electrochem. Soc. 152 (2005), H107. https://doi.org/10.1149/1.1926652.

[313] F. Clabau, X. Rocquefelte, S. Jobic, P. Deniard, M.-H. Whangbo, A. Garcia and T. L. Mercier. Mechanism of phosphorescence appropriate for the long-lasting phosphors Eu^{2+}-doped $SrAl_2O_4$ with codopants Dy^{3+} and B^{3+}. Chem. Mater. 17 (2005), 3904–3912. https://doi.org/10.1021/cm050763r.

[314] T. Aitasalo, J. Hölsä, H. Jungner, M. Lastusaari and J. Niittykoski. Thermoluminescence study of persistent luminescence materials: Eu^{2+}- and R^{3+}-doped calcium aluminates, $CaAl_2O_4$:Eu^{2+},R^{3+}. J. Phys. Chem. B 110 (2006), 4589–4598. https://doi.org/10.1021/jp057185m.

[315] K. V. den Eeckhout, P. F. Smet and D. Poelman. Persistent luminescence in Eu^{2+}-doped compounds: A review. Materials 3 (2010), 2536–2566. https://doi.org/10.3390/ma3042536.

[316] T. Delgado, J. Afshani and H. Hagemann. Spectroscopic study of a single crystal of $SrAl_2O_4$:Eu^{2+},Dy^{3+}. J. Phys. Chem. C 123 (2019), 8607–8613. https://doi.org/10.1021/acs.jpcc.8b12568.

[317] Y. Li, M. Gecevicius and J. Qiu. Long persistent phosphors—from fundamentals to applications. Chem. Soc. Rev. 45 (2016), 2090–2136. https://doi.org/10.1039/c5cs00582e.

[318] K. Takahashi, K. Kohda, J. Miyahara, Y. Kanemitsu, K. Amitani and S. Shionoya. Mechanism of photostimulated luminescence in BaFX:Eu^{2+} (X=Cl,Br) phosphors. J. Lumin. 31–32 (1984), 266–268. ISSN 0022-2313. https://doi.org/10.1016/0022-2313(84)90268-0.

[319] F. Auzel. Upconversion and anti-Stokes processes with f and d ions in solids. Chem. Rev. 104 (2003), 139–174. https://doi.org/10.1021/cr020357g.

[320] B. S. Richards, D. Hudry, D. Busko, A. Turshatov and I. A. Howard. Photon upconversion for photovoltaics and photocatalysis: A critical review. Chem. Rev. 121 (2021), 9165–9195. https://doi.org/10.1021/acs.chemrev.1c00034.

[321] E. L. Cates, A. P. Wilkinson and J.-H. Kim. Visible-to-UVC upconversion efficiency and mechanisms of $Lu_7O_6F_9$:Pr^{3+} and Y_2SiO_5:Pr^{3+} ceramics. J. Lumin. 160 (2015), 202–209. https://doi.org/10.1016/j.jlumin.2014.11.049.

[322] S. N. Misra and K. John. Difference and comparative absorption spectra and ligand mediated pseudohypersensitivity for 4f-4f transitions of Pr(III) and Nd(III). Appl. Spectrosc. Rev. 28 (1993), 285–325. https://doi.org/10.1080/05704929308018115.

[323] M. J. Weber. Chapter 35 Rare earth lasers. In Handbook on the Physics and Chemistry of Rare Earths, pp. 275–315. Elsevier (1979). https://doi.org/10.1016/s0168-1273(79)04008-3.

[324] A. P. Vink, P. Dorenbos, J. T. M. de Haas, H. Donker, P. A. Rodnyi, A. G. Avanesov and C. W. E. van Eijk. Photon cascade emission in SrAlF$_5$:Pr^{3+}. J. Phys. Condens. Matter 14 (2002), 8889–8899. https://doi.org/10.1088/0953-8984/14/38/312.

[325] G. Greuel, T. Juestel, H. Bettentrup, B. Herden and D. Enseling. Red emitting phosphor for plasma display panels and gas discharge lamps. patent (2014).

[326] M. Weil, E. Zobetz, F. Werner and F. Kubel. New alkaline earth aluminium fluorides with the formula (M, N)AlF$_5$ (M, N = Ca, Sr, Ba). Solid State Sci. 3 (2001), 441–453. https://doi.org/10.1016/s1293-2558(01)01165-7.

[327] J. Sommerdijk, A. Bril and A. de Jager. Two photon luminescence with ultraviolet excitation of trivalent praseodymium. J. Lumin. 8 (1974), 341–343. ISSN 0022-2313. https://doi.org/10.1016/0022-2313(74)90006-4.

[328] W. Piper, J. DeLuca and F. Ham. Cascade fluorescent decay in Pr^{3+}-doped fluorides: Achievement of a quantum yield greater than unity for emission of visible light. J. Lumin. 8 (1974), 344–348. ISSN 0022-2313. https://doi.org/10.1016/0022-2313(74)90007-6.

[329] D. Schiffbauer, C. Wickleder, G. Meyer, M. Kirm, M. Stephan and P. C. Schmidt. Crystal structure, electronic structure, and luminescence of Cs$_2$KYF$_6$:Pr^{3+}. Z. Anorg. Allg. Chem. 631 (2005), 3046–3052. https://doi.org/10.1002/zaac.200500316.

[330] Y. Chen. Quantum cutting in Gd$_2$SiO$_5$:Eu^{3+} by VUV excitation. Sci. China, Ser. G 46 (2003), 17. https://doi.org/10.1360/03yg9003.

[331] Y. Liu, J. Zhang, C. Zhang, J. Jiang and H. Jiang. High efficiency green phosphor Ba$_9$Lu$_2$Si$_6$O$_{24}$:Tb^{3+}: Visible quantum cutting via cross-relaxation energy transfers. J. Phys. Chem. C 120 (2016), 2362–2370. https://doi.org/10.1021/acs.jpcc.5b11790.

[332] P. A. Rodnyi, A. N. Mishin and A. S. Potapov. Luminescence of trivalent praseodymium in oxides and fluorides. Opt. Spectrosc. 93 (2002), 714–721. https://doi.org/10.1134/1.1523992.

[333] A. Pirri, R. N. Maksimov, J. Li, M. Vannini and G. Toci. Achievements and future perspectives of the trivalent thulium-ion-doped mixed-sesquioxide ceramics for laser applications. Materials 15 (2022), 2084. https://doi.org/10.3390/ma15062084.

[334] B. M. van der Ende, L. Aarts and A. Meijerink. Near-infrared quantum cutting for photovoltaics. Adv. Mater. 21 (2009), 3073–3077. https://doi.org/10.1002/adma.200802220.

[335] R. J. Reeves, G. D. Jones and R. W. G. Syme. Site-selective laser spectroscopy of Pr^{3+} C4v symmetry centers in hydrogenated CaF$_2$:Pr^{3+} and SrF$_2$:Pr^{3+} crystals. Phys. Rev. B 46 (1992), 5939–5958. https://doi.org/10.1103/physrevb.46.5939.

[336] M. L. Falin, K. I. Gerasimov, V. A. Latypov and A. M. Leushin. Electron paramagnetic resonance and optical spectroscopy of Yb^{3+} ions in SrF$_2$ and BaF$_2$ an analysis of distortions of the crystal lattice near Yb^{3+}. J. Phys. Condens. Matter 15 (2003), 2833–2847. https://doi.org/10.1088/0953-8984/15/17/332.

[337] H. Ishibashi. Mechanism of luminescence from a cerium-doped gadolinium orthosilicate Gd$_2$SiO$_5$ scintillator. Nucl. Instrum. Methods Phys. Res. A 294 (1990), 271–277. https://doi.org/10.1016/0168-9002(90)91843-z.

[338] G. Miersch, D. Habs, J. Kenntner, D. Schwalm and A. Wolf. Fast scintillators as radiation resistant heavy-ion detectors. Nucl. Instrum. Methods Phys. Res. A 369 (1996), 277–283. https://doi.org/10.1016/0168-9002(95)00785-7.

[339] K. W. Krämer, P. Dorenbos, H. U. Güdel and C. W. E. van Eijk. Development and characterization of highly efficient new cerium doped rare earth halide scintillator materials. J. Mater. Chem. 16 (2006), 2773–2780. https://doi.org/10.1039/b602762h.

[340] T. Yanagida, H. Fukushima, G. Okada and N. Kawaguchi. Scintillation properties of Eu:BaFBr crystal. Physica B, Condens. Matter 550 (2018), 21–25. https://doi.org/10.1016/j.physb.2018.08.033.

[341] A. J. Mortlock. Thermoluminescence dating of objects and materials from the south Pacific region. Aust. Archaeol. 9 (1979), 12–17. https://doi.org/10.1080/03122417.1979.12093356.

[342] N. M. Son, L. T. T. Vien, L. V. K. Bao and N. N. Trac. Synthesis of $SrAl_2O_4:Eu^{2+},Dy^{3+}$ phosphorescence nanosized powder by combustion method and its optical properties. J. Phys. Conf. Ser. 187 (2009), 012017. https://doi.org/10.1088/1742-6596/187/1/012017.

[343] R. Mueller-Mach, G. Mueller, M. R. Krames, H. A. Höppe, F. Stadler, W. Schnick, T. Juestel and P. Schmidt. Highly efficient all-nitride phosphor-converted white light emitting diode. Phys. Status Solidi A 202 (2005), 1727–1732. https://doi.org/10.1002/pssa.200520045.

[344] P. Dorenbos. Energy of the first $4f^7 \longrightarrow 4f^65d$ transition of Eu^{2+} in inorganic compounds. J. Lumin. 104 (2003), 239–260. https://doi.org/10.1016/s0022-2313(03)00078-4.

[345] N. Yamashita. Coexistence of the Eu^{2+} and Eu^{3+} centers in the CaO:Eu powder phosphor. J. Electrochem. Soc. 140 (1993), 840–843. https://doi.org/10.1149/1.2056169.

[346] N. Kunkel, H. Kohlmann, A. Sayede and M. Springborg. Alkaline-Earth metal hydrides as novel host lattices for Eu(II) luminescence. Inorg. Chem. 50 (2011), 5873–5875. https://doi.org/10.1021/ic200801x.

[347] P. Pust, A. S. Wochnik, E. Baumann, P. J. Schmidt, D. Wiechert, C. Scheu and W. Schnick. $Ca[LiAl_3N_4]:Eu^{2+}$—A narrow-band red-emitting nitridolithoaluminate. Chem. Mater. 26 (2014), 3544–3549. https://doi.org/10.1021/cm501162n.

[348] J. L. Leaño, M.-H. Fang and R.-S. Liu. Critical review—narrow-band emission of nitride phosphors for light-emitting diodes: perspectives and opportunities. ECS J. Solid State Sci. Technol. 7 (2017), R3111–R3133. https://doi.org/10.1149/2.0161801jss.

[349] U. Nations. The rapid transition to energy efficient lighting: an integrated policy approach. PDF (2013).

[350] H. Amano, N. Sawaki, I. Akasaki and Y. Toyoda. Metalorganic vapor phase epitaxial growth of a high quality GaN film using an AlN buffer layer. Appl. Phys. Lett. 48 (1986), 353–355. https://doi.org/10.1063/1.96549.

[351] S. Nakamura, Y. Harada and M. Seno. Novel metalorganic chemical vapor deposition system for GaN growth. Appl. Phys. Lett. 58 (1991), 2021–2023. https://doi.org/10.1063/1.105239.

[352] S. Nakamura, N. Iwasa, M. S. M. Senoh and T. M. T. Mukai. Hole compensation mechanism of P-type GaN films. Jpn. J. Appl. Phys. 31 (1992), 1258. https://doi.org/10.1143/jjap.31.1258.

[353] S. Nakamura, T. Mukai, M. S. M. Senoh and N. I. N. Iwasa. Thermal annealing effects on P-type Mg-doped GaN films. Jpn. J. Appl. Phys. 31 (1992), L139. https://doi.org/10.1143/jjap.31.l139.

[354] H. Amano, M. Kito, K. Hiramatsu and I. Akasaki. P-type conduction in Mg-doped GaN treated with low-energy electron beam irradiation (LEEBI). Jpn. J. Appl. Phys. 28 (1989), L2112. https://doi.org/10.1143/jjap.28.l2112.

[355] S. Nakamura and T. Mukai. High-quality InGaN films grown on GaN films. Jpn. J. Appl. Phys. 31 (1992), L1457.

[356] S. Nakamura, T. Mukai and M. Senoh. Candela-class high-brightness InGaN/AlGaN double-heterostructure blue-light-emitting diodes. Appl. Phys. Lett. 64 (1994), 1687–1689. https://doi.org/10.1063/1.111832.

[357] K. Przibram. Fluorescence of the bivalent rare earths. Nature 139 (1937), 329. https://doi.org/10.1038/139329b0.

[358] Z. J. Kiss. Energy levels of divalent thulium in CaF_2. Phys. Rev. 127 (1962), 718–724. https://doi.org/10.1103/physrev.127.718.

[359] J. R. Peterson, W. Xu and S. Dai. Optical properties of divalent thulium in crystalline strontium tetraborate. Chem. Mater. 7 (1995), 1686–1689. https://doi.org/10.1021/cm00057a017.

[360] M. Suta and C. Wickleder. Synthesis, spectroscopic properties and applications of divalent lanthanides apart from Eu^{2+}. J. Lumin. 210 (2019), 210–238. https://doi.org/10.1016/j.jlumin.2019.02.031.

[361] O. S. Wenger, C. Wickleder, K. W. Krämer and H. U. Güdel. Upconversion in a divalent rare earth ion: optical absorption and luminescence spectroscopy of Tm^{2+} doped $SrCl_2$. J. Lumin. 94–95 (2001), 101–105. https://doi.org/10.1016/s0022-2313(01)00255-1.

[362] J. Grimm, E. Beurer, P. Gerner and H. Güdel. Upconversion between 4f–5d excited states in Tm^{2+}-doped $CsCaCl_3$, $CsCaBr_3$, and $CsCaI_3$. Chem. Eur. J. 13 (2007), 1152–1157. https://doi.org/10.1002/chem.200600418.

[363] E. P. Merkx, M. P. Plokker and E. van der Kolk. The potential of transparent sputtered $NaI:Tm^{2+}$, $CaBr_2:Tm^{2+}$, and $CaI_2:Tm^{2+}$ thin films as luminescent solar concentrators. Sol. Energy Mater. Sol. Cells 223 (2021), 110944. https://doi.org/10.1016/j.solmat.2020.110944.

[364] V. Bachmann, T. Jüstel, A. Meijerink, C. Ronda and P. J. Schmidt. Luminescence properties of $SrSi_2O_2N_2$ doped with divalent rare earth ions. J. Lumin. 121 (2006), 441–449. https://doi.org/10.1016/j.jlumin.2005.11.008.

[365] H. Lueken. Magnetochemie: eine Einführung in Theorie und Anwendung. Werke. I. Abteilung, Lyrik Und Prosa / Paul Celan. Teubner (1999). ISBN 9783519035305.

[366] F. Cardarelli. Encyclopaedia of Scientific Units, Weights and Measures. Springer (2004). ISBN 9781852336820.

[367] H. A. Höppe, G. Kotzyba, R. Pöttgen and W. Schnick. Synthesis, crystal structure, magnetism, and optical properties of $Gd_3[SiON_3]O$—An oxonitridosilicate oxide with noncondensed $SiON_3$ tetrahedra. J. Solid State Chem. 167 (2002), 393–401. https://doi.org/10.1006/jssc.2002.9677.

[368] J. H. van Vleck. The Theory of Electric and Magnetic Susceptibilities. Clarendon Press, Oxford (1932).

[369] Y. Takikawa, S. Ebisu and S. Nagata. Van Vleck paramagnetism of the trivalent Eu ions. J. Phys. Chem. Solids 71 (2010), 1592–1598. https://doi.org/10.1016/j.jpcs.2010.08.006.

[370] M. Schlipf, M. Betzinger, M. Ležaić, C. Friedrich and S. Blügel. Structural, electronic, and magnetic properties of the europium chalcogenides: A hybrid-functional DFT study. Phys. Rev. B 88 (2013), 094433. https://doi.org/10.1103/physrevb.88.094433.

[371] D. J. Dunlop and O. Özdemir. Rock Magnetism. Cambridge University Press (2001). ISBN 9780521000987.

[372] R. M. Bozorth, B. T. Matthias, H. Suhl, E. Corenzwit and D. D. Davis. Magnetization of compounds of rare earths with platinum metals. Phys. Rev. 115 (1959), 1595–1596. https://doi.org/10.1103/physrev.115.1595.

[373] H. A. Höppe, H. Trill, B. D. Mosel, H. Eckert, G. Kotzyba, R. Pöttgen and W. Schnick. Hyperfine interactions in the 13 K ferromagnet $Eu_2Si_5N_8$. J. Phys. Chem. Solids 63 (2002), 853–859. https://doi.org/10.1016/s0022-3697(01)00239-6.

[374] M. A. Ruderman and C. Kittel. Indirect exchange coupling of nuclear magnetic moments by conduction electrons. Phys. Rev. 96 (1954), 99–102. https://doi.org/10.1103/physrev.96.99.

[375] T. Kasuya. A theory of metallic ferro- and antiferromagnetism on Zener's model. Prog. Theor. Phys. 16 (1956), 45–57. https://doi.org/10.1143/ptp.16.45.

[376] K. Yosida. Magnetic properties of Cu-Mn alloys. Phys. Rev. 106 (1957), 893–898. https://doi.org/10.1103/physrev.106.893.

[377] K. A. McEwen. Chapter 6 Magnetic and transport properties of the rare earths. In Metals, pp. 411–488. Elsevier (1978). https://doi.org/10.1016/s0168-1273(78)01010-7.

[378] P. Wachter. The optical electrical and magnetic properties of the europium chalcogenides and the rare earth pnictides. Crit. Rev. Solid State Sci. 3 (1972), 189–241. https://doi.org/10.1080/10408437208244865.

[379] W. R. L. Lambrecht. Electronic structure and optical spectra of the semimetal ScAs and of the indirect-band-gap semiconductors ScN and GdN. Phys. Rev. B 62 (2000), 13538–13545. https://doi.org/10.1103/physrevb.62.13538.

[380] P. Larson and W. R. L. Lambrecht. Electronic structure and magnetism of europium chalcogenides in comparison with gadolinium nitride. J. Phys. Condens. Matter 18 (2006), 11333–11345. https://doi.org/10.1088/0953-8984/18/49/024.

[381] P. Wachter. Chapter 19 Europium chalcogenides: EuO, EuS, EuSe and EuTe. In Alloys and Intermetallics, pp. 507–574. Elsevier (1979). https://doi.org/10.1016/s0168-1273(79)02010-9.

[382] M. Schlipf, M. Betzinger, C. Friedrich, M. Ležaić and S. Blügel. HSE hybrid functional within the FLAPW method and its application to GdN. Phys. Rev. B 84 (2011), 125142. https://doi.org/10.1103/physrevb.84.125142.

[383] D. B. Ghosh, M. De and S. K. De. Electronic structure and magneto-optical properties of magnetic semiconductors: Europium monochalcogenides. Phys. Rev. B 70 (2004), 115211. https://doi.org/10.1103/physrevb.70.115211.

[384] F. Leuenberger, A. Parge, W. Felsch, K. Fauth and M. Hessler. GdN thin films: Bulk and local electronic and magnetic properties. Phys. Rev. B 72 (2005), 014427. https://doi.org/10.1103/physrevb.72.014427.

[385] B. Díaz, E. Granado, E. Abramof, L. Torres, R. T. Lechner, G. Springholz and G. Bauer. Magnetic ordering and transitions of EuSe studied by x-ray diffraction. Phys. Rev. B 81 (2010), 184428. https://doi.org/10.1103/physrevb.81.184428.

[386] R. Skomski. Permanent magnets: history, current research, and outlook. In Novel Functional Magnetic Materials, pp. 359–395. Springer International Publishing (2016). https://doi.org/10.1007/978-3-319-26106-5_9.

[387] J. Coey. Perspective and prospects for rare Earth permanent magnets. Engineering 6 (2020), 119–131. https://doi.org/10.1016/j.eng.2018.11.034.

[388] M. M. Vopson. Fundamentals of multiferroic materials and their possible applications. Crit. Rev. Solid State 40 (2015), 223–250. https://doi.org/10.1080/10408436.2014.992584.

[389] J. Lyubina. Magnetocaloric materials. In Novel Functional Magnetic Materials, pp. 115–186. Springer International Publishing (2016). https://doi.org/10.1007/978-3-319-26106-5_4.

[390] S. Hirosawa, M. Nishino and S. Miyashita. Perspectives for high-performance permanent magnets: applications, coercivity, and new materials. Adv. Nat. Sci.: Nanosci. Nanotechnol. 8 (2017), 013002. https://doi.org/10.1088/2043-6254/aa597c.

[391] K. J. Strnat and R. M. Strnat. Rare earth-cobalt permanent magnets. J. Magn. Magn. Mater. 100 (1991), 38–56. https://doi.org/10.1016/0304-8853(91)90811-n.

[392] K. Kurima and H. Satoshi. Chapter 208 Permanent magnets. In Handbook on the Physics and Chemistry of Rare Earths, pp. 515–565. Elsevier (2001). https://doi.org/10.1016/s0168-1273(01)32007-x.

[393] K. Strnat, G. Hoffer, J. Olson, W. Ostertag and J. J. Becker. A family of new cobalt-base permanent magnet materials. J. Appl. Phys. 38 (1967), 1001–1002. https://doi.org/10.1063/1.1709459.

[394] K. L. Belener and H. Kohlmann. Reaction pathways of oxide-reduction-diffusion (ORD) synthesis of $SmCo_5$ and in situ study of its hydrogen induced amorphization (HIA). J. Magn. Magn. Mater. 370 (2014), 134–139. https://doi.org/10.1016/j.jmmm.2014.06.066.

[395] B. Das, R. Choudhary, R. Skomski, B. Balasubramanian, A. K. Pathak, D. Paudyal and D. J. Sellmyer. Anisotropy and orbital moment in Sm-Co permanent magnets. Phys. Rev. B 100 (2019), 024419. https://doi.org/10.1103/physrevb.100.024419.

[396] R. Lemaire. Magnetic properties of intermetallic compounds of Co with rare-earth metals or Y (Part I.). Cobalt 32 (Sept. 1966), 132–40.

[397] W. Ostertag and K. L. Strnat. Rare earth cobalt compounds with the A_2B_{17} structure. Acta Crystallogr. 21 (1966), 560–565. https://doi.org/10.1107/s0365110x66003451.

[398] J. Coey and H. Sun. Improved magnetic properties by treatment of iron-based rare earth intermetallic compounds in anmonia. J. Magn. Magn. Mater. 87 (1990), L251–L254. https://doi.org/10.1016/0304-8853(90)90756-g.

[399] T. Iriyama, K. Kobayashi, N. Imaoka, T. Fukuda, H. Kato and Y. Nakagawa. Effect of nitrogen content on magnetic properties of $Sm_2Fe_{17}N_x$ (0<x<6). IEEE Trans. Magn. 28 (1992), 2326–2331. https://doi.org/10.1109/20.179482.

[400] R. Skomski and J. M. D. Coey. Giant energy product in nanostructured two-phase magnets. Phys. Rev. B 48 (1993), 15812–15816. https://doi.org/10.1103/physrevb.48.15812.

[401] M. Sagawa, S. Fujimura, H. Yamamoto, Y. Matsuura and K. Hiraga. Permanent magnet materials based on the rare earth-iron-boron tetragonal compounds. IEEE Trans. Magn. 20 (1984), 1584–1589. https://doi.org/10.1109/tmag.1984.1063214.

[402] U. N. environment programme. Annual Report 2018. Technical report, United Nations (2019).

[403] K. A. Gschneidner, V. K. Pecharsky and A. O. Tsokol. Recent developments in magnetocaloric materials. Rep. Prog. Phys. 68 (2005), 1479–1539. https://doi.org/10.1088/0034-4885/68/6/r04.

[404] H. Johra, K. Filonenko, P. Heiselberg, C. Veje, S. Dall'Olio, K. Engelbrecht and C. Bahl. Integration of a magnetocaloric heat pump in an energy flexible residential building. Renew. Energy 136 (2019), 115–126. https://doi.org/10.1016/j.renene.2018.12.102.

[405] Y. Mozharivskyj, A. O. Pecharsky, V. K. Pecharsky and G. J. Miller. On the high-temperature phase transition of $Gd_5Si_2Ge_2$. J. Am. Chem. Soc. 127 (2004), 317–324. https://doi.org/10.1021/ja048679k.

[406] M. L. Fornasini and S. Cirafici. Crystal structures of Eu_3Ga_2, EuGa, Eu_2In, EuIn and $EuIn_4$. Z. Kristallogr. Cryst. Mater. 190 (1990), 295–304. https://doi.org/10.1524/zkri.1989.190.14.295.

[407] V. Pecharsky, J. K. Gschneidner, A. Pecharsky and A. Tishin. Thermodynamics of the magnetocaloric effect. Phys. Rev. B 64 (2001), 144406. https://doi.org/10.1103/physrevb.64.144406.

[408] N. A. Zarkevich and V. I. Zverev. Viable materials with a giant magnetocaloric effect. Crystals 10 (2020), 815. https://doi.org/10.3390/cryst10090815.

[409] J.-D. Zou. Magnetocaloric and barocaloric effects in a $Gd_5Si_2Ge_2$ compound. Chin. Phys. B 21 (2012), 037503. https://doi.org/10.1088/1674-1056/21/3/037503.

[410] F. Guillou, A. K. Pathak, D. Paudyal, Y. Mudryk, F. Wilhelm, A. Rogalev and V. K. Pecharsky. Non-hysteretic first-order phase transition with large latent heat and giant low-field magnetocaloric effect. Nat. Commun. 9 (2018). https://doi.org/10.1038/s41467-018-05268-4.

[411] N. R. Ram, M. Prakash, U. Naresh, N. S. Kumar, T. S. Sarmash, T. Subbarao, R. J. Kumar, G. R. Kumar and K. C. B. Naidu. Review on magnetocaloric effect and materials. J. Supercond. Nov. Magn. 31 (2018), 1971–1979. https://doi.org/10.1007/s10948-018-4666-z.

[412] V. K. Pecharsky and J. K. A. Gschneidner. Giant magnetocaloric effect in $Gd_5(Si_2Ge_2)$. Phys. Rev. Lett. 78 (1997), 4494–4497. https://doi.org/10.1103/physrevlett.78.4494.

[413] G. J. Liu, J. R. Sun, J. Lin, Y. W. Xie, T. Y. Zhao, H. W. Zhang and B. G. Shen. Entropy changes due to the first-order phase transition in the $Gd_5Si_xGe_{4-x}$ system. Appl. Phys. Lett. 88 (2006). https://doi.org/10.1063/1.2201879.

[414] L. Li, Y. Yuan, Y. Zhang, T. Namiki, K. Nishimura, R. Pöttgen and S. Zhou. Giant low field magnetocaloric effect and field-induced metamagnetic transition in TmZn. Appl. Phys. Lett. 107 (2015). https://doi.org/10.1063/1.4932058.

[415] M.-H. Phan and S.-C. Yu. Review of the magnetocaloric effect in manganite materials. J. Magn. Magn. Mater. 308 (2007), 325–340. https://doi.org/10.1016/j.jmmm.2006.07.025.

www.ingramcontent.com/pod-product-compliance
Lightning Source LLC
Chambersburg PA
CBHW061354210326
41598CB00035B/5981